Adhesion Measurement of Films and Coatings

ADHESION MEASUREMENT OF FILMS AND COATINGS

Editor: K.L. Mittal

CRC Press
Taylor & Francis Group
Boca Raton London New York

CRC Press is an imprint of the
Taylor & Francis Group, an **informa** business

Contents

Adhesion Measurement of Films and Coatings, pp. ix–x
K. L. Mittal (Ed.)
© VSP 1995.

Preface

This book embodies the proceedings of the International Symposium on Adhesion Measurement of Films and Coatings held in Boston, 5–7 December, 1992 under the auspices of Skill Dynamics, an IBM Company. Apropos, the papers from this symposium were earlier published in three issues of the *Journal of Adhesion Science and Technology* as follows: Vol. 7, No. 8 (1993); Vol. 7, No. 12 (1993); and Vol. 8, No. 6 (1994) except the opening article by yours truly. As researchers and technologists evinced considerable interest in acquiring these special issues separately, so we decided to make available a hard-bound book chronicling in one place the acta of this symposium. It should be recorded for historical reasons that the premier symposium on this topic was held in 1976 under the aegis of the American Society for Testing and Materials (ASTM).

Films and coatings are used for a variety of purposes and their adequate adhesion to the underlying substrates, *inter alia*, is of cardinal importance from practical consideration. Concomitantly, the need for techniques for quantitative measurement of adhesion becomes quite patent.

Since the first symposium was held in 1976, there has been brisk activity in devising new ways to measure adhesion or ameliorating the existing repertoire. A legion of techniques, ranging from very mundane to very sophisticated, have been documented in the literature for adhesion measurement of films and coatings. Recently I had the occasion to sift the literature and compile the list of adhesion measurement techniques and, you may believe or not, the final score came to 355 (quite a stupendous number), which are listed in the opening paper in this book. However, it should be mentioned that certain techniques might be listed more than once because of different appellations given to these. It is interesting to note that some of these techniques sound uncouth, primitive and plainly humorous. In spite of this cornucopia of techniques, no single technique has been acceptable to everyone or applicable to all coating-substrate combinations; and this has been the cause for the proliferation of techniques for adhesion measurement. Also there has been a perennial discordance among the people working in this arena as to what exactly is measured when someone uses one of these techniques to measure adhesion.

So in light of the long hiatus since the first symposium coupled with the fact that there was a high tempo of activity and tremendous interest in this topic, the need for another symposium became abundantly evident. Also, most of the people I polled felt that such a symposium was long overdue.

This symposium was planned with the following objectives in mind (i) to bring together the community interested in this topic, (ii) to provide a forum for discussion of latest developments, (iii) to provide an opportunity for cross-pollination of ideas; and (iv) to identify the vexing problems, as well as the techniques which offered good promise and warranted vigorous pursuit. When the announcement of this symposium was sent out, it elicited an excellent response and concomitantly the technical program comprised 51 presentations (overviews as well as original research results). A number of techniques were discussed, a great deal of information was covered, and there were illuminating (not exothermic) discussions throughout the course of this symposium. If comments from the attendees is a barometer of the success of the event, then this symposium was a huge success.

I certainly hope this book will provide bountiful current information on techniques for adhesion measurement of films and coatings, and will be found useful by both veterans and neophytes interested in this subject.

K. L. Mittal

Adhesion Measurement of Films and Coatings, pp. 1–13
K. L. Mittal (Ed.)
© VSP 1995.

Adhesion measurement of films and coatings: a commentary

K. L. MITTAL

92 Saddle Ridge Dr., Hopewell Jct., NY 12533, USA

Abstract—The adhesion of thin films are coatings is of paramount importance in many and diverse technologies; concomitantly, the need to measure adhesion is quite patent. Recently there has been a flurry of activity in devising new or ameliorating the existing techniques for adhesion measurement of films and coatings. As a matter of fact, a cornucopia of methods is available, ranging from very mundane or primitive to very sophisticated. Some techniques have been claimed to measure 'interfacial adhesion.' However, in most situations, strictly speaking, there exists no interface to start with, so what is the significance of interfacial adhesion? Actually what is measured is the so-called practical adhesion, defined as the force or the work required to remove a film or coating from the substrate, irrespective of the locus of failure. Alternatively, practical adhesion can be expressed as the time required for removal or delamination of a film or coating under accelerated stress conditions (e.g. exposure to boiling water, solvents, corrosives, etc.). In a film or coating-substrate combination, the failure will take place at the weakest place and could be interfacial (rare), interphasial, or cohesive. Here the concept and significance of practical adhesion, and its relationship to the fundamental or intrinsic adhesion is discussed. Some comments are made on the locus of failure. A comprehensive list of documented techniques for adhesion measurement of films and coatings is provided.

1. INTRODUCTION

Films and coatings are used in a legion of technologies for a variety of purposes, and irrespective of their intended function these must adhere satisfactorily to the underlying substrate. So the need for quantitative assessment of thin film or coating adhesion is quite patent. Many techniques for adhesion measurement of thin films and coatings [1, 2] have been documented, and even a cursory look at the recent literature will show that there is a tremendous activity in this domain. This proceedings volume is a good testimonial to the brisk activity in the arena of adhesion measurement. Apropos, the difference between a film and coating is: A film is a thin coating and a coating is a thick film; it really is a matter of thickness. There is no universal agreement or standard on the upper limit of thickness of a thin film, but generally speaking thin films are of the order of 100 nm or even less. Many of the adhesion measurement techniques apply to both films and coatings, but some may be more specific to one or the other. In this paper, the terms film and coating will be used interchangeably.

2. WHAT IS ACTUALLY MEASURED?

This question is extremely important as there has been and there still is a great deal of confusion about what exactly is measured when one attempts to measure adhesion of a film or coating. The answer depends on the definition of adhesion.

Adhesion can be manifested in three different forms [3]: (i) Fundamental Adhesion, (ii) Thermodynamic Adhesion, and (iii) Practical Adhesion. Fundamental Adhesion is defined as the summation of all interfacial intermolecular interactions between the contacting materials. If one knows the type of interaction between the film material and the substrate and the number of interacting units per unit area, then one can calculate the fundamental adhesion. Conversely, fundamental adhesion represents the energy required to break chemical bonds at the weakest plane in the film-substrate adhering system under the adhesion measurement conditions used. However, it should be noted that these two forms of fundamental adhesion could be quite different as the former refers to contact formation and the latter represents contact break, and the weakest plane during the disruption of an adhering system may not be where the contact was initially formed. In the following discussion it is the latter form of fundamental adhesion which is relevant. Thermodynamic adhesion signifies the change in free energy when an interface is formed (or separated) and is expressed as $W_A = \gamma_{S1} + \gamma_{S2} - \gamma_{S1S2}$, where W_A is the work of adhesion and γ_{S1} and γ_{S2} represent the surface free energies of material 1 (substrate) and material 2 (film), respectively. γ_{S1S2} is the interfacial free energy. In case of liquid coatings, W_A can be easily determined by $W_A = \gamma_{LV}(1 + \cos\theta)$, where γ_{LV} is the surface free energy of the liquid, and θ is the contact angle of the liquid coating on the substrate.

The practical adhesion signifies the force or the work required to remove or detach a film or coating from the substrate irrespective of the locus of failure (see further discussion regarding this issue in Section 3). Actually this is what is measured when one attempts to measure adhesion by any of the more than 300 techniques. This includes the energy required to deform both the film or coating and the substrate, as well as the energy dissipated as heat or stored in the film or coating, and the component representative of the actual fundamental adhesion. The relationship between the practical adhesion and the fundamental adhesion is expressed as follows:

Practical adhesion = f (fundamental adhesion, other factors).

A myriad of 'other factors' influence the practical adhesion of a coating or film. Some of these are: stress in the film or coating; thickness and mechanical properties of the coating; mechanical properties of substrate; work consumed by plastic deformation and viscous dissipation; mode of failure; mode and rate of applying the force or the energy to detach the film, i.e. the technique used for adhesion measurement, and the parameters of the technique. Let me cite the example of peel test, which is one of the commonly used techniques. For the same film–substrate combination, different angles and rates of peel culminate in different peel strength values; although the fundamental adhesion is expected to be the same irrespective of the angle or rate of peel. This behavior can be easily explained by the above expression in that the first quantity (fundamental adhesion) is the same but the contribution due to 'other factors' is quite different at different angles and different rates. At higher rates of peel, for example for viscoelastic materials, the peel strength is generally higher because of more viscoelastic dissipation of energy.

Also even for the same film–substrate combination, different measurement techniques yield different results. Hull *et al.* [4] published very interesting results in this regard. They studied the effect of thickness of gold film (up to 500 nm) on silicon substrate by peel, pull and scratch tests. They found that peel force decreased with thickness; whereas both pull strength and scratch force increased with thickness but in different manners: linear in the case of scratch test and non-linear when pull test was employed. This clearly shows that different techniques involve different parameters; and, concomitantly, culminate in different practical adhesion values. However, all this can be explained by the expression delineated above.

In a multilayer system, the failure will take place at the weakest plane. Concomitantly, the practical adhesion is the net result of the energy required to break chemical bonds at the weakest place and the contribution due to other factors.

In the discussion above, there has been an implicit assumption that there is an interface to start with and a clear-cut interfacial separation takes place. Strictly speaking, an interface is a mathematical plane or a sharp frontier with no thickness, and the existence of such an interface and a veritable interfacial separation is a rare occurrence. In a situation where one could observe a true interfacial separation, then in that case the fundamental adhesion could be labeled as 'fundamental interfacial adhesion'. However, in most situations, an interphasial separation is the norm (see the next section for further discussion on this topic). So in the case of separation in the interphase or interfacial region, the fundamental adhesion denotes the energy required to break chemical bonds at the weakest place in the interphase. Apropos, some people refer to 'fundamental adhesion' as 'intrinsic adhesion' and the contribution due to other factors as 'extrinsic adhesion.' Intrinsic or fundamental adhesion represents the chemical component as it is dictated by the prevailing chemistry at the weakest place, and the contribution due to other factors constitutes 'non-chemical' or 'mechanical' component.

So it is quite manifest from this discussion that all these techniques measure the cumulative effect of intrinsic or fundamental adhesion and the contribution due to many other factors. One may ask the question: Can one determine fundamental adhesion by making practical adhesion measurement? I do not think so, as it is very difficult, maybe impossible, to quantitate the contribution due to the multitude of non-chemical factors. One can only hope to see increase in practical adhesion by improving fundamental adhesion (by manipulating the interphase, or interface, if it exists) provided no adverse conditions are present (e.g. stresses in the film). However, it would be nice to have a nondestructive and quantitative way to determine fundamental or intrinsic adhesion.

To conclude this section, some comment should be made regarding failure in the bulk of the coating, i.e. cohesive failure. Strictly speaking if the coating fails in the bulk that is really a problem of strength of materials and transcends the purview of adhesion science. As a corollary, in the case of bulk coating failure, attempts to improve the interface or interphase will be futile.

3. LOCUS OF FAILURE

When a film is detached from a substrate, the important question is: Where does the failure take place? It could be at the interface, in the interphase, or in the bulk (called cohesive failure) of the film or the substrate. As mentioned above a true interfacial failure is very uncommon as most often there is no clear-cut interface to start with. Most often failure occurs in an interphase or an interfacial region which has some thickness. Incidentally an interphase could be a single layer or region, or a combination of many regions with differing properties. The interphases are real and they may be present naturally (e.g. oxide on a metal) or are created deliberately (use of intermediate layers or adhesion promoters, surface treatment layers) or are formed by interaction or interdiffusion of the film material with the substrate or by migration of a new component (e.g. plasticizer) from the bulk of one of the adhering materials. It should be kept in mind that interphases have characteristics different from both bulk substrate and bulk coating.

Now the logical question is: If the failure occurs in the interphase, should it be called failure related to adhesion, or is it a bulk failure? This is a moot point and there is actually no easy answer to it. However, most adhesionists will agree that an interphasial failure falls within the purview of practical adhesion. In other words, practical adhesion signifies the force or the work required to detach a film from the substrate if the locus of failure is interfacial, or interphasial. Of course, if there is clear-cut failure in the bulk of the film (i.e. a uniform layer of bulk film material is left behind on the substrate after detachment of the film) that is a patent case of cohesive failure. However, quite often people just measure the force or the energy required to detach a coating without precisely determining the locus of failure, and even a bulk (cohesive) failure can be misconstrued (depending on the technique used to investigate the locus of failure) as interfacial, or interphasial failure. From a pragmatic point of view, the main interest is what sort of force or work a given film–substrate combination can withstand before delamination, irrespective of where it fails. With that in mind, it makes sense to define practical adhesion as the force or the work required to detach a coating or film from the substrate irrespective of the locus of failure. However, to improve practical adhesion, it is imperative to know precisely the locus of failure, so suitable approach can be taken to strengthen the weakest link.

Incidentally, it should be kept in mind that conclusion as to the precise locus of failure depends on the analytical technique used to examine the failed components of an adhering system. If an unaided eye is used to see if there is any film material left on the substrate then one may conclude that there was clearly an interfacial separation even if a thin layer of film was still clinging to the substrate because the human eye cannot see it. On the other hand, examination by sensitive surface spectroscopic analysis techniques will 'see' this thin layer and concomitantly one will conclude that the film failed, which signifies cohesive failure. Or if some other technique (e.g. microscopic) were used one may come to a different conclusion. So it is imperative that one must specify how the failed components were examined when one comments on the locus of failure.

4. ADHESION MEASUREMENT TECHNIQUES

As pointed out in the Introduction, a legion of techniques have been documented in the literature for adhesion measurement of films and coatings. Table 1 provides an alphabetical listing of such techniques. A few comments about this long list of techniques are in order. (i) They range from inexpensive to very sumptuous, and from very primitive to very sophisticated. (ii) Some of these methods are qualitative in nature, so no numerical values can be obtained. (iii) Most of these are mechanical and destructive in nature, and (iv) Just the sheer size of the table shows tremendous interest and activity in the topic of adhesion measurement. It should be noted that certain techniques might be listed more than once because of the different appellations given to these.

Table 1.

Techniques for adhesion measurement/assessment/monitoring of films and coatings

Ablation
Abrasion
Acceleration
Acceleration–Deceleration
Acoustic Emission
Acoustic Microscopy
Adherometer
Adherometer–Integrometer
Adhesive Tape
Angular Scribe-Stripping
Applied Moment
ARCO Microknife
ASTM Tensile Adhesion Method
Automatic Scrape

Balanced-Beam Scrape
Bathroom
Bell Tester
Bend
Bend (180°)
Bend (180°) + Tape
Bending (Three-point, or Four-point)
Bend-Peel Test
Bend Test (ASTM 571-72)
Black Lead Pencil
Blade
Blade Cutting Adhesion Tester

Table 1.

(Continued)

Blister (Constrained, Island)
Blister Peel
Boiling Water
Bolt Tensile
Both Sides Pull
Brenner Nodule Test
Brown and Garnish Crosshatch-Metal Strip Tape Test
Bubble
Buckling
Buffing
Bullet
Burgess Method
Burnishing

Can opener
Capacitance Measurement
Capacity Test
Cathodic Treatment
Centrifugal Hammer
Chisel
Chisel-Knife
Cleavage
Coin Scratch
Compact Tension
Compression
Conical Head Tensile
Conical Mandrel
Conical Mandrel + Tape
Constant Strain (*in-situ* SEM)
Constrained Blister
Continuous Indentation
Creep
Crosscut
Cross-cut Tape Test (ASTM D 3359-78)
Crosshatch
Crosshatch, Impact + Tape
Crosshatch + Tape
Crosshatch Without Tape
Crowfoot Knife Test
Cunningham Wood Cross Adhesion Test
Cupping
Cupping and Indentation
Cutting

Damping (resonator)
Deep Draw
Deformation
Delamination
Diamond Indentation Draw (DID)
Diamond Scratch

Table 1.
(Continued)

___

Die Bond Pull
Dielectrometric
Direct Pull
Disc-On-Disc (DOD)
Dishwasher
Distensibility
Dome
Dot
Double Cantilever Adhesion Test
Double Cantilever Beam
Double Torsion
Draw
Driven Blade Tester
DuPont Sharp Tool
Dynamic
Dynamic Response (based on)

Edge Delamination
Elcometer Adhesion Tester
Electrochemical
Electromagnetic Tensile
Electron Beam (Pulsed)
Electron Spin Resonance
Elongation
Erosion
Exposure + Tape

File
Fingernail
Flat-Wise Tension Test
Flexure
Flexure Spallation Test
Flexure Strain
Floating Image
Fluorescent
Flyer Plate
Ford Motor Co. Crosshatch Tape Test
Four-Point Bend
Fracture Energy
Fracture Mechanics Test
Freeze-Thaw Cycle
Friction (Internal)

Gardner-van Heuckeroth Adhesion Test
General Electric Plug Method
Graham-Linton Edge Test
Gravelometer
Grind-Saw
Grindwheel
Groove

___

Table 1.
(Continued)

Hammering
Hardness
Heating and Quenching
Hesiometer
Hoffman Scratch Tester
Hot Water
Hounsfeld Tensiometer
Hydraulic Adhesiometer
Hydrodynamic
Hydrophil Balance

I-Beam
Ice Pick
ICI Bullet
ICI Gun
$\vec{I} \times \vec{H}$
Impact
Impact Deceleration
Impact + Tape
Impulse
Inboard Wire Peel
Indentation
Indentation-Debonding
Inertia
Inflated membrane
In-situ SEM Constant Strain Method
Interchemical Adherometer
Internal Friction
Internal Stress of Ni film
Interrupted Bend Test
Inverse Ollard Method
Inverse peel
Inverted Blister
Ion-Migration
Island Blister

Jacquet Method

Knife
Konig Knife-Wedge
Koole Chisel

Lamb Waves (use of)
Lap Shear
Laser Ablation
Laser Acoustic Test
Laser Beam Holography
Laser Spallation
Liquid Jet
Liquid Wedge
LSRH-Revetest

Table 1.
(Continued)

Mandrel
Mechanical Resonance Method
MEEM
Membrane (Inflated)
Meredith and Guminski Chisel
Meredith-Guminski Adhesion Test
Mesle 'Can Opener'
Metal Stamping
Microindentation
Micro-scratch
Microtribometer
Microwedge scratch
Modified Ollard
Modified Pull Test
Moment

Nailhead
Nailhead Lead Tension Test
Nano-indentation
Napkin Ring
New Jersey Zinc Co. Test
New York Club Chisel
New York Club Tensile Method
NMP (*N*-methyl pyrrolidone)
Nodule
Normalized Sticking Tape
Notch
Nucleation

OEMS
Ollard Method
Ollard (Modified) Method
Olson Ball + Tape
Orange Peel Meter
Outboard Wire Peel

Parallel Gap Welding
Parallel Scratch
Parking Lot
Particle (Solid) Erosion
Pascoe Torque
Pass Test
Peel
Peel with Spatula
Pen Knife
Pencil
Pencil Hardness
Pendulum
Pendulum Scratching
Photoacoustic Pulse

Table 1.

(Continued)

Photothermal Radiometry
Pin-Pull
Ploughing
Plug Pull
Pneumatic Adhesion Tester
Pocket Knife
Pocket Scrape
Portable Pull Tester
Pre-cut Scrape
Pressure Cooker
Pressure Sensitive Tape
Princeton Adhesion and Scratch Tester
Pulling-down
Pull-off (Schmidt, Hoffman)
Pulsed Electron Beam
Pulsed Laser Beam
Push-in
Push-out

Q-Meter
Q-Tip
Quad Sebastian Tester

Raman Frequency Shift
Raman-Scratch
Razor Blade
Resonator Damping
Resonance Measurement (based on)
Reverse Impact
Reverse Impact + Tape
Revetest (LSRH)
RFL (British Motor Driven)
Ribbon Lead Shear Test
Ribbon Peel Test
Ring-Shear
Rivet
Rod and Ring
Rolling With Slip
Rondeau Scratch Tester
Rossman Chisel
Rub
Russian Method

Salt Bath
Sand Erosion
Sandwich Pull-off
Saw
Scalpel
Scanning Acoustic Microscope

Table 1.

(Continued)

Sclerometric
Score + Salt Spray
Scotch® Tape
Scrape + Solvent Wash
Scraping
Scratch
Scratchmaster
Scribe
Scribe-Grid
Sebastian Tester
Self-delamination Method
Separation Method
Shear
Shear Stress Deformation
Shockwave
Simple Cut + Tape
Single Cantilever Adhesion Test
Single Edge Notched Test
Single Pass Pendulum Scratching
Soldered-Wire Tension-Peel
Solvent
Spall
Spiral Gut
Springscale Pull-off Test
Squashing
Squeezing in Compression
Stiffness
Stoneley Waves (use of)
Strain, Constant (*in-situ* SEM)
Stretch Deformation
Stretching
Stud Pull Test
Stylometer
Stylus
Surface Acoustic Wave Sensor
Surface and Interfacial Cutting Method
Swab
Sward Adhesion Tester

Taber Scratch-Shear
Tape
T-Bend + Tape
TC Peel
Tear Test
Tensile
Tensile Extension
Tensile Shear
Thermal
Thermal Cycling

Table 1.

(Continued)

Thermal Gradient Adhesion Meter
Thermocompression Bended Peel
Thermoreflectance
Three-Point Bending
Three-Point Flexure
Threshold Adhesion Failure
Thumbnail
Tipple
Tooke Inspection Gage
Topple
Torque Wrench
Torsion Balance
Torsion Elcometer Adhesion Tester
Transient Joule Heating
Twisting
Twisting Cork
Twisting-off

Ultracentrifugal
Ultrasonic Pulse-Echo
Ultrasonic Resonance
Ultrasonic Surface Wave
Ultrasonic Vibration
Ultrasound (use of)
Undercutting
Uniaxial Compression
Uniaxial Tension

Van Laar Scratch Test
Vibratory
Voltage-Cyclic Technique
VTT Scratch Test
V(z) Curve Method

Water (boiling)
Wedge Bend
Wedge Bend + Tape
Wedge Insertion
Weight-fall
Westinghouse Scratch Meter
Weyerhaeuser Paint Adhesion Tester
Whirling Ball
Window Adhesion Test
Wire Bend
Wire Peel
Wolf Adhesion Chisel Test
Wrapping

X-Cut Tape Test (ASTM D3359-78)
X-ray Diffraction

5. CONCLUSIONS

1. There exists a plethora of techniques for practical adhesion measurement of films and coatings; however, there is no single technique which will be acceptable to everyone or will be applicable to all coating-substrate combinations. Apropos, it would be highly desirable to have a nondestructive and quantitative way to assess fundamental or intrinsic adhesion, which signifies the energy required to break bonds exclusively at the weakest plane in an adhering system.

2. For relative purposes, any of these techniques can be used. In other words, any of these techniques will rank coating-substrate samples in a series, or discriminate cases of poor practical adhesion.

3. While reporting practical adhesion values, all the parameters which can influence the results obtained using a particular technique must be specified.

4. Along with practical adhesion values, one should also comment on the locus of failure and how it was determined.

5. The best test for practical adhesion measurement is the one that simulates usage stress conditions as closely as possible.

REFERENCES

1. K. L. Mittal, (Ed.), *Adhesion Measurement of Thin Films, Thick Films and Bulk Coatings*, STP No. 640. American Society for Testing and Materials, Philadelphia (1978).
2. K. L. Mittal, *J. Adhesion Sci. Technol.* 1, 247–259 (1987). This paper presents a selected bibliography on adhesion measurement of films and coatings. Note: An update of this bibliography is in preparation.
3. K. L. Mittal, in: ref. 1, pp. 5–17
4. T. R. Hull, J. S. Colligon and A. E. Hill, *Vacuum* 37 (3/4), 327–330 (1987).

Adhesion Measurement of Films and Coatings, pp. 15–39
K. L. Mittal (Ed.)
© VSP 1995.

Adherence* failure and measurement: some troubling questions

SHERMAN D. BROWN

*Department of Materials Science and Engineering, University of Illinois at Urbana-Champaign,
1304 West Green Street, Urbana, IL 61801, USA*

Revised version received 4 January 1994

Abstract—It is pointed out that many methods used to determine the adherence of films and coatings to their substrates are inadequate. Sometimes, they are misleading. Key test conditions must appropriately simulate the conditions of service under which adherence failure may be brought about. It is indicated that the test environment and the rate and mode of stress application are among the important factors to consider. The nature of adherence failure is discussed against a background of multibarrier fracture kinetics. A brief review of multibarrier fracture kinetics as it applies to adherence failure and testing is given. Some evidence is cited to show that competitive failure mechanisms operate in many cases of adherence failure, and that which mechanism dominates depends on the conditions of failure.

Keywords: Adherence; adhesion; fracture kinetics; subcritical failure processes; coatings.

1. INTRODUCTION

It has already been pointed out by earlier writers that the terms *adherence* and *adhesion*, as used apropos of coating–substrate systems, have many semantic difficulties (e.g. [1, 2]). For instance, *adherence* was recognized as a term with broad meaning by Andrews [1] in his treatment of porcelain enamel coatings on a metallic substrate. He reported that in his time enamellers considered adherence to mean the '... resistance of the enamel to mechanical damage by impact, torsion, bending or heat shock.' At times, adherence was taken to mean '... the actual attraction of the enamel and the metal to each other.' Andrews also stated that '... the most common acceptance of the term is that adherence involves the resistance of the enamel coating to mechanical damage and whether the enamel comes off the metal leaving it clean or leaving variable degrees of broken glass retained in contact with the metal.' Meanings cited by Andrews for coating–substrate adherence were not limited to enamel–metal systems. Rather, they have been used in connection with various types of coating–substrate combinations; e.g. ceramic and/or metallic coatings applied to a metallic substrate by thermal spray methods [3], and ceramic coatings put down by chemical vapor deposition [4]. In the case cited last, the authors used the terms adherence and adhesion interchangeably.

Mittal [2] dealt with the term *adhesion* by dividing it into three categories: namely, (1) *basic or fundamental adhesion*, (2) *thermodynamic or reversible adhesion*, and

---

*The rationale for use of the word 'adherence' is given in the text.

(3) *experimental or practical adhesion. Basic or fundamental adhesion* was '... related to the nature and strength of the binding forces between two materials in contact with each other.' Mittal cited as types of bonding included in his definition of basic adhesion the following: ionic, covalent, coordinate, metallic, hydrogen and van der Waals forces. It was pointed out that '... this basic definition of adhesion is not very helpful as it is not possible either to calculate the magnitude or to measure such adhesion forces in practical systems.'

Thermodynamic or reversible adhesion was defined in terms of the reversible work of adhesion, W_{AB}, defined by equation (1),

$$W_{AB} = \gamma_A + \gamma_B - \gamma_{AB},\qquad(1)$$

in which γ_A and γ_B represent, respectively, the specific surface free energies of substances A and B, and γ_{AB} represents the interfacial specific free energy. In some cases, this definition is not useful. If at least one of the phases involved is a liquid, however, equation (1) may be useful. That is, surface tensions of liquids and contact angles generally are easily measured. Thus, if θ is the contact angle and γ_B is taken to be the specific surface free energy of the liquid phase, then

$$W_{AB} = \gamma_B(1 + \cos\theta).\qquad(2)$$

Mittal identified *experimental or practical adhesion* with terms such as bond strength or adhesion strength for those cases in which measurement is made by some method in which the maximum force per unit area required to separate a coating from its substrate is determined. When measurement is made by a method which yields results in terms of the work per unit area required to cause coating–substrate separation, terms such as work of adhesion or energy of adhesion are used. Mittal was careful to point out that '... experimental values of adhesion may not have direct relevance to the basic or fundamental adhesion ...'. This is because processes unrelated to the simple separation of the coating from its substrate ordinarily act within the system during testing to confound the result. An important consequence of this is the fact that different methods for measuring adhesion often yield significantly different values for coating–substrate systems that are essentially the same. This problem is a matter of central importance in the present paper.

Some final comments as regards the terms adherence and adhesion are now appropriate: Most dictionaries treat the terms as being synonymous in the technical context (e.g. [5–7]). However, I use adherence more broadly than the term adhesion. *Adherence includes but is not limited to adhesion. Adhesion failure* is taken to mean failure that is restricted to the coating–substrate interface. *Adherence failure* also includes failure that may be partially cohesive in character. It is not infrequently observed that some portions of a crack front in a coating–substrate system move through the interface while other portions are moving through the coating and/or the substrate. In some instances, there may be multiple cracks moving within the same time frame to make up the major crack front, some at the interface and some within the near-interface bulk phases. In this paper, such complicated failure is called adherence failure. Adhesion failure is taken to be that which occurs strictly at the coating–substrate interface or within the interfacial region.

In the late 1950s, adherence testing of ceramic or vitreous coatings on metallic substrata was typically done by methods that left much to be desired. In one test, for example, a known weight (often a ball bearing) was dropped at a right angle from some predetermined height onto the exposed surface of a coating that was to be checked (e.g. [8]). If no fracture occurred in the coating, owing to the impact of the weight, its adherence to its substrate was deemed satisfactory. On the other hand, if fracture was observed, the area of coating dislodged from the substrate in the locality of the point of impact was taken as a crude measure of the lack of adherence of the coating to its substrate. One engineer that I met used a silver dollar for the weight in his test. In a somewhat similar approach, a ball bearing was pushed against the coating surface under a predetermined static load (e.g. [9]). Again, if no fracture occurred, the coating–substrate adherence was considered adequate. Fracture of the coating was interpreted as an indication of unsatisfactory adherence. Such tests often yield confusing results that provide little insight. Nevertheless, some who used them were convinced of their validity. It was about this time that various pull-off tests were first advanced. Many other kinds of tests to measure adherence then came along; moreover, the understanding of adherence increased. Even so, problems regarding adherence and its measurement remain unresolved as the present paper is written.

Quantitative determination of the adhesion between coatings (or films) and their substrates is a matter of considerable practical importance. If coatings or films fail to adhere adequately to their substrates during service, functions for which they are applied may not be achieved. It is often necessary that adherence testing be included in quality control procedures. Furthermore, scientific investigation into the nature of coating–substrate adherence and the development of strategies to improve it require its accurate and meaningful measurement. More than 200 different test methods have been advanced for one or more of these purposes [1, 2, 10–16]. Some of these are inadequate on fundamental grounds. In addition to being fraught with the kinds of complications that distinguish practical adhesion from basic or fundamental adhesion (see above), some methods may not provide any valid measurement of the adherence. Or the conditions under which some tests are made may hopelessly confound the results. Other methods appear to be valid, but only within very limited contexts. Troubling questions and problems remain unresolved. Some of these are as follows:

(1) Testing often is limited to ordinary temperatures. Service may occur at elevated temperatures. How does temperature affect adherence? Do adherence tests made at ordinary temperatures adequately represent adherence failure conditions at elevated temperatures?

(2) Service conditions often include environmental factors of a chemical nature (e.g. moisture or sulfur oxides) that may affect adherence. Nevertheless, adherence testing conditions may not take adequate cognizance of such factors.

(3) The adhesion of a coating to its substrate may increase [17] or decay [18–21] with time, following coating application, depending on the nature of the system and the conditions imposed during storage and/or service. Moreover, the stress level to which a given coating–substrate system is exposed during coating application, storage, and/or service may affect the direction of such change and the rate at which it occurs (e.g. [19–21]).

(4) A key question is this: 'Is coating–substrate failure ever truly interfacial in character?' This matter needs to be considered carefully [1, 2, 13, 22–24]. Stress level

and loading rate have been observed to affect the path taken by the major crack during the failure process and, hence, the measured adherence [19–21, 25]. Much testing of adherence fails to take these facts into account.

(5) Some methods used for testing adherence apply stress or energy unevenly to the coating–substrate interface in an unacceptable measure. One important consequence of this in some cases is that the stress or energy density at the locus of failure initiation is not really known. Hence, the value calculated for the adherence from the data obtained, using the conventional formulas for the case, may have an unacceptable, and often unrecognized, error associated with it.

(6) Methods used for testing adherence (e.g. [1, 2, 10–13, 15, 16]) yield results that seem to differ importantly as regards their fundamental meanings. For instance, tensile pull-off test results have dimensions of M/Lt^2 whereas data obtained by the impact deceleration test or the widely used peel test have dimensions of M/t^2. The scratch test yields data with dimensions of force, ML/t^2. L, M, and t represent the dimensions length, mass, and time, respectively. Other tests may give data with still different fundamental dimensions. Another problem: The literature dealing with measurement of adhesion and/or adherence reveals confusion in this regard on the part of too many workers; i.e. dimensions cited with data too often are incorrect.

(7) Are the samples tested truly representative of the coating–substrate system that will see service? This matter is complicated by the fact that adherence failure — even that which occurs altogether at the coating–substrate interface — follows extreme value statistics [19, 26]. Can proof testing be used?

(8) If the coating–substrate interface is not smooth but irregular, stresses at the crack front during adhesion failure may be more complex than assumed. This could lead to misinterpretation.

(9) Undetected residual stresses within the coating–substrate system may affect adherence and confound adherence testing.

It is the intent of this paper to discuss the questions and problems in a fashion that will stimulate further in-depth study of adherence testing and the interpretation of its results. Detailed treatment of some issues is beyond its scope; earlier authors have discussed them (e.g. [2, 11, 24]). Moreover, an extensive bibliography [13] and critical reviews (e.g. [2, 10–12, 15, 16, 27]) of most methods used for adhesion and/or adherence testing have already been published and will not be dealt with extensively here. Testing methods will be described only to the extent necessary to facilitate presentation of the ideas broached here.

2. DISCUSSION

2.1. Testing at ordinary temperatures vs. service at elevated temperatures

That temperature can affect coating–substrate adherence seems quite obvious. The mechanical properties of most materials change with temperature, albeit not always for the same reasons. If a coating is intended to serve as a thermal barrier on, say, an interior surface of a jet engine or rocket thrust chamber, its mechanical properties including the coating–substrate adherence are likely to vary widely from those measured at ordinary temperatures. Moreover, stresses that derive from temperature gradients and differences in thermal expansion properties will develop in service that are not taken into account in typical adherence tests. These stresses will affect the coating–substrate adherence. In some instances, more or less ductile bond coats can be used to dissipate to some extent the stresses developed. Even so, the pertinent question remains: 'How should coating–substrate systems be tested for adherence if their service is to be at elevated temperatures?'

Another relevant factor that comes into play as a consequence of testing temperatures being markedly different from service temperatures stems from the fact that adherence failure often involves thermally activated rate processes. Thus, reactions that contribute to adherence failure may proceed at significantly different rates at service temperatures if those temperatures differ in substantial measure from those associated with the test conditions. This underscores the need, in some cases at least, of testing coating–substrate adherence at temperatures characteristic of the projected service.

Many tests used to measure coating–substrate adherence require the use of an adhesive (typically an epoxy or a cyanoacrylate) to attach a load-transmitting fixture to the free surface of the coating or film. This is done so that a measured tensile or shear stress can be applied to the coating–substrate system. The applied stress on the coating at the time of failure is then taken as the measure of the adherence. Examples of such tests include most of the direct pull-off methods (e.g. [28, 29]), the moment or topple method [30–32], lap shear tests [33–35], the napkin ring test [36], and certain fracture mechanics tests adapted to the testing of adherence; e.g. the double cantilever beam configuration [15, 37–47], the double torsion test [15, 38, 39, 41, 48–50], and the four-point composite bend test [19, 51]. Figures 1 through 7 respectively provide schematic representations of (1) a direct pull-off method, (2) the moment or topple method, (3) a lap shear test, (4) the napkin ring test, (5) the double cantilever beam configuration, (6) the double torsion test, and (7) the four-point composite bend test. Use of an organic adhesive for attachment of the fixture required for load transmission to the coating–substrate system limits such tests to ordinary or only moderately elevated temperatures. Yet, service conditions may often involve temperatures that are much higher, well beyond the capability of the adhesive used. In some instances, metallic brazes have been used, rather than organic adhesives, to connect the aforementioned fixture to the coating–substrate system in an attempt to extend the range of testing to higher temperatures (e.g. [52]). Even so, the test temperatures possible in such cases are often well below those met with in service.

There are other difficulties that emerge as regards the use of organic adhesives or brazes as just described. For instance, the characteristics of the adhesive affect significantly the measured value of the coating–substrate adherence for at least some ceramic–metal systems [19]. This effect is observed even though (1) there is no penetration through the coating by the adhesive and (2) failure does not occur at all near the glue line. Thermal expansion/contraction coefficients of brazes are such, compared with ceramics, that when they are used to attach a load transmission fixture to the coating surface in ceramic–metal, coating–substrate systems, they tend to set up residual stresses that seriously affect the measured values of adherence. Most of the reported adherence data that were obtained using an organic adhesive or braze for attachment failed to take these matters into account.

It is true that the maximum stresses which tend to cause coating failure by spallation may occur at comparatively modest temperatures during the heat-up portion of the

S. D. Brown

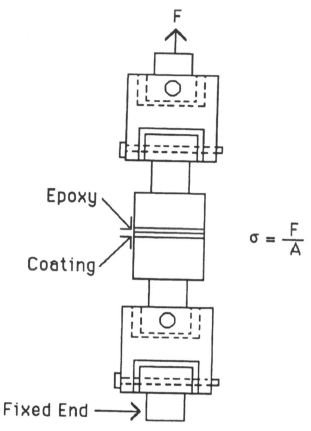

$$\sigma = \frac{F}{A}$$

Figure 1. Direct tensile pull-off test (ASTM C633-79). This illustration appeared previously [15] and is reprinted here by permission of the American Society of Mechanical Engineers.

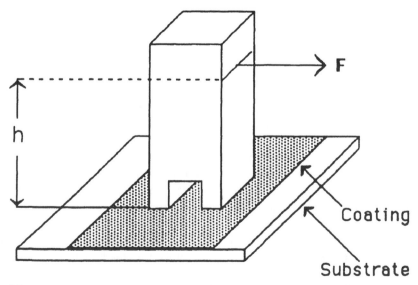

Figure 2. Moment or topple adhesion test. This illustration appeared previously [15] and is reprinted here by permission of the American Society of Mechanical Engineers.

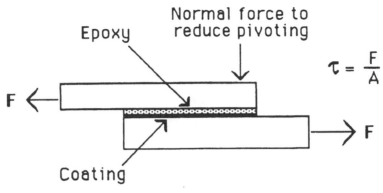

Figure 3. Lap shear test. This illustration appeared previously [15] and is reprinted here by permission of the American Society of Mechanical Engineers.

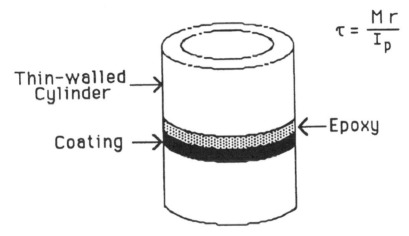

Figure 4. Napkin ring test. This illustration appeared previously [15] and is reprinted here by permission of the American Society of Mechanical Engineers.

service cycle. It is also true that the maximum stresses tending to cause tensile cracking of the coating may occur, again at comparatively modest temperatures, during the cooling portion of the cycle. This is particularly the case if the system is quenched. Nevertheless, it is also a fact that in many instances, significant stresses may occur at coating–substrate interfaces owing to temperature gradients that exist at the maximum service temperature. For instance, heat may be removed from the back surface of a substrate by a coolant, and this would result in a temperature gradient across the coating–substrate system. In the absence of any mechanism provided to avoid and/or reduce thermal stresses (e.g. grading the coating with substrate metal to minimize thermal expansion mismatch, or use of a highly ductile bond layer between the coating and substrate), this temperature gradient may lead to failure. Mechanical strains may be generated in substrates during service at high temperatures by factors such as impact, pressure and pressure fluctuations, and vibration. As pointed out before, material properties such as elastic modulus, strength, and toughness can change, sometimes dramatically, as the temperature is increased. It is likely, therefore, that adherence will change as well. The key question here is this: 'How can adherence be determined

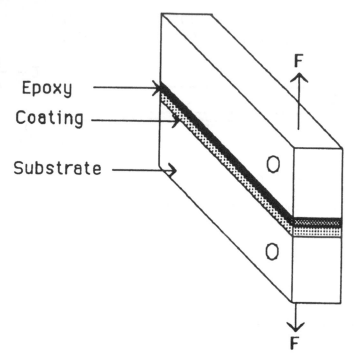

Figure 5. Double cantilever beam (DCB) configuration. This illustration appeared previously [15] and is reprinted here by permission of the American Society of Mechanical Engineers.

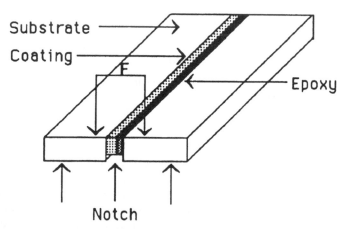

Figure 6. Double torsion test. This illustration appeared previously [15] and is reprinted here by permission of the American Society of Mechanical Engineers.

reliably at the high temperatures typical of those encountered under some conditions of service?'

2.2. Possible methods for testing at elevated temperatures

There are adherence testing methods that do not rely upon adhesives or brazes for attachment. Examples are the modified Ollard test [10–12, 15, 16, 53], the rod and

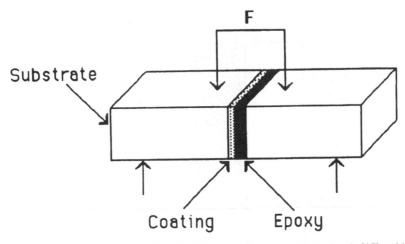

Figure 7. Four-point composite bend test (FPCB). This illustration appeared previously [15] and is reprinted here by permission of the American Society of Mechanical Engineers.

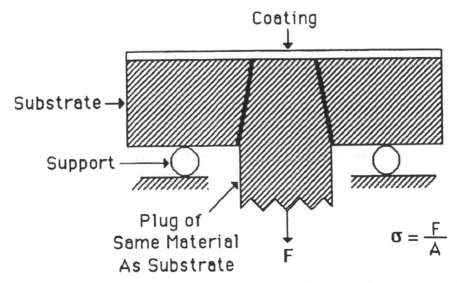

Figure 8. Modified Ollard test. This illustration appeared previously [15] and is reprinted here by permission of the American Society of Mechanical Engineers.

ring test [15, 16, 54, 55], the impact deceleration method [15, 16, 25, 56–59], the scratch test (e.g. [11, 15–17, 60–65]), and various strain methods that rely on the response of the coating to bending, stretching, or twisting of the substrate to provide an indication of the coating–substrate adherence [10, 66–68]. Figures 8 through 11 respectively provide schematic representations of (1) the modified Ollard test, (2) the rod and ring test, (3) an impact deceleration test, and (4) the scratch test. In many respects, such methods are marked improvements over those that must use an organic adhesive or a metallic braze if the coating or film must serve at high temperatures. Even so, there are troubling problems, as described below.

Adherence failure that involves crack growth apparently conforms to extreme value statistics [19, 26]. An important consequence of this fact is that the specimen size must

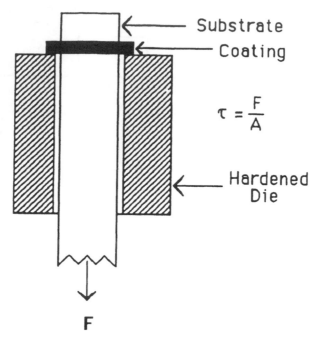

Figure 9. Rod and ring test. This illustration appeared previously [15] and is reprinted here by permission of the American Society of Mechanical Engineers.

be sufficiently large that a representative flaw size distribution is obtained. For example, if the tested area in a pull-off test is too small, scatter in the data becomes an important problem [28]. In the modified Ollard test, the coating tends to fail at the edges of the plug if the ratio of the interface test area diameter to the coating thickness becomes too large. In some cases, the required representative flaw size distribution simply cannot be obtained in the modified Ollard test. Another consequence of the fact that adherence failure conforms to extreme value statistics is that substantial numbers of specimens may need to be tested to achieve meaningful data. Typically, 25–50 specimens are used (e.g. [69]); however, some investigators have used 100 or more (e.g. [19]). Most users may be reluctant to conduct the number of tests required except, perhaps, to establish a sound basis for proof testing.

In some methods, a uniform lateral distribution of the applied stress over the portion of the coating–substrate interface tested is most difficult to achieve. For example, avoidance of bending moments that confound results can be a major problem in the rod and ring test. The clearance between the rod and die also must be kept to a close tolerance, otherwise thin coatings may slip through the die and/or application of force to the coating could be uneven.

The impact deceleration test has considerable potential. Nevertheless, it is not a simple test. It is not sufficient to measure only the translational velocity and mass of the coating just separated from its substrate by impact with the target [25]. The rotational velocity must also be determined. Moreover, adherence failure is abrupt in this instance and may not involve crack growth of the same kind as that associated with less rapid methods. Is the test, then, an adequate simulation of the service conditions? This question also relates to tests such as the laser spallation method [70].

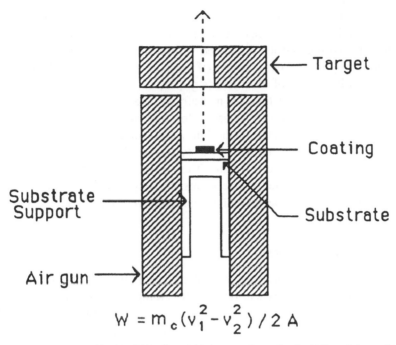

$$W = m_c(v_1^2 - v_2^2)/2A$$

Figure 10. Impact deceleration test. This illustration appeared previously [15] and is reprinted here by permission of the American Society of Mechanical Engineers.

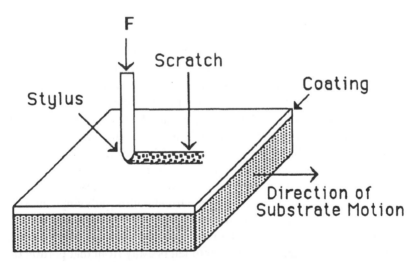

Figure 11. Scratch test. This illustration appeared previously [15] and is reprinted here by permission of the American Society of Mechanical Engineers.

The widely used and accepted scratch test does not depend on any adhesive or braze. In some of its more advanced forms, it is highly sophisticated (e.g. [62–65]). Nevertheless, the method as commonly used has many serious problems. It is altogether unsuitable for measuring the adhesion in some types of coating–substrate systems (e.g. porcelain enamels and bulk coatings such as those produced by thermal spray techniques). Of course, if it is desired to measure only coating cohesion, the scratch test may be useful. However, even for cases in which many deem the test acceptable for the measurement of adhesion, an excessive fraction of the force applied to move the stylus in scratching the coating is often dissipated within the coating by cracking (if the coatings is brittle) or by plastic deformation (if the coating is ductile). This can seriously confound the results. Some of its critics (e.g. [15, 23, 61]) have concluded on rational grounds that it is an invalid method for measuring adhesion. There are, of course, some instances in which the method may be useful for qualitative, comparative purposes. Furthermore, a recent variation on the method may hold some promise [62–65]. It consists of a frictional force measurement made by a stylus that vibrates. The amplitude changes when the coating peels from the substrate. The extent to which the method can be used to measure adhesion at elevated temperatures remains to be explored. Moreover, its appropriate application may be limited to comparatively thin coatings with relatively smooth substrata.

2.3. Some miscellaneous problems

It is widely recognized that adherence may change with time if the environment to which the coating–substrate system is exposed during storage and/or service can access the interfacial region [17–21, 27, 71, 72]. In most instances, corrosion at the interface weakens adherence. For example, autoclaved, plasma-sprayed alumina coatings on 316L stainless steel or Ti-6Al-4V ELI substrates were subjected to *in vivo* aging in Sprague–Dawley rats for periods up to 29 weeks [18, 71]. Adherence, measured by the standard tensile pull-off test [28, 29], degraded markedly with time of exposure to the *in vivo* environment. Average adherence losses of 64% and 63% were observed for the specimens having substrates of 316L stainless steel and Ti-6Al-4V ELI, respectively.

Another, related problem is the fact that corrosion of elements (e.g. iron, copper) within a coating can generate stresses that impair adherence. For instance, a protective ceramic coating, graded with a metallic substrate as a means of reducing the effects of thermal expansion mismatch, may be invaded during service at an elevated temperature by corrosive species that can react with the metallic particles used to effect grading. If the volume of the solid reaction products exceeds that of the solid reactants, compressive stresses may be generated within the coating. This has resulted in coating spallation.

In other instances, even a breach in the coating or film the size of a pin prick can admit sufficient amounts of corrosive species from the environment to significantly degrade the substrate strength and/or coating–substrate adhesion within a distance of several centimeters from the breach [73]. For example, a hermetic film [polyvinyl chloride (PVC)] was applied to a glass substrate (a rod) that had been etched with hydrofluoric acid (HF) to strengthen it prior to application of the film. Subsequently, the film was deliberately punctured with a pin at a location several centimeters away from that portion of the substrate to which stress sufficient to cause fracture eventually was applied. A substantial (approximately 690 MPa) weakening of the substrate was noted within a few hours of the introduction of the breach. Presumably, moisture from the environment had passed through the breach and then diffused along the PVC-glass interface to flaws that earlier had been blunted by the HF etching. Assuming that residual tensile stresses were sufficiently proximate to some of the said flaws, the water could then react with the glass at the flaw tips to enhance their acuities and thus weaken the substrate. In the absence of the breach, no such weakening effect was observed. Admittedly, the adherence of the

PVC film to the glass was not measured in this instance. Nevertheless, it is difficult to believe that the film–substrate adherence was not similarly affected, given the propensity of water to react with glass surfaces and interfaces. Treatments used to provide hermetic seals that prevent corrosive elements from gaining access to coating–substrate interfaces may provide barriers that are vulnerable to minor mechanical damage. For instance, sealing treatments are sometimes given to anodic spark deposited films or coatings put down by a thermal spray method because it is recognized that these kinds of coatings tend to be porous. If the hermetic seal is breached, even to a minor extent, interfacial weakness may develop within comparatively short periods of time owing to the subsequent invasion of corrosive species from the environment.

The point is that the chemical characteristics of the storage and/or service environments to which a coating–substrate system is subjected can have a dramatic effect on its adherence. This is not new information. Yet, it is often found that adherence testing reported in the literature fails to take cognizance of this important fact.

One problem associated with the tensile testing of brittle materials, and this includes the tensile pull-off methods frequently used apropos of ceramic–metal coating–substrate systems, is that of improper alignment of the specimen [74]. Such adherence measurement methods as the nail-head lead tension test [75], various peel tests, and the lap shear test may also suffer from this problem. A key consequence of improper alignment is that the stress at the locus of failure is not known with sufficient accuracy. The problem becomes particularly severe when stress corrosion enters into the picture [19]. Inasmuch as many adherence tests are performed in ordinary air environments (which usually are moist), this problem is one that deserves consideration. If the coating–substrate system must serve in some corrosive environment, then that environment should be simulated appropriately in the testing effort (e.g. [19–21, 26, 49, 50, 71]). Moreover, measures need to be taken to ensure that the test and setup selected permit the stress at the locus of failure to be known with sufficient accuracy.

2.4. *Crack paths during adherence failure*

It is commonly observed, at least in some coating–substrate systems, that adherence failure yields a substrate-side fracture surface that exhibits patches of the coating as well as bare areas of the substrate. Under some circumstances, patches of substrate material may be found adhering to the coating–side fracture surface of a coating or film that has been removed from its substrate. Scrutiny of the fracture surface of systems that apparently have fractured cleanly at the interface of a coating–substrate system not infrequently reveals very thin films or patches from the coating remaining on the substrate. Such observations raise questions regarding the mechanism of adherence failure. Moreover, an old question is resurrected: 'Is coating–substrate failure ever truly interfacial in character?' This question has been actively argued (e.g. [22–24]).

Useful insight into the nature of adherence failure can be gained from investigation of subcritical crack growth in ceramic–metal coating–substrate systems. For instance, in systems consisting of plasma-sprayed alumina coatings on either 316L stainless steel or Ti-6Al-4V ELI metallic substrates, the paths taken by cracks during subcritical adherence failure were found to be dependent on the magnitude of the applied stress and/or the rate of crack propagation [19–21, 26, 49, 50]. Note that these systems were exposed to chemically active environments (namely, water and aqueous solutions) that could access the interfacial region. Figures 12 through 15 show static fatigue data obtained by

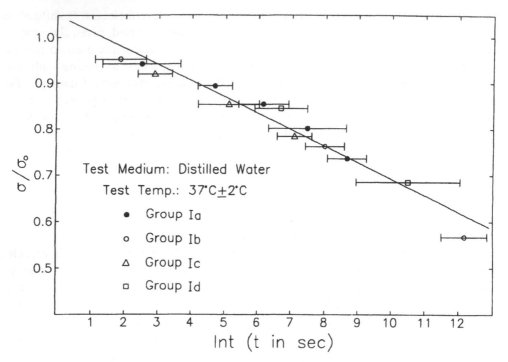

Figure 12. Static fatigue results for plasma-sprayed alumina-coated Ti-6Al-4V ELI substrates exposed to distilled water at $37 \pm 2\,^\circ$C [19, 20]. The four-point composite bend test was used to obtain the data. This illustration appeared previously [20] and is reprinted here by permission of the American Ceramic Society.

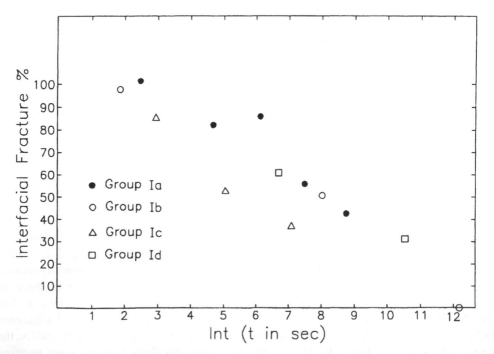

Figure 13. Time to failure dependence of the percent interfacial fracture associated with the static fatigue data of Fig. 12 [19, 20]. This illustration appeared previously [20] and is reprinted here by permission of the American Ceramic Society.

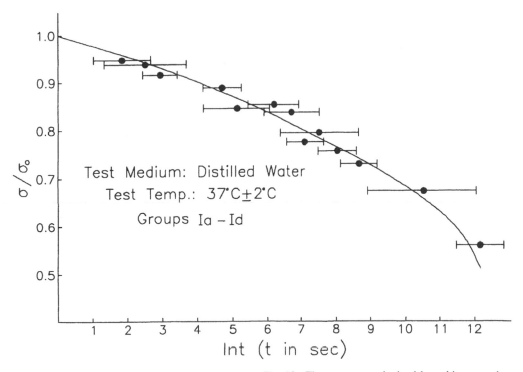

Figure 14. Optimal fit curve for the data shown in Fig. 12. The curve was obtained by cubic regression analysis [19, 21]. This illustration appeared previously [21] and is reprinted here by permission of Elsevier Sequoia S.A.

Ferber [19] for plasma-sprayed alumina coatings on Ti-6Al-4V ELI substrates. The four-point composite bend test (see above) was used for the purpose. Note that in all cases the environment was distilled water at $37 \pm 2°C$, and that all four figures represent the same set of data (a total of 76 specimens). These data show trends that are similar to those revealed by other data taken at different temperatures, in a range of different aqueous test media, and/or for alumina-316L stainless steel coating–substrate systems. Rapidly propagating cracks tended to move mostly along the ceramic–metal interface; slower cracks, through the ceramic coating, roughly parallel to the coating–substrate interface (cf. Figs 12 and 13). Cracks having intermediate propagation rates left visible evidence of having propagated stochastically both along the interface and through the coating. Post-failure examinations of the fracture surfaces, interpreted within the framework of multibarrier fracture kinetics (see below), indicated that competitive fracture processes were operative at all propagation rates. Another clue to the operation of competitive fracture processes is that when the velocity of the crack that acts to separate the coating or film from its substrate is plotted against the applied stress (or the stress intensity factor), the more rapid process within a given regime is most often seen to dominate (cf. Figs 13, 14, and 15). Careful examination of Figs 12 and 14 reveals that the data points in the two figures are the same; the lines, of course, are different. The optimal fit of the data (indicated by the curved line in Fig. 14) was obtained by cubic regression. Figure 15 was derived from Fig. 14 by the method of Fuller [76]. It is interesting that the slope of the line in Fig. 15 increases for failure that occurs predominantly at the coating–substrate interface, and that it is steeper than that portion which pertains

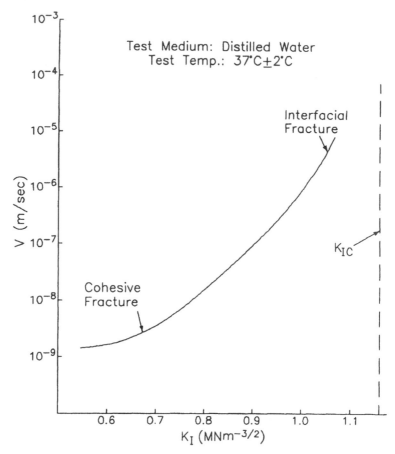

Figure 15. $V - K_I$ curve obtained from the static fatigue data of Figs 12 and 14 by Ferber [19] using the approach of Fuller [76]. This illustration appeared previously [21] and is reprinted here by permission of Elsevier Sequoia S.A.

primarily to cohesive failure. This fact will be revisited shortly apropos of the activation volume. It has interesting implications as regards the mechanism.

Of course, if the coating–substrate bonds are sufficiently weak compared with the bonds within the coating or substrate, the crack will be expected to move only along the interface. In such instances, evidence for other, competitive processes would not be seen because the process at the interface would dominate under all conditions. Nevertheless, coating–substrate separation that is absolutely clean, with no fracture occurring either within the coating or within the substrate, is rarely, if ever, achieved.

Evidence of competitive fracture processes is also observed in some polyphase solids. What is observed in such systems may provide insights that can apply to coating–substrate systems. For example, processes that participate in intergranular fracture, on the one hand, and those that lead to transgranular fracture, on the other, are often seen to be competitive. The fracture of MDF (macro-defect-free) cement is a case in point [77, 78]: Comparatively slow crack growth usually results in intergranular failure. That is, the crack tends to pass around the ceramic grains of the cement and through the polyvinyl alcohol phase. As the crack propagation rate is increased, a gradual transition from intergranular to transgranular failure occurs. At comparatively high crack growth rates, failure is predominately transgranular. Numerous other polyphase and/or polycrystalline materials exhibit similar behavior. However, there are systems and conditions in which one mode of failure or the other predominates regardless of the strain rate (e.g. [79]). If phase changes are initiated as the crack passes through a particular region of the solid, this further complicates the picture. Nevertheless, the fact that competing processes operate in many instances is clear. It is interesting that within the aforementioned transition zone, portions of a single crack front may exhibit both intergranular and transgranular fracture, and that this pattern may change with time as the crack propagates through the material. This suggests that either or both of the following conditions may exist: namely, (1) the character and/or properties of the material encountered by the crack as it moves through the solid are not uniform, and (2) the energy available to drive the crack propagation is not uniform in time or space along the crack front. Nonuniformity in the concentrations of environmental species along the length of the crack is also expected to influence the paths taken by various segments of a crack front. Fractographic evidence for this was found in connection with crack propagation in glasses [80].

Admittedly, the MDF cement case just cited does not involve a coating–substrate system of any kind. The example nevertheless provides additional evidence that the paths taken by cracks propagating through a material of more than a single phase may depend on the crack velocity (which is a function of the applied stress), among other factors. It is quite evident from the data in Figs 12–15 that competitive adherence failure processes operated in the thermal spray coating–substrate system that was investigated. The MDF cement example is helpful because it suggests that competitive failure processes may occur in various other systems that involve two or more solid phases. In fact, data reported by Oh *et al.* [81] on subcritical stress-corrosion crack velocities along plain and chemically etched glass–copper interfaces (their Fig. 7) gave clear evidence to me that both sequential and competitive adherence failure processes were active in each case. How prevalent such effects may be in the adherence failure of ceramic–metal, coating–substrate systems other than those mentioned here is not yet known. Even so, this places additional emphasis on the importance of adequate simulation of service conditions in adherence testing. It also raises some most interesting questions of a scientific nature. For instance, is the local energy density uniform along the front of a propagating crack, as has often been assumed, even when the applied stress is constant? Can variations of material character and/or properties be determined quantitatively from microanalysis of the fracture surfaces of specimens, and then related to local variations in the fracture process? It can be shown that the slope of the $\ln V$ vs. σ (or $\ln V$ vs. K_I) curve is proportional to the *activation volume* [82]. Here v and σ represent the crack front velocity and the applied stress, respectively; and K_I is the mode I stress intensity factor. What is implied, then, by the fact that the slopes of curves that represent failure at or very near coating–substrate interfaces are often greater than those that represent cohesive failure within the corresponding coatings (e.g. [19–21])? In the context of the fracture processes that occur during adherence failure, what occurs physically when the activation volume is increased so markedly? Does it mean that patches of bonds are broken as the crack front advances near the interface whereas single bonds are broken when the crack moves through the coating at lower values of K_I? These questions deserve investigation. Again, are the often simplistic interpretations of adherence data, which stem from the usual, so-called practical methods of testing, sufficient?

2.5. Multibarrier fracture kinetics

Many of the questions associated with adherence failure in coating–substrate systems, and the growth of cracks at and/or near solid–solid interfaces have been addressed very well in the context of fracture mechanics (e.g. [52, 83–85]). However, theory based strictly upon fracture mechanics omits aspects of the failure process that in some instances are important. This is particularly so in connection with subcritical failure. The problem is not unique to coating–substrate systems; failure of some bulk materials and fibers requires insight beyond that provided by fracture mechanics alone. Some workers have merged fracture mechanics with rate process theory in response to this need (e.g. [86–100]). In some of its later versions, the theory of multibarrier rate processes [101–103] has been melded with fracture mechanics for the purpose inasmuch as much of the data obtained indicated clearly the operation of both sequential and competitive rate processes. I have suggested the use of the term *multibarrier fracture kinetics* for the hybrid theory that has resulted [93–100] and continues in its development.

Earlier investigators of delayed failure and/or subcritical growth in glasses and ceramics recognized the need for including stress-dependent activation energies in their considerations (e.g. [86–89]). The later studies of Wiederhorn (e.g. [90–92]) provided data that in addition clearly indicated the operation of both sequential and competitive rate processes. For instance, Wiederhorn found three regions in his $\log V$ vs. K_I curves. Within region I (i.e. that just above the threshold stress intensity factor), a stress corrosion process controlled the rate at which the crack advanced. As K_I was increased, v increased until region II became dominant. Within region II, the crack velocity was controlled by the rate at which corrosive species (water in the case of Wiederhorn's data) could reach the crack tip. In cases that involved ceramics or glasses, this generally required diffusion of the corrosive species (usually water) through the gas phase in the crack and/or along the exposed surfaces at and near the crack tip. Data obtained by Williams and Nelson [104], in which gaseous hydrogen induced subcritical cracking of Ti-5Al-2.5Sn alloy, indicated that the diffusion path of the hydrogen was in the bulk metal phase vicinal to the crack tip. When the crack growth became sufficiently rapid that the rate was controlled by the diffusion of the corrosive species to the crack tip, region II became prominent. That is, the slower of the two processes — stress corrosion at the crack tip and transport of the corrosive species to the crack tip — dominated. *This is typical of sequential reactions in which the slowest step (or series of steps) controls the overall rate.*

In the case of region III, data obtained by various investigators uniformly indicated that the processes active in region III were in *competition* with those in regions I and II. That is, once K_I became sufficiently large that the transition from region II to region III behavior occurred, it was the *more rapid set of processes that dominated.* In Figs 14 and 15, it is quite clear that the data indicate a transition from the less rapid to the more rapid set of processes. In other words, the processes involved in cohesive failure are slower than those involved in interfacial failure. The two sets of processes are competitive.

Figures 16 through 18 provide additional solid evidence that the subcritical adherence failure of the plasma-sprayed alumina coating/Ti-6Al-4V ELI substrate system, discussed in connection with Figs 12–15, involved chemical rate processes. The fact that the failure process has a significant temperature dependence is shown quite amply in Fig. 16. Figure 17 shows Arrhenius plots of the same data as those illustrated in Fig. 16.

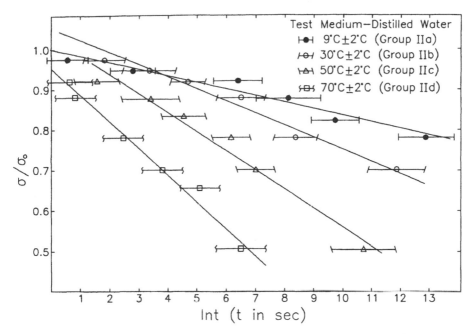

Figure 16. Temperature dependence of static fatigue failure of plasma-sprayed alumina-coated Ti-6Al-4V ELI substrates exposed to distilled water [19, 20]. This illustration appeared previously [20] and is reprinted here by permission of the American Ceramic Society.

Figure 18 is a plot of the activation energy vs. the ratio of the applied stress, σ, to that obtained for 'instant' failure, σ_0. Note that the results for the plasma-sprayed alumina on 316L stainless steel system are included in Fig. 18 for purposes of comparison. There are at least two important facts to glean from Figs 17 and 18: namely,

(1) The activation energies associated with the adherence failure are of such a magnitude that they can reasonably be associated with chemical reactions including the chemically assisted rupture of highly stressed bonds at the tips of cracks within the surfaces of solid phases and/or at solid–solid interfaces.

(2) There is a significant stress dependence of the activation energy. This is compatible with the findings of certain earlier investigators (e.g. [86–92]).

The data in Fig. 18 raise some interesting questions: Why are the activation energies associated with the specimens having 316L stainless steel substrates significantly less than those with the Ti-6Al-4V ELI substrates for all values of σ/σ_0? Also, what causes the deviation from straight-line behavior for σ/σ_0 values less than about 0.7? These questions have not yet been answered definitively. However, it is believed that the effects of residual stresses that derive from differences in substrate properties and possibly the behavior of the epoxy adhesive used for attachment of the load transmission rod to the coating surface may be responsible. In any case, these questions do not erase the basis provided for the validity of melding fracture mechanics and the theory of rate processes.

2.6. Selection of adherence testing methods

It is widely realized that adherence test methods must be selected on the basis of the situation at hand, e.g. on the basis of the nature of the coating–substrate system to be

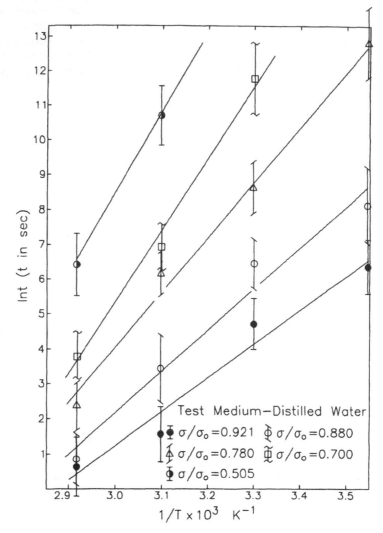

Figure 17. Arrhenius plots for the static fatigue failure of plasma-sprayed alumina-coated Ti-6Al-4V ELI substrates exposed to distilled water [19, 20]. This illustration appeared previously [20] and is reprinted here by permission of the American Ceramic Society.

tested. No universal test is available. Yet, it is troubling that the various tests that have been devised yield results that are, in many instances, basically different by nature: that is, the results achieved by some tests may differ markedly in their fundamental dimensions from those obtained by other methods. This fact alone implies certain incompatibilities amongst the methods. Moreover, attempts to reconcile the differences on any quantitative basis have seldom, if ever, succeeded. This suggests that various kinds of tests may yield different kinds of information. A tensile pull-off test, for example, will not provide the same kind of information that a scratch test will. Nor can a laser spallation method yield the same kind of information that, say, a four-point bend test does (and vice versa). The challenge, then, is to ascertain just what the test results obtained in any given case really mean. Also, there is a question that should always be addressed: namely, 'Does the test provide failure conditions sufficiently similar to

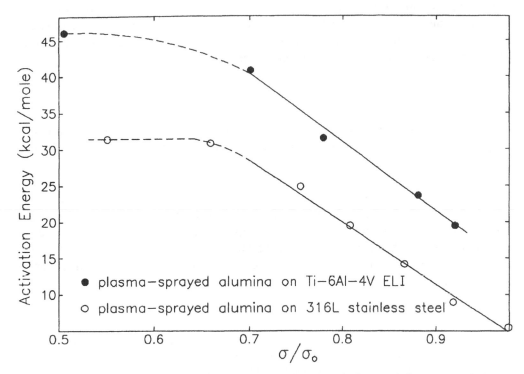

Figure 18. Stress dependencies of the activation energies for the static fatigue of plasma-sprayed alumina coatings on Ti-6Al-4V ELI and 316L stainless steel substrata [19, 20]. This illustration appeared previously [20] and is reprinted here by permission of the American Ceramic Society.

those that the system will encounter in service?' Failure processes involved in different tests may not be the same. Therefore, a major question that must be considered is this: 'What changes occur in the rate processes and fracture mechanics pertinent to adherence failure owing to differences in the nature of the tests?' The situation also poses basic questions regarding the very meaning of adherence.

Residual stresses are seldom distributed evenly over a coating–substrate interface [105, 106], nor are applied stresses [74, 105]. Even when coating–substrate systems have been annealed in some fashion, localized residual stresses may vary. Therefore, as a crack propagates through the interfacial region of a coating–substrate system, some segments of the crack front can be expected to be moving more slowly than others at any given moment; moreover, some portions will move along the interface while others will spread through either the coating or the substrate near the interface. The paths assumed by various segments of the crack front may change with time as the crack propagates. Those available paths of least resistance and/or those offering the greatest motivation will be taken by the crack. Regions within the coating–substrate interfacial space that have pores and cracks, or weakly bonded grains, will offer less resistance to the movement of the crack. Moreover, those regions with which comparatively large residual tensile stresses are associated will be more susceptible to the entry of some segment of the crack front than those having compressive or significantly smaller tensile stresses. Another complication is the fact that substrate roughness affects significantly the character and distribution of stresses at the coating–substrate interface. Finally, some segments of the crack front may be more accessible to crack-promoting environmental agents (e.g. moisture) than others. All of these factors and others must be considered when a fracture kinetics approach is taken to explain the observed phenomena. This is anything but a simple task. Nevertheless, improved and much needed insights into the nature of adherence and its measurement can be expected to result.

3. CONCLUSION

Testing of adhesion is not yet well understood notwithstanding the considerable progress that has been made since the 1950s in the pertinent art and science. Many different tests have been devised; however, there is considerable confusion. Results from some of the most widely used test methods do not have the same dimensions, and this suggests that data from various tests may have different meanings. In a practical sense, it is vital that the test conditions simulate well the expected service conditions. Otherwise, the value of the test is brought seriously into question. Service temperature, loading, and environmental chemistry are some of the factors that must be considered. Fracture kinetics is a promising theoretical framework within which to examine both adhesion and adherence failure.

REFERENCES

1. A. I. Andrews, *Porcelain Enamels*, pp. 517–579. The Garrard Press, Champaign, IL (1961).
2. K. L. Mittal, *Electrocomp. Sci. Technol.* **3**, 21–42 (1976).
3. S. W. Bradstreet, Private commuinication to S. D. Brown (1965).
4. D. O'Neil, J. H. Selverian and S. F. Wayne, *Development of Adherent Ceramic Coatings to Reduce Contact Stress Damage of Ceramics*. Final Report No. ORNL/Sub/89-95915/2. U.S. Department of Energy, Oak Ridge National Laboratories, Martin Marietta Energy Systems, Inc., Oak Ridge, TN (1992).
5. J. Stein and L. Urdang (Eds), *The Random House Dictionary of the English Language*, p. 18. Unabridged Edition. Random House, New York (1970).
6. D. B. Guralnik (Ed.), *Webster's New Dictionary*, p. 9. Second Concise Edition. Simon and Schuster, New York (1982).
7. F. de M. Vianna, A. D. Steinhardt *et al.* (Eds), *Roget's II, The New Thesaurus*, p. 16. Houghton Mifflin, Boston, MA (1980).
8. *ASTM Designation C284-51T; ASTM Standards Pt. 3*, 854–857 (1955); as cited in ref. 1.
9. *ASTM Designation C313-55*; adopted (1955), modified (1959); *ASTM Standards* **3**, 834–842 (1955); as cited in ref. 1.
10. D. Davies and J. A. Whittaker, *Metall. Rev.* **12**, 15–26 (1967).
11. K. L. Mittal (Ed.), *Adhesion Measurement of Thin Films, Thick Films and Bulk Coatings*, ASTM STP 640, pp. 5–16. American Society for Testing and Materials, Philadelphia, PA (1978).
12. C. C. Berndt and R. McPherson, *Aust. Weld. Res.* **6**, 75–85 (1979).
13. K. L. Mittal, *J. Adhesion Sci. Technol.* **1**, 247 (1987).
14. K. L. Mittal, Private communication to S. D. Brown (1993).
15. B. A. Chapman, H. D. DeFord, G. P. Wirtz and S. D. Brown, in: *Technology of Glass, Ceramic, or Glass–Ceramic to Metal Sealing*, W. E. Moddeman, C. W. Merten and D. P. Kramer (Eds), MD-Vol. 4, pp. 77–87. American Society of Mechanical Engineers, New York (Copyright 1987).
16. S. D. Brown, B. A. Chapman and G. P. Wirtz, in: *Thermal Spray Technology, New Ideas and Processes*, D. L. Houck (Ed.), pp. 147–157. ASM International, Metals Park, OH (1989).
17. P. Benjamin and C. Weaver, *Proc. R. Soc. London, Ser. A* **254**, 163–176 (1960).
18. S. D. Brown, J. L. Drummond, M. K. Ferber, D. P. Reed and M. R. Simon, in: *Mechanical Properties of Biomaterials, Advances in Biomaterials, Vol. II*, G. W. Hastings and D. F. Williams (Eds), pp. 249–264. John Wiley, Chichester (1980).
19. M. K. Ferber, PhD Thesis, University of Illinois, Urbana, IL (1981).
20. M. K. Ferber and S. D. Brown, *J. Am. Ceram. Soc.* **64**, 737–743 (Copyright 1981).
21. S. D. Brown, *Thin Solid Films* **119**, 127–139 (Copyright 1984).

22. J. J. Bikerman, in: *Adhesion and Cohesion*, P. Weiss (Ed.), pp. 36–45. Elsevier, Amsterdam (1962).
23. J. J. Bikerman, in: *Adhesion Measurement of Thin Films, Thick Films and Bulk Coatings*, ASTM STP 640, K. L. Mittal (Ed.), pp. 30–40. American Society for Testing and Materials, Philadelphia, PA (1978).
24. R. J. Good, in: *Adhesion Measurement of Thin Films, Thick Films and Bulk Coatings*, ASTM STP 640, K. L. Mittal (Ed.), pp. 18–29. American Society for Testing and Materials, Philadelphia, PA (1978).
25. B. A. Chapman, G. P. Wirtz and S. D. Brown, in: *Proceedings of a Symposium on Ceramic Thin and Thick Films, Ceramic Trans. 11*, B. V. Hiremath (Ed.), pp. 271–278. American Ceramic Society, Westerville, OH (1990).
26. M. K. Ferber and S. D. Brown, in: *Fracture Mechanics of Ceramics, Vol. 6*, R. C. Bradt, A. G. Evans, D. P. H. Hasselman and F. F. Lange (Eds), pp. 523–544. Plenum Press, New York (1983).
27. D. M. Mattox, in: *Deposition Technologies for Films and Coatings, Developments and Applications*, R. F. Bunshah *et al.* (Eds), pp. 63–82. Noyes Publications, Park Ridge, NJ (1982).
28. S. D. Brown, *Room-Temperature Adhesive Strength Tests of Various Flame-Sprayed and Flame-Plated Ceramic Coatings*. Progress Report No. 20-374. Jet Propulsion Laboratory, California Institute of Technology, Pasadena, CA (1959).
29. *ASTM Standard C633-79, Standard Test Method for Adhesion of Cohesion Strength of Flame-Sprayed Coatings. 1992 Annual Book of the ASTM Standards, Vol. 02.05*, pp. 663–667 (1992).
30. K. Kuwahara, H. Hirota and N. Umemoto, in: *Adhesion Measurement of Thin Films, Thick Films and Bulk Coatings*, ASTM STP 640, K. L. Mittal (Ed.), pp. 198–207. American Society for Testing and Materials, Philadelphia, PA (1978).
31. D. W. Butler, *J. Phys. E* **3**, 979–980 (1970).
32. D. W. Butler, C. T. H. Stoddart and P. R. Stuart, in: *Aspects of Adhesion 6*, D. J. Alner (Ed.), pp. 54–63. University of London Press, London (1971).
33. G. P. Anderson, S. J. Bennett and K. L. DeVries, *Analysis and Testing of Adhesive Bonds*. Academic Press, New York (1977).
34. J. W. Guinn, W. H. Griswold and S. G. Vermilyea, *J. Prosthet. Dent.* **48**, 551–554 (1982).
35. D. S. Lin, *J. Phys. D* **4**, 1977–1990 (1971).
36. N. A. DeBruyne, in: *Adhesion and Cohesion*, P. Weiss (Ed.), pp. 46–64. Elsevier, Amsterdam (1962).
37. S. M. Wiederhorn, A. M. Shorb and R. L. Moses, *J. Appl. Phys.* **39**, 1569–1572 (1968).
38. A. G. Evans, in: *Fracture Mechanics of Ceramics, Vol. 1*, R. C. Bradt, D. P. H. Hasselman and F. F. Lange (Eds), pp. 17–48. Plenum Press, New York (1974).
39. C. C. Berndt and R. McPherson, *Mech. Eng. Trans. Eng. Aust.* **ME6**, 53–58 (1980).
40. C. C. Berndt and R. McPherson, in: *Proc. Conf. Aust. Fracture Group*, pp. 100–105. DSTO (Department of Defence), ARL Melbourne, Australia (1980).
41. C. C. Berndt and R. McPherson, in: *Proc. 9th Int. Thermal Spraying Conf. 1980*, pp. 310–316. Nederlands Instituut voor Lasteniek, The Hague, Netherlands (1980).
42. C. C. Berndt and R. McPherson, in: *Surfaces and Interfaces in Ceramic and Ceramic–Metal Systems, Materials Science Research, Vol. 14*, J. A. Pask and A. G. Evans (Eds), pp. 619–626. Plenum Press, New York (1981).
43. C. C. Berndt, in: *Proc. 6th Int. Conf. Fracture, Vol. 4*, S. R. Valluri *et al.* (Eds), pp. 2545–2552. Pergamon Press (1984).
44. S. Mostovoy and E. J. Ripling, *J. Appl. Polym. Sci* **10**, 1351–1371 (1966).
45. P. F. Becher and W. L. Newell, *J. Mater. Sci.* **12**, 90–96 (1977).
46. W. D. Bascon and J. L. Bitner, *J. Mater. Sci.* **12**, 1201–1410 (1977).
47. W. D. Bascom, P. F. Becher, J. L. Bitner and J. S. Murday, in: *Adhesion Measurement of Thin Films, Thick Films and Bulk Coatings*, ASTM STP 640, K. L. Mittal (Ed.), pp. 63–81. American Society for Testing and Materials, Philadelphia, PA (1978).
48. A. G. Evans, *J. Mater. Sci.* **7**, 1137–1146 (1972).
49. R. F. Lowell, M.S. Thesis, University of Illinois, Urbana, IL (1980).
50. J. H. Enloe, M.S. Thesis, University of Illinois, Urbana, IL (1981).
51. G. Elssner, W. F. Barisch and R. Pabst, in: *Mechanical Properties of Biomaterials, Advances in Biomaterials, Vol. II*, G. W. Hastings and D. F. Williams (Eds), pp. 265–274. John Wiley, Chichester (1980).
52. C. A. Andersson, in: *Fracture Mechanics of Ceramics, Vol. 6*, R. C. Bradt, A. G. Evans, D. P. H. Hasselman and F. F. Lange (Eds), pp. 497–509. Plenum Press, New York (1983).

53. J. W. Dini and H. R. Johnson, *Met. Finish.* **75**, 42–46, 48–51 (1977).
54. J. W. Dini and H. R. Johnson, in: *Adhesion Measurement of Thin Films, Thick Films and Bulk Coatings*, ASTM STP 640, K. L. Mittal (Ed.), pp. 305–326. American Society for Testing and Materials, Philadelphia, PA (1978).
55. R. J. Dent, J. D. Preston, J. P. Mofta and A. Caputo, *J. Prosthet. Dent.* **47**, 59–62 (1982).
56. W. D. May, N. D. P. Smith and C. I. Snow, *Nature* **179**, 494 (1957).
57. W. D. May, N. D. P. Smith and C. I. Snow, *Paint Manuf.* **27**, 233–224 (1957).
58. W. D. May, N. D. P. Smith and C. I. Snow, *Trans. Inst. Met. Finish.* **34**, 369–382 (1957).
59. S. D. Brown and M. K. Ferber, *J. Vac. Sci. Technol.* **A3**, 2506–2508 (1985).
60. J. Ahn, K. L. Mittal and R. H. MacQueen, in: *Adhesion Measurement of Thin Films, Thick Films and Bulk Coatings*, ASTM STP 640, K. L. Mittal (Ed.), pp. 134–157. American Society for Testing and Materials, Philadelphia, PA (1978).
61. J. Oroshnik and W. K. Croll, in: *Adhesion Measurement of Thin Films, Thick Films and Bulk Coatings*, ASTM STP 640, K. L. Mittal (Ed.), pp. 158–183. American Society for Testing and Materials, Philadelphia, PA (1978).
62. S. Baba, A. Kikuchi and A. Kinbara, *J. Vac. Sci. Technol.* **A4**, 3015 (1986).
63. S. Baba, A. Kikuchi and A. Kinbara, *J. Vac. Sci. Technol.* **A5**, 1860 (1987).
64. A. Kinbara, S. Baba and A. Kikuchi, *J. Adhesion Sci. Technol.* **2**, 1–10 (1988).
65. A. Kinbara, A. Kikuchi, S. Baba and T. Abe, *J. Adhesion Sci. Technol.* **7**, 457–466 (1993).
66. R. M. Jarvinen, T. Mantyla and P. Kettinen, *Thin Solid Films* **114**, 311–317 (1984).
67. B. Y. Ting, W. O. Winer and S. Ramaligam, *Trans. ASME, J. Tribol.* **107**, 472–477 (1985).
68. B. Y. Ting, S. Ramalingam and W. O. Winer, *Trans. ASME, J. Tribol.* **107**, 478–482 (1985).
69. J. C. Wurst, M.S. Thesis, University of Dayton, Dayton, OH (1968).
70. J. L. Vossen, in: *Adhesion Measurement of Thin Films, Thick Films and Bulk Coatings*, ASTM STP 640, K. L. Mittal (Ed.), pp. 122–133. American Society for Testing and Materials, Philadelphia, PA (1978).
71. J. L. Drummond, PhD Thesis, University of Illinois, Urbana, IL (1979).
72. J. Black, R. A. Latour, Jr, and B. Miller, *J. Adhesion Sci. Technol.* **3**, 65–67 (1989).
73. G. Y. Onoda, Jr, Private communication to S. D. Brown (1966).
74. A. Rudnick, C. W. Marshall, W. H. Duckworth and B. R. Emrich, *The Evaluation and Interpretation of Mechanical Properties of Brittle Materials*, AFML-TR-67-316. DCIC 68-3. Air Force Materials Laboratory, OH (1968).
75. G. J. Ewell, in: *Adhesion Measurement of Thin Films, Thick Films and Bulk Coatings*, ASTM STP 640, K. L. Mittal (Ed.), pp. 251–268. American Society for Testing and Materials, Philadelphia, PA (1978).
76. E. Fuller,. Paper No. 1-B-80, presented in the Basic Science Division, 82nd Annual Meeting of the American Ceramic Society, Chicago, IL (1980).
77. R. E. Robertson, Private communication to S. D. Brown (1991).
78. S. K. Holzgraefe and S. D. Brown, Unpublished results.
79. J. Lankford, in: *Fracture Mechanics of Ceramics, Vol. 5*, R. C. Bradt, A. G. Evans, D. P. H. Hasselman and F. F. Lange (Eds), pp. 625–637. Plenum Press, New York (1983).
80. T. A. Michalske, J. R. Varner and V. D. Frechette, in: *Fracture Mechanics of Ceramics, Vol. 4*, R. C. Bradt, D. P. H. Hasselman and F. F. Lange (Eds), pp. 639–649. Plenum Press, New York (1978).
81. T. S. Oh, J. Rodel, R. M. Cannon and R. O. Ritchie, *Acta Metall.* **36**, 2083–2093 (1988).
82. S. D. Brown, Unpublished results (1993).
83. P. F. Becher, W. L. Newell and S. A. Halen, in: *Fracture Mechanics of Ceramics, Vol. 3*, R. C. Bradt, D. P. H. Hasselman and F. F. Lange (Eds), pp. 463–471. Plenum Press, New York (1978).
84. Z. Suo, *Scripta Metall. Mater.* **25**, 1011–1016 (1991).
85. J. W. Hutchinson and Z. Suo, in: *Advances in Applied Mechanics, 29*, J. W. Hutchinson and T. Y. Wu (Eds), pp. 63–191. Academic Press, Boston (1992).
86. P. Gibbs and I. B. Cutler, *J. Am. Ceram. Soc.* **34**, 200–206 (1951).
87. R. J. Charles and W. B. Hilig, in: *Symposium on the Mechanical Strength of Glass and Ways of Improving It*, pp. 511–527. Union Scientifique Continentale du Verre, Charleroi, Belgium (1962).
88. W. B. Hillig and R. J. Charles, in: *High-Strength Materials*, V. F. Zackay (Ed.), pp. 682–705. John Wiley, New York (1965).
89. S. N. Zhurkov, *Int. J. Fract.* **1**, 311–323 (1965).
90. S. M. Wiederhorn, *J. Am. Ceram. Soc.* **50**, 407–414 (1967).
91. S. M. Wiederhorn, *Int. J. Fract.* **4**, 171–177 (1968).

92. S. M. Wiederhorn, *J. Am. Ceram. Soc.* **55**, 81–85 (1972).
93. S. D. Brown, in: *Environmental Degradation of Engineering Materials*, M. R. Louthan, Jr, and R. P. McNitt (Eds), pp. 141–149. Virginia Tech. Printing Department, Balcksburg, VA (1977).
94. S. D. Brown, in: *Fracture Mechanics of Ceramics, Vol. 4*, R. C. Bradt, D. P. H. Hasselman and F. F. Lange (Eds), pp. 597–621. Plenum Press, New York (1978).
95. S. D. Brown, *J. Am. Ceram. Soc.* **61**, 367–368 (1978).
96. A. S. Krausz, *Int. J. Fract.* **14**, 5–15 (1978).
97. S. D. Brown, *J. Am. Ceram. Soc.* **62**, 515–524 (1979).
98. A. S. Krausz, *J. Eng. Fract. Mech.* **11**, 33–42 (1979).
99. R. M. Thomson, *J. Mater. Sci.* **15**, 1014–1026 (1980).
100. A. S. Krausz and J. Mshana, *Int. J. Fract.* **19**, 277–293 (1982).
101. F. H. Ree, T. S. Ree, T. Ree and H. Eyring, in: *Advances in Chemical Physics, Vol. 4*, I. Prigogine (Ed.), pp. 1–66. Wiley-Interscience, New York (1962).
102. H. Eyring, T. S. Ree, T. Ree and F. M. Wanlass, in: *Symposium on the Transient State (Spec. Publ. No. 16)*, pp. 1–22. The Chemical Society, London (1962).
103. H. Eyring and E. M. Eyring, in: *Modern Chemical Kinetics*, H. H. Sisler and C. A. VanderWerf (Eds), Ch. 4. Reinhold, New York (1962).
104. D. P. Williams and H. G. Nelson, *Metall. Trans. A* **3**, 2107–2113 (1972).
105. A. G. Evans, G. B. Crumley and R. E. Demaray, *Oxid. Met.* **20**, 189–212 (1983).
106. L. C. Cox, in: *Residual Stress in Design, Process and Materials Selection*, W. B. Young (Ed.), pp. 109–115. ASM International, Metals Park, OH (1987).

Adhesion Measurement of Films and Coatings, pp. 41–70
K. L. Mittal (Ed.)
© VSP 1995.

Measurement of adhesion for thermally sprayed materials

C. C. BERNDT* and C. K. LIN

*The Thermal Spray Laboratory, Department of Materials Science and Engineering,
SUNY at Stony Brook, Stony Brook, NY 11794-2275, USA*

Revised version received 21 May 1993

Abstract—Thermally sprayed coatings have a distinctive microstructure which can be described as '*a three-dimensional layered structure of discs which are interlaced to form a material of composite nature*'. The coatings are normally greater than 25 μm in thickness and can thus be described as bulk coatings. The minimum microstructural detail would be a single splat (often described as a lamella), which is about 5 μm in thickness and up to 80 μm in diameter. This paper focuses on methods used to define and measure the adhesion of coatings or deposits formed by thermal spray technology. The properties distinguished include those of strength and toughness. Measurements such as the tensile adhesion (according to ASTM C633) and double cantilever beam (DCB) tests will be addressed to illustrate the relevance (if any) of such methods to present industrial practice. Acoustic emission studies have also assessed a function termed as the 'crack density function', i.e. a product of the number of cracks and crack size. Other measuring methods applied to this technology include micro-hardness and scratch testing. The former technique has demonstrated that the material properties of coatings are anisotropic, and the latter method is being considered within the biomedical industry to assess the adhesion of hydroxyapatite to orthopedic prostheses. These techniques, among others, may be used for both fundamental understanding of coating performance (i.e. life prediction and cracking mechanisms) and as tests for quality control.

Keywords: Acoustic emission; adhesion; adhesion measurement; coating failure; degradation; double cantilever beam test; fracture mechanics; indentation; lifetime modeling; scratch test; thermally sprayed coatings.

1. INTRODUCTION

1.1. Formation and structure of thermally sprayed coatings

A variety of thermal spray processes are available to deposit thick coatings for a broad range of applications [1–3]. The processes are essentially similar in that a material is heated up by a gaseous medium and simultaneously accelerated and projected onto the substrate. The family of thermal spray processes includes, among other processes, flame spraying, plasma spraying, vacuum plasma spraying (also called low pressure plasma spraying), high velocity oxygen fuel, arc metallization, and detonation gun spraying. The prime distinctions between these spraying processes are the temperature of the process and the velocity of the thermal source used in the process. These process variables control the nature of the materials that can be sprayed. The techniques also differ with regard to their process economics; that is, factors such as the cost of the equipment, the cost of

---

*To whom correspondence should be addressed.

the feedstock materials and other consumables such as gases, grit blast media, etc. that may limit the viability of a particular process.

Thermal spray technology is not limited to coating substrates but now also encompasses the manufacture of net shapes [4], which can be considered as very thick coatings [up to, say, 1″ (25 mm) in thickness] stripped from the substrate (or 'forming tool') and then used directly as an engineered material [5]. The availability of thermal spray processes enables the production of materials of varying composition and structure–the so-called 'functionally gradient materials'.

Several conferences have been devoted to thermal spray and such proceedings [6–8] cover many of the processing variables associated with this technology. In addition, a quarterly journal called the *Journal of Thermal Spray Technology* is now available that covers the complete engineering and scientific arenas of this technology.

Thermally sprayed coatings consist of a layered structure that is highly anisotropic such that individual splats are oriented parallel to the substrate surface [9]. A microstructural cross-section of a thermally sprayed material is illustrated in Fig. 1, where unmelted particles are embedded within the layered structure and also pores exist either totally inside or between each layer [10]. After formation of the coating, the first property criterion for this coating system is 'How well is the coating adhered to the substrate' and it therefore follows 'How can we evaluate this adhesion strength, especially for coatings which are in service?' The intent of this paper is to describe and then discuss methods of

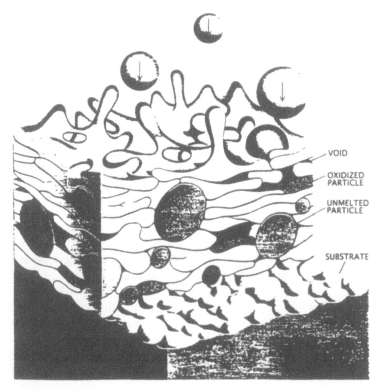

Fig. 1. Microstructure of thermally sprayed coatings [10].

measuring the 'adhesion strength' of thermally sprayed coatings. A strong intent is to always relate such measurements to the ultimate application(s) of the coatings.

1.2. Rationale for measuring adhesion

According to Mittal [11], adhesion can be expressed in various ways. For example, 'basic adhesion' signifies the interfacial bond strength and is the summation of all intermolecular or interatomic interactions. The result of an adhesion test is called 'practical adhesion' and is a function of basic adhesion and many factors—all of which represent the work required to detach a film or coating from a substrate. Many theories or mechanisms for adhesion have been proposed [12, 13]; however, none is fulfilled by all situations and there is no adhesion test available which satisfies all requirements. Therefore the best test method often becomes the one that simulates practical stress conditions.

The term 'adhesion' requires special definition for the purposes of thermal spray coatings. For example, the American Society for Testing and Materials (ASTM) [14] states adhesion as '*the state in which two surfaces may consist of valence forces or interlocking forces or both*'. The theories of adhesion between two materials, in general terms, include mechanisms based on diffusion, mechanical interlocking, electrostatic attraction, physical adsorption, chemical bonding, and weak boundary layers [15]. However, the above global definition cannot be universally applied to thermal spray materials since these coatings can be considered as 'composites' at the microstructural level. Thus, bonding mechanisms for forming an integral coating or net shape will also be complex and may involve adhesion processes which are exclusive of those established for classical joining technology. The basic bonding mechanisms that have been defined for thermal spray coatings can be categorized into three major groups [16] (Table 1).

The specific thermal spraying process will influence the microstructure of the coating and, therefore, it can be inferred that the adhesion strength of the deposit will vary. For example, the high velocity oxygen fuel (HVOF) technique produces a very dense microstructure with porosity typically less than 2% compared to the less than 5% porosity, at best, for a flame sprayed or an atmospheric plasma sprayed material. Thus, factors affected by the spray parameters, including the size and distribution of porosity, oxide content, residual stresses, and macro- or micro-cracks, have an important influence on the performance and eventual failure of the coating system.

Table 1.
The basic bonding mechanisms of thermally sprayed coatings

Mechanical interlocking (anchoring)
Metallic bonds (metal–metal bonds) • dispersion forces • chemisorption and epitaxy • diffusion
Chemical bonds • intermetallic compounds

The service failure mode(s) can be described as interfacial, cohesive, or of mixed interfacial/cohesive failures.* This service failure mode must also be reflected by any testing method which seeks to perform a meaningful quality control test of coatings. It is thus preferable that any laboratory test induces the observed service failures, otherwise they will be of limited application to engineering design. Useful measurements on adhesion strength and interpretation of the test results to predict the service life of coatings or net shapes are the most challenging problem for thermal spray scientists.

1.3. Application of fracture mechanics to adhesion measurement

The adhesion of thermally sprayed coatings is not only an interfacial problem of the individual lamella within a coating, but also concerns examination of the integrity of the interface between the substrate and coating, residual stresses, crack population, and pore size and distribution. Fracture mechanics [17, 18] considers the energy required to initiate or propagate cracks and evaluates the adhesion of the coating system in terms of 'fracture toughness, [19–21]. Four-point bending methods, single-edge notched specimens, double cantilever beam tests, indentation techniques, and other measurements have been employed to assess the adhesion of coating systems [22–24].

The purpose of all these methods is fundamentally the same when they are expressed from the viewpoint of fracture mechanics. The experimental proposition is to establish the equilibrium condition where the elastic energy provided by an external force (as defined by the geometry of the specimen and the applied load) is balanced by the propagation of a stable crack. One form of this energy balance criterion derives the strain-energy release rate, G (in J/m^2), and is defined as:

$$G = \frac{\partial(W - U)}{\partial A} \tag{1}$$

where W is the work done by external forces (in J), U is the elastic energy stored in the system (in J), and A is the crack area (in m^2).

It is convenient to write G as

$$G = \frac{F^2}{2B} \frac{dC}{dL} \tag{2}$$

where F is the force required to extend a crack (in N), L is the crack length (in m), B is the thickness (in m), and C is the compliance (in m/N).

The strain-energy release rate can be related to the fracture toughness by

$$K = \sqrt{\frac{EG}{1 - \nu^2}} \tag{3}$$

where E is the elastic modulus (in MPa) and ν is Poisson's ratio (dimensionless).

---

*Thermal spray engineers should recognize that the term 'interfacial' is used in preference to 'adhesive' to avoid confusion with an 'adhesive' which is used to join materials (i.e. adhesive will be used as a noun rather than as an adjective).

When G exceeds a critical value, G_c, crack propagation occurs and failure of the coating system results. The evaluation of K assumes that both the elastic modulus and Poisson's ratio of the material are known. The physical representation of the above equations is that the change in compliance of the specimen controls the energy input into the creation of new fracture surfaces. Thus, the corollary of this theory is that the crack path and rate of crack growth can be controlled very precisely by appropriate design of the specimen.

Thus, 'mechanical adhesion' can be evaluated in terms of adhesion strength or fracture toughness. Such measurements of coating–substrate adhesion have been reviewed recently and are listed in Table 2 [25–27]. However, test methods for thermally sprayed coatings are restricted. The techniques to be discussed in this paper include the double cantilever beam test, acoustic emission technology, microhardness assessment, and the scratch test. The tensile adhesion test (TAT) is not covered in detail but will be referenced.

Table 2.
Methods that can be used to determine coating–substrate adhesion

Qualitative	Quantitative
Mechanical methods	
Scotch tape test	Direct pull-off method
Abrasion test	Laser spallation test
Bend and scratch test	Indentation test
	Ultracentrifugal test
	Scratch test
Non-mechanical methods	
X-ray diffraction	Thermal method
	Nucleation test
	Capacitance test

2. ADHESION MEASUREMENTS

2.1. Methods

One difficulty with any mechanical property assessment of coatings is how to attach a loading device to the coating without influencing the property that is being measured. Some investigators [28] have approached this problem by manufacturing a pin or disk that could be removed from a mating component (Fig. 2). The surface of this assembly was thermal sprayed and then the pin or disk, depending on the assembly used, removed. The force at fracture was used to find a parameter termed 'the adhesion strength' of the coating. The corresponding shear test is performed by spraying the outside diameter of a cylinder or the head of a pin (Fig. 3) [29]. One potential difficulty with the above test is that the duplex nature of the specimen assembly may influence the coating quality since the heating and cooling behavior of the deposit will be affected by the interface between the two components. Another feature of both the tensile and the shear tests is that the fracture mode is ambiguous since mixed mode failure often occurs.

Other geometries that do not need an adhesive are illustrated in Figs 4 [30] and 5 [31]. These tests have not been widely implemented outside of their

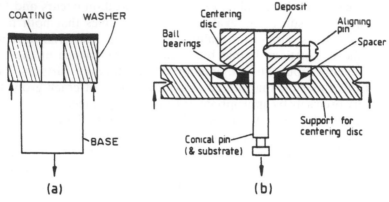

Fig. 2. (a) Specimen for determining adhesion strength; (b) centering device and specimen in cross-section [28].

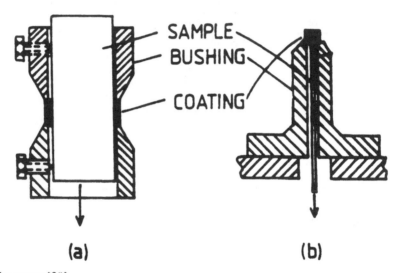

Fig. 3. Shear tests [29].

laboratories of origin. The method of spraying two adjoining conical parts is expected to incorporate a failure which has a large shear component, whereas the other shear test may exhibit either shear failure parallel to the substrate surface or shear failure through the coating thickness.

The adhesion properties of coatings were investigated on the microscopic level [32] by shearing individual particles from the substrate surface (Fig. 6). This study considered that adhesion to the substrate arose from interactions across the particle–substrate interface and a 'strength of growth rate constant' (K') was defined as

$$K' = -\frac{1}{t} \ln\left(1 - \frac{N(t)}{N_0}\right) \tag{4}$$

where t is the time, $N(t)$ is the number of atoms that react during the interaction

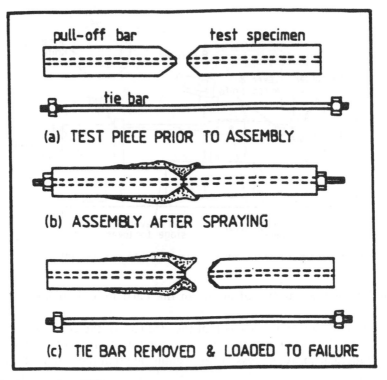

Fig. 4. Adhesion test piece [30].

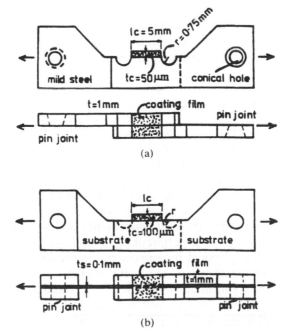

Fig. 5. (a) Shear stress deformation; (b) critical shear stress of adhesion [31].

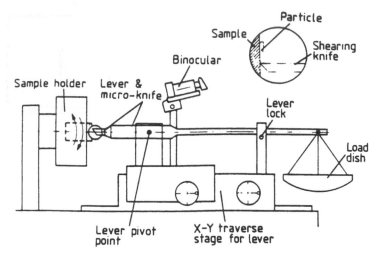

Fig. 6. An instrument for measuring the adhesion strength of the particles [32].

time, and N_0 is the number of atoms in the particle and substrate that are in contact.

The upshot of this analysis was that the strength of the coating at some inter-action time of 't' was compared with the maximum strength of the coating at the end of the thermal spray process. It was established that the extent of the particle/interface reaction increased with both increasing particle pressure and temperature, and this agreed with the experimental observations that coating adhesion also increased under these conditions. Further theoretical work [33] has treated adhesion as a stochastic process which depends on the formation characteristics of the first monolayer of material. The coating buildup is treated as a statistical process involving input data from the thermal spray processing parameters such as the relative motion between the torch and substrate (i.e. the spray pattern), the velocity and temperature distribution of the thermal source, and the particle size.

2.2. Application of adhesion measurements

The reason for performing adhesion measurements can be brought into focus by examining how such experiments are used to ascertain the utility of coatings. For example, bond strength measurements allow optimization of different grit blast media and the angle of grit blasting, as well as establishing the best coating thickness for aluminum coatings (Fig. 7) [34]. Other workers [35] have optimized ceramic compositions and plasma spraying parameters with respect to strength. Thus, the historical basis for these measurements is significant. In recent years, with the adoption of design of experiment methods [36], bond strength measurements along with other physical measurements (roughness, porosity, etc.) have been used to establish specifications to select coatings.

The strength of the specimen may not be the same as the strength of the engineered coating. It can also be difficult to establish how process-induced residual stresses influence the strength measurements. The adhesion measurement is now taken as a control parameter that can be used as a guide to optimize

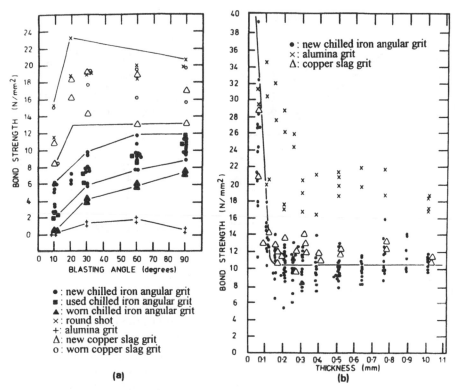

● : new chilled iron angular grit
■ : used chilled iron angular grit
▲ : worn chilled iron angular grit
× : round shot
+ : alumina grit
△ : new copper slag grit
○ : worn copper slag grit

(a)

● : new chilled iron angular grit
× : alumina grit
△ : copper slag grit

(b)

Fig. 7. Variation of the bond strength of aluminum coatings with (a) the blasting angle and (b) the coating thickness [34].

the many process parameters that are involved during thermal spraying. Often the tensile adhesion test is performed as a quality control test and numerous references can be found in thermal spray conference proceedings [6–8].

2.3. Fracture mechanics approach

2.3.1. Overview. The fracture mechanics approach to evaluate crack propagation is based on defining adhesion in terms of a stress intensity factor K or strain-energy release rate G. Methods of measurement include the double cantilever beam (DCB) test; the double torsion test; the bending (three-point or four-point), single-edge notched test; and the compact tension test. Among these, the DCB method allows multiple fracture toughness readings by testing a single specimen and will be discussed in the following section.

The fracture mechanics mode of failure is also a material property that can not only be controlled (to some degree), but must be quantified for engineering design purposes. Thus, a major justification for a fracture mechanics test is that a mode I (tensile) or a mode II (shear) test can be performed and the response of pre-existing flaws ascertained. There is also the possibility of carrying out mixed mode tests that may better replicate a variety of service conditions that the coating may experience throughout its lifetime.

The double torsion test [37, 38] has been applied to thermally sprayed coatings (Fig. 8). The prime advantage of this test is that there is no need to measure the

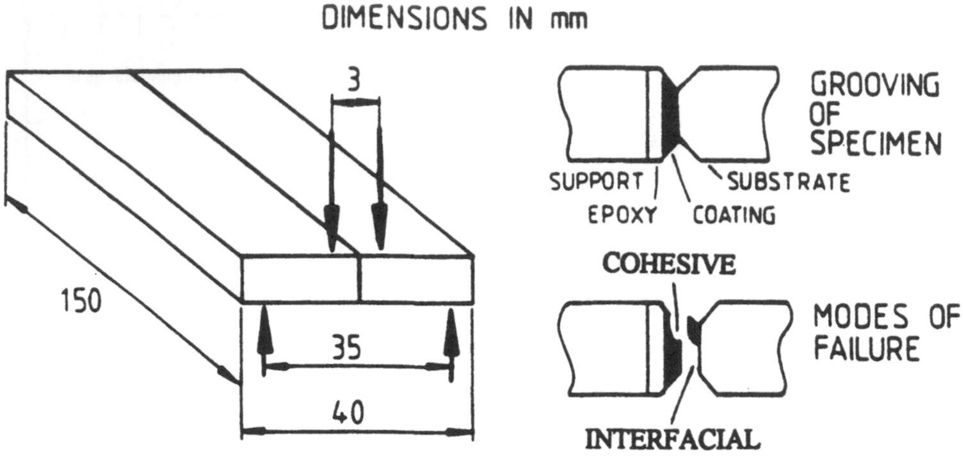

DIMENSIONS IN mm

Fig. 8. Double torsion geometry and test configuration.

crack length because cracking occurs at a constant strain-energy release rate or stress intensity factor. Specimen manufacture involves incorporating the coating into an arrangement so that a torque can be applied to the crack front. The short-comings of the test are that mode I cracking, where the crack front is orthogonal to the crack propagation direction, was never verified despite several attempts. It also appeared that both arms of the double torsion specimen were deflecting at a constant angle along their entire length. Both these conclusions are reflected in the fracture surface morphology as indicated in the inset diagram (on the bottom right) of Fig. 8 where a mixed mode of failure is observed.

The four-point bending test has also been applied to measure the fracture toughness of ceramics joined with metals [39]. The configuration is quite simple and one measurement can be obtained from each test.

2.3.2. Double cantilever beam (DCB) test. A major advantage of the DCB test is that it may have wide applicability to the design engineer; however, this ideal comes at the expense of complex specimen preparation and more sophisticated experimental techniques than the quality control departments of thermal spray shops may be prepared to undertake.

Many configurations of this test are available (Fig. 9) and an example of the technique on a thick film conductor of alumina is shown in Fig. 10 [40]. The applied moment method was used in this case since the crack length, a difficult property to measure for these opaque coatings, was not required. These studies are analogous to those on thermal spray coatings (Fig. 11) [41], where interfacial and cohesive modes of failure were distinguished in terms of fracture toughness. Another DCB geometry arrangement is illustrated in Fig. 12 [42], where a single contoured arm was used.

The general nature of the force vs. displacement requirement of a DCB experiment is shown in Fig. 13. The elastic energy of the specimen arms controls the amount of energy which is transferred to the creation of new crack faces within the locus of fracture. Thus, crack growth, as indicated by the dashed line in Fig. 13, will continue until a stable condition is reached and crack propagation

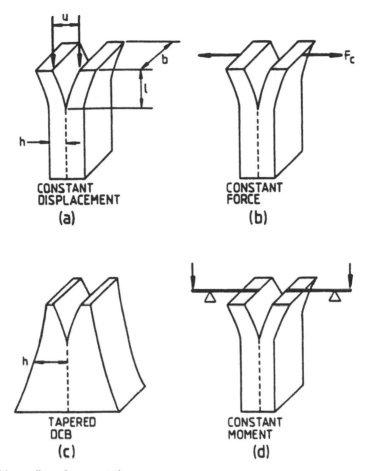

Fig. 9. Double cantilever beam test pieces.

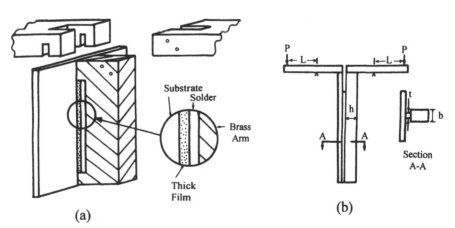

Fig. 10. Applied moment DCB specimen for measuring the adhesion of thick films. (a) Specimen preparation; (b) testing arrangement [40].

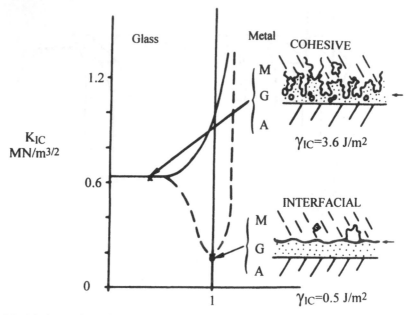

Fig. 11. Model of stress intensity profile through the film and substrate [41]. (In the detailed figure, M is metal, G is glass, and A is alumina.)

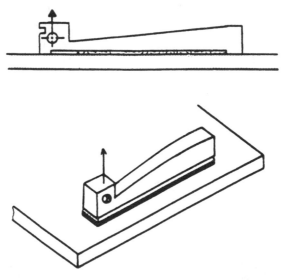

Fig. 12. Specimen with DCB geometry in which one arm is contoured [42].

stops at that point. The new slope of the force–displacement curve (i.e. the compliance) indicates the magnitude of the new crack length, and the area enclosed within the force–displacement curve is related to an energy transfer (from the DCB arms to the crack) during crack growth. Thus, the versatility of the DCB method is that the energy input into the coating can be controlled by altering the elastic modulus and/or geometry of the DCB arms.

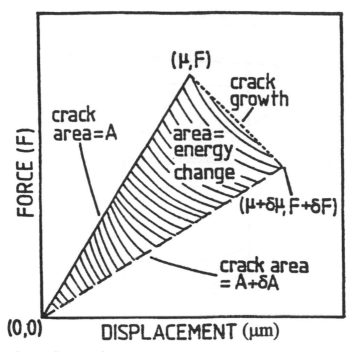

Fig. 13. Energy change when a crack grows.

It is also necessary to verify that the arms of the DCB are indeed bending and that there is no rigid movement of the adherends (Fig. 14a). It was established that true bending of the DCB arms did take place as indicated in Fig. 14c, rather than the ideal situation as depicted in Fig. 14b. The composite DCB geometry which incorporated an adhesive joint does not directly obey the Mostovoy formulation [43] but a modified equation that incorporated displacement at the crack-tip and deformation beyond the crack-tip fitted reasonably well to the theory (Fig. 15).

2.3.3. DCB test for thermally sprayed materials. In the basic DCB test on coatings (Fig. 16), a tension force is applied to the specimen assembly and displacement is measured by an extensometer placed on the arms [22]. When cracking initiates, as noticed by load decreasing with extension increasing, the DCB is unloaded. Several loading/unloading sequences are performed until complete failure of the specimen, at which point the morphology can be examined by optical microscopy and scanning electron microscopy (SEM). Either one of the three failure modes, i.e. interfacial, cohesive, and mixed, can be observed and both inter- and trans-lamellar cracking is exhibited. The locus of fracture can be controlled by grooving the edges of the DCB as shown in the inset (right-hand side) of Fig. 16. Another feature of the DCB method is that the method is not limited by the strength of the adhesive, as is the case for tensile adhesion tests but since it is a fracture mechanics test, it is only necessary for the fracture toughness of the adhesive to be greater than that of the coating. This is a relatively easy condition to satisfy.

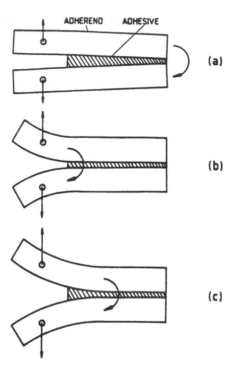

Fig. 14. Bending modes of a DCB specimen. (a) Rigid arms; (b) cantilever beams built in at the crack-tip; (c) rotation of cantilever beams at a point beyond the crack-tip.

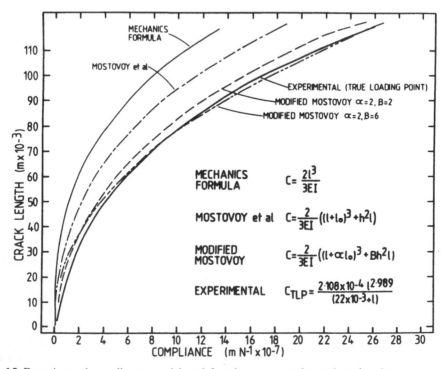

Fig. 15. Experimental compliance–crack length functions compared to various theories.

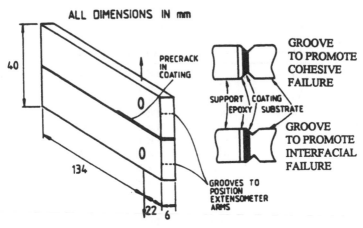

Fig. 16. DCB specimens used for adhesion measurement. The inset shows the grooving procedure to promote either interfacial or cohesive failure.

2.3.4. G_{IC} determination. The critical strain-energy release rate, G_{IC}, is determined from equation (2). The compliance values, dC/dL, are determined from the displacement at the loading points (C_{LP}), which differs from the values recorded by the extensometer which is termed the crack opening displacement (C_{COD}). An experimentally determined calibration curve is necessary to calculate dC/dL [22, 44]. Several laboratories have used this technique [45, 46] to generate G_{IC} data. The results are compiled with fracture indentation measurements in Table 3. The mean G_{IC} values for ceramic coatings exhibit a large range but generally they can be considered to lie below ~ 100 J/m², whereas G_{IC} values for metallic coatings are greater than 100 J/m².

Table 3a.
G_{IC} values (in J/m²) for thermally sprayed alumina coatings[a]

Material	Mean	SD	Method	Note	Ref.	Appendix
Al_2O_3	12	2	DCB	IF	22	Metco 101
Al_2O_3	16.1	5.3	DCB	CF	23	Metco 105
Al_2O_3	21	5	DCB	CF	22	Metco 101
Al_2O_3	31.6	9.3	DCB	CF	23	Sealed
Al_2O_3	34	9	DT	CF	22	Metco 101
Al_2O_3	49.8	14.0	DCB	CF	23	PC-WAF
Al_2O_3	52.7	14.5	DCB	CF	23	As sprayed
Al_2O_3	58	16	DCB	Int. BC Fail.	22	Metco 101
Al_2O_3	78	29	DT	IF	22	Metco 101
Al_2O_3	209	—	Ind.	Load: 47 N	23	—
Al_2O_3	189	—	Ind.	Load: 98 N	23	—
Al_2O_3	97.6	—	Ind.	Load: 147 N	23	—
Al_2O_3–2.5 wt% TiO_2	15.9	5.6	DCB	CF	23	—
Al_2O_3–2.5 wt% TiO_2	218	—	Ind.	Load: 73.5 N	23	—
Al_2O_3–2.5 wt% TiO_2	268	—	Ind.	Load: 147 N	23	—
Al_2O_3–2.5 wt% TiO_2	301	—	Ind.	Load: 98 N	23	—
Al_2O_3–40% TiO_2	48.7	15.8	DCB	IF	23	—

[a] For data given in K, equation (3) is used to convert K into G by assuming E = 48 GPa: $\nu = 0.25$.

SD: standard deviation; IF = interfacial failure; CF = cohesive failure; Int. BC Fail.: interfacial bond coat failure; Ind.: indentation.

Table 3b.

G_{IC} values (in J/m^2) for thermally sprayed zirconia coatings[a]

Material	Mean	SD	Method	Note	Ref.
ZrO_2–10 CeO_2	11.4	3.6	DCB	IF	21
ZrO_2–15 CeO_2	74.1	22.9	DCB	IF	21
ZrO_2–15 CeO_2	125.4	51.1	DCB	CF	21
ZrO_2–6 Y_2O_3	49.9	16.7	DCB	IF	21
ZrO_2–6 Y_2O_3	95.5	39.4	DCB	CF	21
ZrO_2–6 Y_2O_3	148.3	52.3	DCB	Mixed	21
ZrO_2–20 Y_2O_3	30.0	10.4	DCB	IF	21
ZrO_2–20 Y_2O_3	69.7	31.8	DCB	CF	21
ZrO_2–20 Y_2O_3	43.2	19.6	DCB	Mixed	21
ZrO_2–8 Y_2O_3	11.1	1.3	Ind.	Load: 50 N	68
ZrO_2–8 Y_2O_3	10.4	1.3	Ind.	Load: 100 N	68
ZrO_2–8 Y_2O_3	16.0	1.0	Ind.	Load: 50 N	68
ZrO_2–8 Y_2O_3	21.2	1.2	Ind.	Load: 100 N	68
ZrO_2–8 Y_2O_3	5.7	0.9	Ind.	Load: 50 N	68
ZrO_2–8 Y_2O_3	3.8	0.5	Ind.	Load: 100 N	68
ZrO_2–8 Y_2O_3	5.3	0.5	Ind.	Load: 50 N	68
ZrO_2–8 Y_2O_3	5.8	0.7	Ind.	Load: 100 N	68
YSZ	24.0	6.4	Ind.	Load: 588 N	65

[a] For data given in K, equation (3) is used to convert K into G by assuming $\nu = 0.25$. Mixed: mixed mode failure.

Table 3c.

G_{IC} values (in J/m^2) for thermally sprayed coatings of some metals and ceramics[a]

Materials	Mean	SD	Method	Note	Ref.
Mild Steel	116	21	DCB	IF	22
Mild Steel	261	71	DCB	CF	22
Ni–Al	319	95	DCB	CF	22
Cr_2O_3	43.0	5.2	Ind.	Load: 9.8–98 N	64
Spinel	47.0	12.7	Ind.	Load: 294 N	65
Ti	6.79	0.34	4-P Bending	With stress	39
Ti	2.98	0.19	4-P Bending	Without stress	39
Ni–20% Al	362	16	DCB	CF	23

[a] For data given in K, equation (3) is used to convert K into G by assuming $\nu = 0.25$.

2.4. Acoustic emission

2.4.1. Background. 'Acoustic emission (AE) is a term describing a class of phenomena whereby transient elastic waves are generated by the rapid release of energy from localized sources within a material.' [47]. The energy usually arises from one or more sources [48], which include phase transformation, plastic deformation, corrosion, and crack initiation and growth [49]. An AE event is detected by a piezoelectric transducer when energy is released from the material. The output is amplified and then features of the AE signal such as the ring down count, rise time, and/or pulse height are subjected to analysis. Often a multi-channel system is used to examine different energy levels and the signal may also be digitized or integrated for an energy analysis.

Special interest lies in formulating crack initiation and growth criteria which are based on the microstructural design of coatings. This is important because microstructural features can be quantitatively determined [50] by image analysis methods and leads to the conclusion that the microstructure of coatings can be 'controlled' by the thermal spray process. AE technology has been combined with fracture mechanics measurements [51–53] or thermal tests [54, 55] to characterize coating properties. AE technology has been applied to better understand failure mechanisms and to predict lifetimes [56–58]. It has also been applied to quality control and in-service monitoring.

2.4.2. Crack density function. A thermal spray coating has a very rich micro-structure, and both macro- and micro-cracking, among other sources, can release energy during coating service. A difficulty is that the AE response is over-whelming with regard to acquiring data and this often limits the AE method to be a qualitative technique since individual AE response-to-coating morphology correlations cannot be made. However, quantitative AE analysis may still be possible through a test procedure which is combined with calibration [59, 60]. For example, studies on thermal barrier coatings that were subjected to heat cycling changed from a systematic response to a stochastic response during a certain thermal cycle. This was considered to be indicative of a change in cracking response from micro- to macro-cracking since the change in AE response was correlated to the observation of the formation of large cracks (Fig. 17) [61].

The record of AE response for coatings will be a combination of all possible noise origins, and a 'crack density function' (CDF) [62] which incorporates both the number of cracks and the size of cracks has been proposed. It is found that macrocracking events tend to occur at low values of the CDF. Figure 18 shows an example of a CDF analysis for two coatings that were prepared to exhibit different behaviors. The essential details are that one sample (indicated by the filled-in parts of the histogram) exhibited a lower frequency of the CDF function and this is indicative of a lower degree of cracking, in terms of both the number of cracks and the size of cracks (since the CDF incorporates these physical charac-teristics of coatings). At failure and after failure (Figs 18b and 18c), the frequency of these events increased.

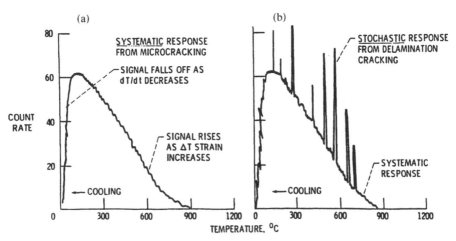

Fig. 17. Schematic diagram of AE effects. (a) Typical cooling cycle; (b) failure during cooling cycle.

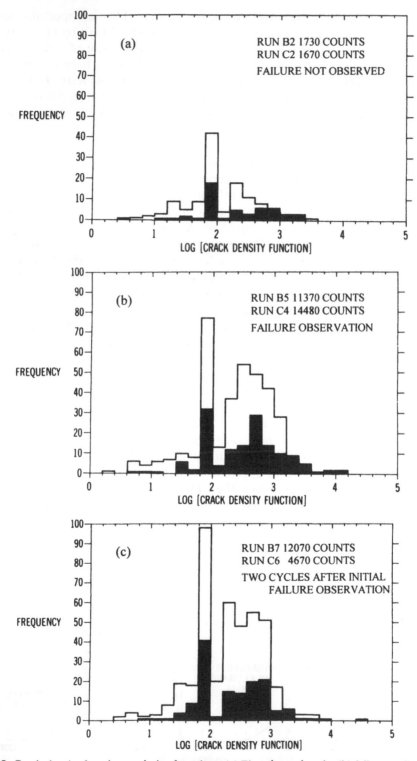

Fig. 18. Crack density function analysis of coatings. (a) First thermal cycle; (b) failure cycle; (c) two cycles after failure.

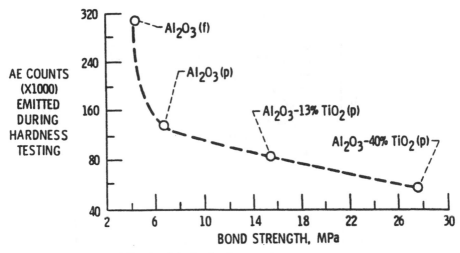

Fig. 19. Bond strength vs. AE emitted during hardness testing.

AE methods have also been use in conjunction with mechanical property measurements. For instance, it has been established that the number of AE counts emitted during a hardness test increases as the density of the material decreases. Figure 19 indicates a number of processes and materials where it is generally understood that flame spraying (indicated by 'f') will produce a less dense deposit than plasma spraying ('p') and also that additions of titania to alumina increase the deposit density. Thus, an intuitive interpretation is that the most dense material exhibits the least cracking behavior and this is reflected in a lower AE response.

Similar correlations have been proposed on the basis of AE measurements performed during tensile adhesion tests (TATs). The AE count accumulated during a TAT is graphed with respect to the so-determined bond strength in Fig. 20. The coatings which incorporate metallic constituents exhibit an activity lower than that of the non-metallic coatings (at equivalent bond strengths). Therefore, a physical interpretation of the mechanical response is that the metallic materials allow more plastic deformation than the ceramic materials. This simple explanation relates well to the general understanding of bulk material behavior, i.e. ceramics are more brittle than metals. However, one caution is that such correlations between bulk material properties and thermally sprayed materials may not be correct since these thick coatings are formed by a rapid solidification process.

The purpose of this discussion is to show that adhesion and cracking mechanisms are symbiotic material properties that can be linked by AE processes. Thus, in a very broad sense, a study and understanding of cracking mechanisms will lead to real improvements in maximizing the adhesion of coatings. For example, consider relating the AE response during a TAT to the cracking and deformation behavior of the specimen assembly. Figure 21 is a schematic diagram which indicates that the strain in the bond coat (the metallic constituent) is always greater than the ceramic strain, although the absolute extension in the ceramic layer is greater than that of the bond coat. The overall view is that cohesive failure occurs by many microcracks throughout the material, whereas interfacial failure has a lower density of cracks. The other implication

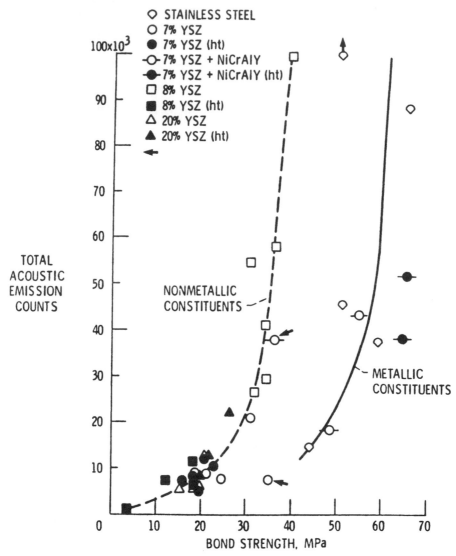

Fig. 20. Bond strength vs. AE emitted during tensile testing.

from Fig. 21 is that cracking during a TAT always begins at the edges of the material and that flaws in this region may dictate the so-determined strength value since they are the weakest link.

2.5. Microhardness assessment

2.5.1. Indentation fracture toughness. Indentation techniques are often used as surface characterization tests. The hardness, implying the resistance of a material to permanent indentation, is measured by a sharp or blunt indenter. Specific crack patterns can be observed [63] in the material at certain loads on the indenter. The indentation method allows the fracture toughness of materials [64–66] and, in particular for thermal spray coatings, the properties of the interface between the

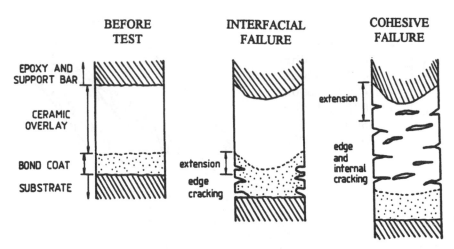

Fig. 21. Schematic diagram to illustrate the coating deformation response during the interfacial and cohesive failure modes.

coating and substrate to be measured. The coating toughness has been measured by the Vickers indentation test [67, 68], where it is necessary to measure the indent diagonals and the crack lengths produced during the test.

The results of several workers have been summarized in Table 3. The indentation fracture toughness measurements tend to be greater than those obtained from DCB tests. Aspects of these measurements that lead to critical discussion are mainly based on the application of the indentation theory to thermal spray coatings since these materials are highly anisotropic. Thus, obtaining a symmetrical crack pattern during any test is never assured, since the coating microstructure has many features that influence their formation and propagation, and therefore the so-determined values are often highly variable. For example, the Weibull modulus of an alumina coating is 0.5 and the modulus of an alumina–titania coating is 0.8. The other main point, as will be discussed in the following section, is that coatings are highly anisotropic throughout their thickness and thus randomly placed indentation fracture mechanics tests would not be expected to have consistent values.

2.5.2. Anisotropy of thermally sprayed coatings. The microstructure of thermally sprayed coatings is a mix of lamellae, pores of varying geometry, and oxides, etc. It is recognized that the coating is not homogeneous and microhardness measurements can be used to examine any anisotropy. The mean values of microhardness and the distributions of data sets across the coating thickness change with respect to the test position [69]. Hence, characterizing thermally sprayed coatings by only their hardness is of limited value, since hardness depends on the precise location of the measurement. However, the microhardness measurements can quantify the material property variation in the specimen if the Weibull modulus is determined. In this fashion it was found that the variability of surface properties is greater compared to the properties throughout the specimen cross-section. The morphology of microhardness indents also changes and this feature can be used to study the variation in homogeneity and stress concentration within the specimen.

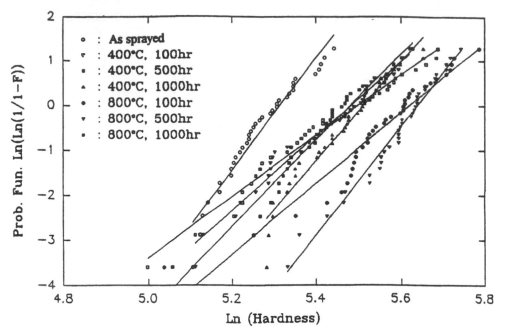

Fig. 22. Weibull plots for microhardness data for as-sprayed and aged samples within the bond coat.

As shown in Fig. 22, the Weibull modulus fluctuates and suggests that the data distributions are a reflection of microstructural changes in the bond coat. For example, the variation in 'm' may imply the formation and distribution of oxides and crack networks.

2.5.3. Interpretation to lifetime. Microhardness measurements have been used to monitor coating behavior after thermal treatments [70]. For example, a series of thermal barrier coatings (a NiCoCrAlY intermetallic bond coat and Ce-stabilized zirconia layer) aged at different times and temperatures (at 400 and 800°C for 100, 500, and 1000 h) were tested for microhardness to assess any material property changes [71]. The major failure mechanisms of thermal barrier coatings are oxidation within the bond coat and cracking due to thermal expansion mismatch within the coating system. This can be reflected by the hardness variation and, in the future, a failure model and lifetime prediction may be based on the analysis of such data.

Temperature effects were noticed in the coating systems. For ceramic coatings, microcracks produced by thermal expansion mismatch may have different sizes and densities. These cracks will be responsible for the variation of mechanical properties. At the same time, the oxide film surrounding the lamellae within the bond coat should have different thicknesses according to the oxidation kinetics at various temperatures. These oxide films, though they may contribute to the increase of microhardness, decrease the adhesion strength of the bond coat. Schematic illustrations of mechanisms that cause coating variation are presented in Fig. 23. Non-monotonic response of hardness and the low Weibull modulus imply complex processes such as stress relaxation, growth of oxides within the bond coat, and phase change.

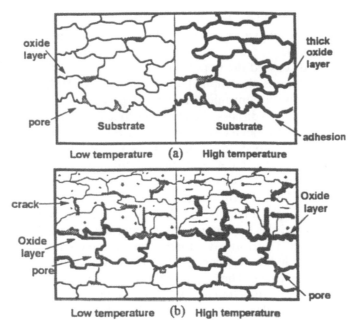

Fig. 23. Physical model of (a) the bond coat–substrate interface and (b) the bond coat–ceramic coating interface.

Thermal cycling of TBCs not only deteriorates the strength of the material at and near the bond coat interface, but also reduces the reliability of the ceramic thermal barrier layer. Microhardness measurements can be used to quantify the material property variation in the specimen. Weibull modulus values obtained in the study show that the reliability of the ceramic coating decreases from 10.5 before cycling to 5.5 after thermal cycling. The wide scatter in the hardness data indicates the variable nature of stress concentration at the test location.

Thus, increasing the reliability of coatings by controlling the variable nature of the material properties requires that the mechanical response throughout coatings be precisely quantified. Microhardness has been selected since this has been used by many authors not only to characterize specific coatings, but also to compare coatings formed from different feedstock materials, spray processs, and process parameters, as well as many other properties.

2.6. Other methods

2.6.1. The scratch test. The scratch test, originally studied by Benjamin and Weaver [72], is often used to characterize thin hard coatings, such as TiN and TiC [73, 74]. In this test, a loaded Rockwell C diamond stylus is drawn across the coating surface under either constant or gradually increasing load. In one variation of the test, the AE is also monitored during the scratching procedure so that the critical load 'L_C' for failure can be measured. The failure morphology is examined by optical microscopy or scanning electron microscopy. If interfacial, cohesive, or a mixed failure mode is observed, then L_C is used as a qualitative value of coating–substrate adhesion.

Three contributions to coating loss have been identified for the scratch adhesion test [75–78], these being an elastic/plastic indentation (a ploughing component), and internal stress component, and a tangential frictional stress (an adhesion component).

Bull *et al.* [76] discussed the importance of frictional drag and suggested that under certain limitations the applied load, together with the scratched cross-sectional area, can be a convenient means of predicting the adhesion of thin coatings. Sekler *et al.* [77] discussed techniques to determine the critical load, and the failure modes in the scratch test were recently reviewed by Bull [78].

The scratch test has been applied in the evaluation of thermally sprayed coatings [52, 79–82]. The major problem for utilization is that all the theories were developed based on thin coatings and may no longer be appropriate for bulk coatings. Das *et al.* [52], in studying plasma sprayed yttria-stabilized zirconia (YSZ) coatings, proposed a method for the determination of the critical load and discussed the effect of the loading rate dL/dt and the scratch table speed dx/dt. Beltzung *et al.* [81, 82] performed scratch tests on the cross-section of alumina-based coatings. A half-cone-shaped fracture was formed as the indenter approached the free surface. The height of this cone can be related to the cohesive strength or intrinsic fracture toughness of the coating. Interfacial cracking may also occur and can be utilized as a measure of the adhesion strength.

2.6.2. Some tests not covered. There are many other methods that can be employed to evaluate adhesion in the qualitative or quantitative sense. The following are some examples:

(1) wear tests [83, 84], which are related to the interfacial or cohesive strength of coatings;
(2) thermal tests [85, 86] during thermal cycling and thermal shock protocols that influence the adhesion strength during heating and cooling processes;
(3) shear tests [87, 88], which may best reflect the in-service conditions;
(4) modified short bar [89] and crack-opening displacement methods [90] for fracture toughness tests.

There is still no ideal adhesion test which can satisfy all requirements. Modifications of existing techniques and designing new methods can further improve adhesion tests.

3. DEGRADATION AND FAILURE OF THERMALLY SPRAYED COATINGS

3.1. Failure mechanisms

Thermally sprayed coatings have been used widely in applications varying from biomaterials to thermal barrier coatings [91]. Coating failure can occur by one or more mechanisms such as surface damage (e.g. wear or corrosion), elastic or plastic deformation, fracture, etc. The degradation or failure of coatings is fundamentally related to the decrease of adhesion and cohesion strength, both of which cause spallation. Fundamental studies on failure mechanisms, especially for TBCs, have been discussed by NASA [92].

Thermal barrier coatings, usually comprising a metallic bond coat and a ceramic coating, endure detrimental thermal and chemical environments. General failure modes include thermal–mechanical ceramic failure, oxidation bond coat failure, hot corrosion, erosion, and fatigue [93]. The thermal variations and inelastic strain due to interfacial oxidation, which leads to crack propagation and coating spallation, should exacerbate coating failure. Meanwhile, phase transformation and bond coat plasticity (or pseudo-elasticity) may also contribute to these mechanism(s). It has been suggested that failure is the result of slow crack growth and microcrack link-up within the ceramic which takes place in the ceramic layer close to the bond coat interface.

Chang *et al.* [94] used finite element analysis to calculate the stress field within a hypothetical wavy interface and found that radial stress would promote crack propagation. The stress owing to the thermal expansion mismatch can be estimated as [95]

$$\sigma_{x,y} = \Delta\alpha(T_{\text{cool}} - T_{\text{hot}})\frac{E_c}{(1 - \nu_c)} \qquad (5)$$

where $\Delta\alpha$ is the difference in thermal expansion coefficients between the substrate and coating, T_{cool} is the lower temperature to which a coated specimen cooled, T_{hot} is the upper temperature of a stress-free state, E_c is the elastic modulus of the ceramic, and ν_c is Poisson's ratio of the ceramic.

A representation of the thermal–mechanical properties resulting from the coating splat structure is shown in Fig. 24. It has generally been recognized that the coating failure is 'time-at-temperature' dependent, especially for oxidation. Macro-and micro-cracking will decrease the adhesion strength of the system and thus appropriate interpretation of the data obtained from adhesion measurements may be beneficial to improving the reliability and durability of the coatings.

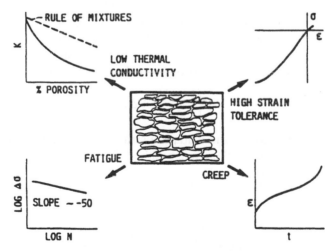

Fig. 24. Schematic representation of thermal–mechanical properties resulting from the coating splat structure of a plasma-sprayed thermal barrier coating. The diagram represents the influence of microstructure on the thermal conductivity, the stress *vs.* strain response, fatigue, and creep properties [93].

3.2. Lifetime modeling

A thermal barrier coating life model was proposed by Miller [95] based on the assumption that oxidation is the 'single most important factor' that limits coating lifetime. An oxidation-based model was used to calculate the cycles to failure as a function of heating cycle duration. The coating life can be expressed by the oxidative weight gain and oxidation-induced strain as

$$\sum_{N=1}^{N_f} \left[\left(1 - \frac{\varepsilon_r}{\varepsilon_f} \right) \left(\frac{W_N}{W_C} \right)^m + \frac{\varepsilon_r}{\varepsilon_f} \right]^b = 1 \tag{6}$$

where N is the number of cycles, N_f is the number of cycles to failure, ε_r is the radial strain, ε_f is the failure strain, W_N is the oxidative weight gain at cycle N, W_C is the critical weight gain which would lead to failure in a single cycle, m is the relationship between effective strain and weight gain, and b is the subcritical crack-growth exponent.

The NASA-sponsored HOST program contributed more effort to model the TBC life [96–98]. Hillery et al. [96] used time-dependent, nonlinear finite element analysis to model the stress and strain within the coating system and expressed the life model as

$$\Delta\varepsilon_{RZ} + 0.4\Delta\varepsilon_R = 0.121 \, N_f^{-0.486} \tag{7}$$

where $\Delta\varepsilon_{RZ}$ is the shear strain range, $\Delta\varepsilon_R$ is the normal strain range, and N_f is the number of cycles to failure.

Strangman et al. [97] expressed the TBC life as a function of bond coat oxidation, zirconia transformation, and damage due to molten salt deposits. The empirical equation was

$$\text{TBC life} = \frac{(t^{0.25} + 0.181)\text{MTBREF}}{\{\exp[-0.015 \, T + C_1]\}^{-1} + \{\exp[-0.041 \, T + C_2]\}^{-1}} \tag{8}$$

where MTBREF is the multi-temperature burner rig experience factor; T is the temperature (in K); t is the time; and C_1 and C_2 are constants.

DeMasi et al. [98] considered the fatigue performance and expressed the relationship between strain and the oxide layer thickness, δ, as

$$\Delta\varepsilon_f = \Delta\varepsilon_{fo} \left(\frac{1 - \delta}{\delta_c} \right)^c + \Delta\varepsilon_i \left(\frac{\delta}{\delta_c} \right)^d \tag{9}$$

where $\Delta\varepsilon_f$ is the strain, $\Delta\varepsilon_{fo}$ is the oxidation strain, $\Delta\varepsilon_i$ is the inelastic strain, δ is the oxide thickness at a particular cycle number, δ_c is the critical oxide thickness, and c and d are constants.

Recently, Meier et al. [99] studied TBC deposited by the electron beam PVD fatigue life model and gave

$$N = \left[\left(\frac{\Delta\varepsilon_{ff}}{\Delta\varepsilon} \right) \left(1 - \frac{\delta}{\delta_c} \right)^c + \left(\frac{\delta}{\delta_c} \right)^c \right]^b \tag{10}$$

where N is the cyclic life, $\Delta\varepsilon_{ff}$ is the furnace failure strain, $\Delta\varepsilon$ is the strain range, δ is the oxide thickness at a particular cycle number, δ_c is the critical oxide thickness, and b and c are constants ($b = 7.64$, $c = 1.0$).

The lifetime modeling of active in-service engineering components is a more complex problem and further effort is in progress.

4. CONCLUDING REMARKS

The measurement of the adhesion of thermally sprayed materials is, at least on the conceptual level, a routine operation. The tensile adhesion method as detailed in ASTM C633 is simple and often used in industry for ranking different coatings. However, the major shortcoming of this test is that it does not promote any understanding of coating performance, i.e. how coatings can be designed to be more functional. Thus, the present paper has addressed other methods based on fracture mechanics and mechanism-based studies.

The design of experiments with regard to material property optimization of coatings is another area of intense effort. Experimental protocols which are based on Taguchi and response surface methodology allow engineers efficient and viable ways to optimize the process parameters [100, 101]. Such statistical methods are executed to discriminate key parameters which induce the variation of coating properties. The signal-to-noise ratios of the processing parameters are derived from such studies [102]. The Taguchi method does have some short-comings and limitations [103] but it is a simple and powerful process control procedure.

The coating should be considered as part of the overall component system and therefore current trends are to design the coating as an integral part of the component assembly rather than as an add-on to the substrate. Whereas the property of coating adhesion to the substrate is of principal interest, there is still no single measurement which can satisfy all the requirements for determining material properties. Standardization of measurements, which may be achieved by improving existing experimental techniques or by the combination of two or more techniques, will aid future coating development. Finally, a coating design (i.e. both microstructural and mechanical engineering designs) which is based on lifetime modeling is the critical information that should be forthcoming from any test method. Such designs will increase the knowledge-base and understanding of thermal sprayed materials and coatings so that their reliability and application will grow.

Acknowledgements

We wish to thank the Alcoa Foundation for supporting parts of this work related to examining failure mechanisms of alumina. We thank Paul S. Fussell and Roger Kaufold as the managers of our program.

REFERENCES

1. T. N. Rhys Jones, *Surface Coat. Technol.* **43/44**, 402–415 (1990).
2. K. T. Scott and R. Kingswell, in: *Advanced Surface Coatings*, D. S. Rickerby and A. Mattews (Eds), pp. 217–243. Chapman and Hall, New York (1991).
3. J. H. Zatt, *Annu. Rev. Mater. Sci.* **13**, 9–42 (1983).
4. L. E. Weiss, F. B. Prinz, D. A. Adams and D. P. Siewiorek, *J. Thermal Spray Technol.* **1**, 231–237 (1992).
5. K. Neufuss, B. Kolman, J. Dubsky and P. Chraska, *Proc. AustCeram 92*, Melbourne, 16–21 August pp. 124–129 (1992).
6. C. C. Berndt (Ed.), 1992 International Thermal Spray Conference and Exposition, 28 May– 5 June, *Proceedings: Thermal Spray: International Advances in Coatings Technology*, Orlando, FL. ASM International, Materials Park, OH (1992).
7. T. F. Bernecki (Ed.), 1991 National Thermal Spray Conference and Exposition, 4–10 May, *Proceedings: Thermal Spray Coatings: Properties, Processes and Applications*, Pittsburgh, PA. ASM International, Materials Park, OH (1992).

8. S. Blum-Sandmeier, H. Eschnauer, P. Huber and A. R. Nicoll (Eds), *2nd Plasma-Technik-Symposium, Proceedings,* Plasma-Technik AG, Wohlen/Switzerland, Häfliger Druck AG, Wettingen (1991).
9. R. McPherson, *Surface Coat. Technol.* **39/40**, 173–181 (1989).
10. H. Herman, *Sci. Am.* **256**, 112–117 (1989).
11. K. L. Mittal (Ed.), in: *Adhesion Measurement of Thin Films, Thick Films, and Bulk Coatings,* ASTM STP640, pp. 5–17. American Society for Testing and Materials, Philadelphia, PA (1978).
12. K. L. Mittal (Ed.), in: *Adhesion Measurement of Thin Films, Thick Films, and Bulk Coatings,* ASTM STP640, American Society for Testing and Materials, Philadelphia PA (1978).
13. R. L. Patrick (Ed.), *Treatise on Adhesion and Adhesive,* Vol. 1. Marcel Dekker, New York (1967).
14. D 907-91b, *Terminology of Adhesives,* American Society for Testing and Materials, Philadelphia, PA (1991).
15. J. Comyn, *Int. J. Adhesion Adhesives* **10**, 161–165 (1990).
16. A Matting and H.-D. Steffens, *Metallwiss. Tech.* **17**, 583–593, 905–922, 1213–1230 (1963). Available under Translations Register Index No. 72-14247-13H.
17. M. D. Thouless, *Mater. Res. Soc. Symp. Proc.* **119**, 51–62 (1988).
18. B. R. Lawn and T. R. Wilshaw (Eds), *Fracture of Brittle Solids.* Cambridge University Press, Cambridge (1975).
19. C. C. Berndt and R. McPherson, *Trans. Int. Eng.* **6**, 53–58 (1981).
20. C. C. Berndt, in: *Advances in Fracture Research,* Vol. 4, S. R. Valluri, D. M. R. Taplin, P. Rama Rao, J. F. Knott and R. Dubey (Eds), pp. 2545–2552. Pergamon Press, Oxford (1984).
21. G. N. Heintze and R. McPherson, *Surface Coat. Technol.* **34**, 15–23 (1988).
22. C. C. Berndt, Ph.D. Thesis, Monash University, Australia (1980).
23. P. Ostojic, Ph.D. Thesis, Monash University, Australia (1986).
24. K. L. Mittal, *J. Adhesion Sci. Technol.* **1**, 247–259 (1987).
25. D. S. Rickerby, *Surface Coat. Technol.* **36**, 541–557 (1988).
26. S. J. Bull and D. S. Rickerby, *Br. Ceram. Trans. J.* **88**, 177–183 (1989).
27. P. R. Chalker, S. J. Bull and D. S. Rickerby, *Mater. Sci. Eng.* **A140**, 583–592 (1991).
28. B. A. Lyashenko, V. V. Rishin, V. G. Zil'berberg and S. Yu. Sharivker, *Sov. Powder Metall. Meth. Ceram.* **8**, 331–334 (1969).
29. B. M. Zakharov, M. G. Trofimov, L. I. Guseva, Y. I. Golovkin, A. A. Kononov and V. V. Vinokurova, *Sov. Powder Metall. Met. Ceram.* **9**, 925–929 (1970).
30. W. E. Stanton, in: *Proc. 7th Int. Metal Spraying Conf.,* 10–14 Sept. 1973. pp. 157–164, 312–314. The Welding Institute, Cambridge (1974).
31. T. Suhara, K. Kitajima and S. Fukada, in: ref. 30, pp. 179–184.
32. N. N. Rykalin, *Pure Appl. Chem.* **48**, 179–194 (1976).
33. V. E. Belashchenko and Y. B. Chernyak, in: ref 6, pp. 433–437.
34. R. L. Apps, *Chem. Eng.* **292**, 769–773 (1974).
35. V. Wilms and H. Herman, *Thin Solid Films* **39**, 251–262 (1976).
36. T. J. Steeper, D. J. Varacalle, Jr., G. C. Wilson, W. L. Riggs (II), A. J. Rotolico and E. Nerz, in: ref. 6, pp. 415–420.
37. J. O. Outwater and D. J. Gerry, *J. Adhesion* **1**, 290–298 (1969).
38. J. A. Kies and A. B. J. Clark, in: *Fracture 1969, Proceedings of the 2nd International Conference,* Brighton, P. L. Platt (Ed.), Chapman and Hall, London (1969).
39 S. J. Howard and T. W. Clyne, *Surface Coat. Technol.* **45**, 333–342 (1991).
40. P. F. Becher and W. L. Newell, *J. Mater. Sci.* **12**, 90–96 (1977).
41. P. F. Becher, W. L. Newell and S. A. Halen, in: *Fracture Mechanics of Ceramics,* R. C. Bradt, D. P. H. Hasselman and F. F. Lange (Eds), Vol. III, pp. 463–471. Plenum Press, New York (1978).
42. W. D. Bascom and J. L. Bitner, *J. Mater. Sci.* **12**, 1401–1410 (1977).
43. S. Mostovoy, P. B. Crosley and E. J. Ripling, *J. Mater.* **2**, 661–681 (1967).
44. G. N. Heintze and R. McPherson, *Surface Coat. Technol.* **36**, 125–132 (1988).
45. G. N. Heintze and R. McPherson, in: 1987 National Thermal Spray Conference and Exposition, 14–17 Sept., *Proceedings: Thermal Spray: Advances in Coatings Technology,* Orlando, FL, D. L. Houck (Ed.), pp. 271–275. ASM International Materials Park, OH (1988).
46. P. Ostojic and R. McPherson, *J. Am. Ceram. Soc.* **71**, 891–899 (1988).

47. M. J. Noone and R. L. Mehan, in: *Fracture Mechanics of Ceramics*, R. C. Bradt, D. P. H. Hasselman and F. F. Lange (Eds), Vol. 1, pp. 201–229. Plenum Press, New York (1974).
48 *Acoustic Emission*, STP505, American Society for Testing and Materials, Philadelphia PA (1972).
49. J. R. Matthews (Ed.), *Acoustic Emission*, Gordon and Breach, New York (1983).
50. T. C. Nerz, J. E. Nerz, B. A. Kushner, A. J. Rotolico and W. L. Riggs, in: ref. 6, pp. 405–414.
51. L. C. Cox, *Surface Coat. Technol.* **36**. 807–815 (1988).
52. D. K. Das, M. P. Srivastava, S. V. Joshi and R. Sivakumar, *Surface Coat. Technol.* **46**, 331–345 (1991).
53. M. M. Mayuram and R. Krishnamurphy, in: ref. 6, pp. 711–715.
54. N. Iwamoto, M. Kamai and G. Ueno, in: ref. 6, pp. 259–265.
55. H. Nakahira, Y. Harada, N. Mifune, T. Yogoro and H. Yamane, in: ref. 6, pp. 519–524.
56. H. L. Dunegan, *Prevention of Structural Failure*, pp. 86–113, ASM, Materials Park, OH (1975).
57. T. Tsuru, A. Sagara and S. Haruyama, *Corrosion* **43**, 703–707 (1987).
58. F. Bordeaux, C. Moreau and R. G. Saint Jacques, *Surf. Coat. Technol.* **54/55**, 70–76 (1992).
59. C. C. Berndt, *J. Mater. Sci.* **24**, 3511–3520 (1989).
60. I. G. Scott (Ed.), *Basic Acoustic Emission*. Gordon and Breach, New York (1991).
61. C. C. Berndt and R. A. Miller, *Thin Solid Films* **119**, 173–184 (1984).
62 C. C. Berndt, in: *Proc. Thermal Barrier Coatings Workshop*, 21–22 May, pp. 127–137. NASA Lewis Research Center, Cleveland, OH (1985).
63. B. Lawn and R. Wilshaw, *J. Mater. Sci.* **10**, 1049–1081 (1975).
64. C. Richard, J. Lu, J. F. Flavenot, G. Beranger and F. Decomps, in: ref 6, pp. 11–16.
65. J. G. Binner and R. Stevens, *Trans. Br. Ceram. Soc.* **83**, 168–172 (1984).
66. H. Nayeb-Hashemi and C. A. Tracy, *Exp. Mech.* No. 12, 366–372 (1991).
67. G. K. Beshish, C. W. Florey, F. J. Worzala and W. J. Lenling, *J. Thermal Spray Technol.* **2**(1), 35–38 (1993).
68. R. Dal Maschio, V. M. Sgavo, L. Bertamini and E. Galvanetto, in: ref. 6, pp. 947–951.
69. C. C. Berndt, J. Karthikeyan, R. Ratanarj and Yang Da Jun, in; ref. 7, pp. 199–203.
70. C. C. Berndt, J. Ilavsky and J. Karthikeyan, in: ref. 6, pp. 941–946.
71. C. K. Lin and C. C. Berndt, in: 1993 National Thermal Spray Conference and Exposition, 7–11 June, *Proceedings: Thermal Spray Coatings: Research, Design and Application*, Anaheim, CA. C. C. Berndt and T. F. Bernecki (Eds.), pp. 561–568. ASM International, Materials Park, OH (1993).
72. P. Benjamin and C. Weaver, *Proc. R. Soc. London, Ser. A* **254**, 163 (1960).
73. A. J. Perry, J. Valli and P. A. Steinmann, *Surface Coat. Technol.* **36**, 559–575 (1988).
74. C. Julia-Schmutz and H. E. Hintermann, *Surface Coat. Technol.* **48**, 1–6 (1991).
75. P. J. Burnett and D. S. Rickerby, *Thin Solid Films* **154**, 403–416 (1987).
76. S. J. Bull, D. S. Rickerby, A. Matthews, A. Leyland, A. R. Pace and J. Valli, *Surface Coat. Technol.* **36**, 503–517 (1988).
77. J. Sekler, P. A. Steinmann and H. E. Hintermann, *Surface Coat. Technol.* **36**, 519–529 (1988).
78. S. J. Bull, *Surface Coat. Technol.* **50**, 25–32 (1991).
79. C. W. Anderson and K. H. Heffner, in: ref. 6, pp. 695–704.
80. M. Gudge, D. S. Rickerby, R. Kingswell and K. T. Scott, in: 1990 National Thermal Spray Conference and Exposition, 20–25 May, *Proceedings: Thermal Spray Research and Application*, Long Beach, CA. T. F. Bernecki (Ed.), pp. 331–337. ASM International, Materials Park, OH (1991).
81. E. Lopez, F. Beltzung and G. Zambelli, *J. Mater. Sci. Lett.* **8**, 346–348 (1989).
82. F. Beltzung, G. Zambelli, E. Lopez and A. R. Nicoll, *Thin Solid Films* **181**, 407–415 (1989).
83. E. Lugscheider, P. Jokiel, G. Purshe, O. Roman and K. Yushchenko, in: ref. 6, pp. 647–651.
84. K. Furukubo, S. Oki and S. Gohda, in: ref. 6, pp. 705–709.
85. R. C. Hendricks and G. McDonald, *Assessment of Variations in Thermal Cycle Life Data of Thermal Barrier Coated Rods*, NASA TM-81743 (1981).
86. H.-D. Steffens and Fischer, *Surface Coat. Technol.* **32**, 327–338 (1987).
87. S. J. Grisaffe, *Analysis of Shear Bond Strength of Plasma-Sprayed Alumina Coatings on Stainless Steel*, NASA TN D3113 (1965).
88. H. Grützner and H. Weiss, *Surface Coat. Technol.* **45**, 317–323 (1991).
89 K. K. Schweitzer, M. H. Zeihl and Ch. Schwaminger, *Surface Coat. Technol.* **48**, 103–111 (1991).

90. M. J. Filaggi and R. M. Pilliar, *J. Mater. Sci.* **26**, 5383–5395 (1991).
91. R. A. Miller, *Surface Coat. Technol.* **30**, 1–11 (1987).
92. Proc. Thermal Barrier Coatings Workshop, NASA Lewis Research Center, Cleveland, OH, 21–22 May, 1985.
93. R. A. Miller, *J. Eng. Gas Turbines Power* **111**, 301–305 (1989).
94. G. C. Chang, W. Phucharoen and R. A. Miller, *Surface Coat. Technol.* **30**, 13–28 (1987).
95. R. A. Miller, *J. Am. Ceram. Soc.* **67**, 517–521 (1984).
96. R. V. Hillery, B. H. Pilsner, R. L. McKnight, T. S. Cook and M. S. Hartle, *Thermal Barrier Coating Life Prediction Model*, Final Report, NASA CR-180807 (1987).
97. T. E. Strangman, J. Neumann and A. Liu, *Thermal Barrier Coating Life Prediction Model Development*, Final Report, NASA CR-179648 (1987).
98. J. T. DeMasi, M. Ortiz and K. D. Sheffler, *Thermal Barrier Coating Life Prediction Model Development*, Phase 1 Final Report, NASA CR-182230 (1989).
99. S. M. Meier, D. M. Nissley and K. D. Sheffler, *Thermal Barrier Coating Life Prediction Model Development*, Phase II Final Report, NASA CR-189111 (1991).
100. G. Taguchi and S. Konishi, *Taguchi Methods, Orthogonal Arrays and Linear Groups*, American Supplier Institute, Dearborn, MI (1987).
101. P. Ross, *Taguchi Techniques for Quality Engineering*, McGraw-Hill, New York (1988).
102. G. E. P. Box, *Technometrics* **30**, 1–17 (1988)
103. S. Bisgaard, in: ref. 80, pp. 661–667.

Adhesion Measurement of Films and Coatings, pp. 71–86
K. L. Mittal (Ed.)
© VSP 1995.

Adhesion measurement of thin metal films by scratch, peel, and pull methods

AKIRA KINBARA[1,*] and ICHIHARU KONDO[2]

[1] *Department of Applied Physics, The University of Tokyo, Tokyo 113, Japan*
[2] *Nippondenso Co. Ltd, Kariya, Aichi Prefecture 448, Japan*

Revised version received 29 March 1993

Abstract—Scratch, peel, and pull methods for adhesion measurement were applied to deposited thin film/solid substrate combinations. Scatter in the experimental data was observed and its origin is discussed. The interface between the thin film and the substrate was investigated by transmission electron microscopy (TEM), energy dispersive spectroscopy (EDS), and Auger electron spectroscopy (AES), and the correlation of the interface structure with the adhesion strength was investigated. Ion bombardment and heat treatment were carried out to enhance the adhesion. Accumulation of bombarding gas ions at the interface was observed and the role of ion bombardment in improving adhesion is considered.

Keywords: Adhesion; thin films; ion bombardment; scratch test; peel test; pull test.

1. INTRODUCTION

Adhesion of thin films to substrates is generally evaluated through processes of detaching the films from the substrates. Many methods for detaching films have been developed. Although very sophisticated methods such as the Lorentz force method [1], the laser spallation method [2], etc. have been devised, they are not really of routine practical use. More conventional but significant methods will be discussed here. These methods have been covered by Mittal [3, 4] and here we classify them into the following three categories according to the physical quantities obtained by them:

(1) Force method
 Measured quantity = force (unit N or J/m)
 e.g. scratch method.

(2) Energy method
 Measured quantity = energy per unit area (unit N/m or J/m^2)
 e.g. peel method.

---

*To whom correspondence should be addressed.

(3) Stress method
   Measured quantity = force per unit area (unit N/m^2 or J/m^3)
   e.g. pull/topple method.

The relationship between the adhesion measurement results obtained by the various methods is not clear. In addition, the data scatter obtained in adhesion measurement is usually large; so, it is not easy to determine the value of the adhesion strength with high reproducibility.

In spite of the difficulty in evaluating adhesion, adhesion enhancement or improvement techniques have been developed and are widely used in a variety of industries.

In this paper, we present experimental data on the adhesion measurement for several combinations of thin films and substrates obtained by the various methods as well as the scatter in the data. A detailed discussion on the relationship between the interface structure and the adhesion strength is also given.

2. EXPERIMENTAL

As a force measurement technique, the scratch test is one of the most popular methods. This method is essentially based on the measurement of the load to peel off the film by scratching. It can be applied to almost any kind of thin film/substrate combination and is applicable even where strong adhesion exists. The principle of the scratch test is very simple. The scratch pattern which indicates the detached area is observed by an optical or electron microscope. One well-known method is the acoustic detection method [5]. We have used a vibrating stylus, which enables us to electronically detect the scratch peeling. The details of this method are described elsewhere [6]. Our experimental results presented here were obtained mainly by this method. The scratch method is widely used in laboratories and industries but the relationship between values obtained by the scratch method and those obtained by other methods still remains an unsolved problem.

The peel method has been used for measurement of the adhesion energy, where adhesion energy is the energy required to detach the unit area of the film from the substrate in a quasi-static process. However, the method to peel off the film is not easy. The so-called Scotch tape test is generally used for the 'yes–no' decision. This is not adequate for obtaining quantitative results. The wedge insertion method developed by Obreimoff [7] and Yoon [8] is preferred instead, but insertion of the wedge is often not possible, particularly when the adhesion is strong. In the present paper, wedge insertion data in the case of weak adhesion films are given.

Another method for peeling has been developed for the evaluation of the interfacial free energy using the internal stress generated in thin Ni films [9]. We have observed that the internal stress in a vacuum-deposited Ni thin film is almost constant (approximately 0.5 GPa) and quite reproducible. The relationship between the internal stress × Ni film thickness and the Ni film thickness is shown in Fig. 1. We deposited Ni films on the sample films. So, by depositing Ni films on the sample films, we can apply various shear stresses to the sample films under the Ni films by changing the

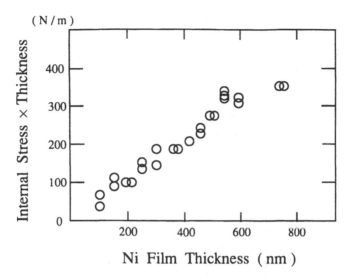

Figure 1. Internal stress generated in Ni films × film thickness as a function of the film thickness.

thickness of the Ni films. This method was applied to the adhesion measurement of Ti thin films deposited on Si substrate and was found to be effective.

The adhesion strength has also been measured by the pull or topple test [10, 11]. Although devices to carry out this test are commercially available, we have developed our own pull and topple test machine and have applied it to samples prepared in our laboratory. The machine, in principle, consists of a usual adhered rivet and a pull force measurement system. Two types of rivets were prepared. One is a circular type for the pull test, with a diameter of 5 mm. The other is a square type pillar for the topple test. The length of one side and the height of the pillar are 7 mm and about 50 mm, respectively. They are adhered to the samples by an epoxy adhesive. About 70 samples of gold films on glass substrates were prepared under the same conditions and divided into two groups. One group was used for the pull test and the other for the topple test. Actually, the difference between the results obtained by the pull and the topple methods cannot be distinguished within experimental errors. Hence, we will regard the topple method as the same as the pull method hereafter.

3. RESULTS

It is well known that the results on the adhesion measurement show considerable scatter even for carefully prepared samples; hence, we measured the adhesion for many samples prepared under the same conditions in order to obtain an average value. All the data are indicated as a histogram. Figure 2 shows the adhesion strength vs. frequency results on vacuum-deposited gold films on glass substrates measured by the pull method. More than 200 samples were examined. The results show scatter but not necessarily random scatter. The histogram was rearranged and a Weibull plot was drawn (Fig. 2, inset). The horizontal axis shows ln (applied stress F) and the vertical

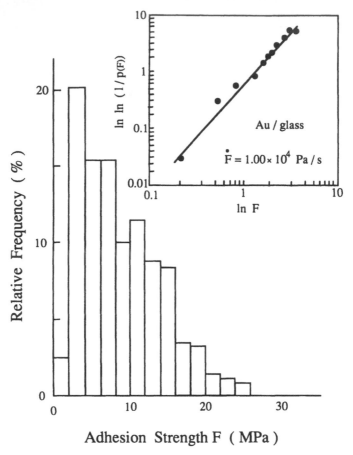

Figure 2. Histogram showing scatter in the adhesion strength, as measured by the pull method, of Au films on glass substrates. A Weibull plot is also shown.

axis shows $\ln \ln[1/p(F)]$, where $p(F)$ is the percentage of undetached samples. The rate of the stress increase was 1.00×10^4 Pa/s. As shown in this figure, good linearity was found. The slope is about 2 and hence, the relation.

$$p(F) = \exp(-F^2), \tag{1}$$

is obtained, where F values are expressed in units of 10 MPa. The condition $p(0) = 1$ is clearly satisfied.

The cumulative integral distribution function $D(F)$ for the fracture strength of solids is generally written as

$$D(F)(= 1 - p) = 1 - \exp\left[-F^{b+1}(b+1)\right]. \tag{2}$$

So, relation (1) seems to be a special case of (2), where $b = 1$, and it is suggested that the detaching process of the films from the substrate is similar to the fracture process of materials.

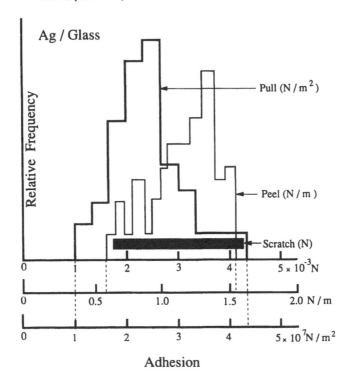

Figure 3. Adhesion of Ag films on glass substrates measured by the pull, peel, and scratch methods.

The scatter in the data is not only present in the pull test, but is common to every method. As mentioned before, the adhesion is evaluated by three quantities of different dimensions. We measured the adhesion of Ag films on glass substrates by these three different methods. Histograms of the peel and the pull tests, and the range of the data scatter in the scratch test are shown in Fig. 3. Each method reveals considerable scatter. Figure 3 suggests that the average values in the three methods are quite different while the spread in the data does not differ from each other, and therefore these three methods may have a correlation, but not necessarily a linear correlation. Perhaps, we can select any one of these methods to find the order of the adhesion strength, i.e. each is valid for relative comparison.

It should be noted that the scatter in the experimental data is not only attributed to the method of the measurement, but it should also be attributed to the actual scatter in the adhesion strength. We scratched films by a stylus at constant loads and observed partial detachment [11, 12]. In other words, even by scratching at a constant load, only some part of the film is detached and the rest is still attached. We scratched a Ag [11] and a MgO [12] film by a graphite and a diamond stylus at constant loads and evaluated the fraction of detached film. The detached fraction was found to increase with increasing applied load, but for complete detachment we needed a considerably larger amount of load than the load required to initiate detachment. These results seem to suggest a non-uniformity of the adhesion strength and the scatter in the data reflect partly the non-uniformity of the adhesion strength itself.

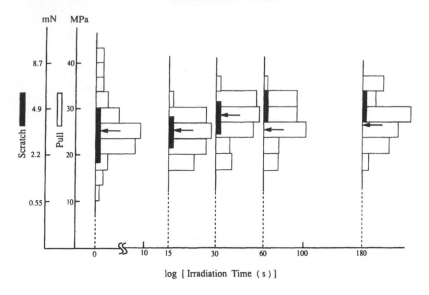

Figure 4. Adhesion of Ag films on Ar ion-bombarded glass substrates. The irradiation time was varied.

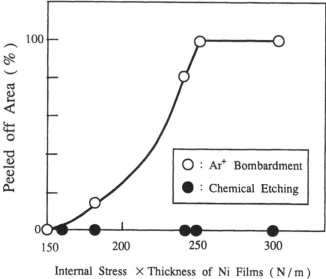

Figure 5. Percentage of peeled-off area of Ti films on Si substrates. The internal stress generated in Ni films is applied to Ti films to detach them.

In order to investigate the effect of ion bombardment, we bombarded the substrate before film deposition. As a typical example of the ion bombardment effect, Ag films on glass substrates were investigated. The substrates were placed in an Ar ion plasma generated by 5 kV AC voltage. The supplied power density was approximated to be 5 kW/m^2 at the maximum. Adhesion was measured by the pull and the scratch methods. We also observed scatter in the data and partial detachment; the results are shown in Fig. 4. The irradiation time was varied as shown in the figure. The effect of the bombardment on this film/glass combination was not so remarkable. As another example [9], the results of the adhesion energy of sputter-deposited Ti thin films on Si(100) surfaces are shown in Fig. 5. The Si surfaces were bombarded by Ar ions prior to Ti film deposition. The acceleration voltage and the power of Ar ions were 400 V and 6 kW/m^2, respectively. The number of Ar atoms included in the Ti films estimated by the Rutherford back-scattering (RBS) method was about 5.9×10^{18} atoms/m^2. The adhesion was evaluated by the internal stress generated in Ni films formed on the Ti films, as mentioned before. When the shearing force per unit length (shear stress $\times$ Ni film thickness) originated by the internal stress was only 150 N/m (J/m^2), detachment was not observed but the detached fraction increased monotonically with increasing shearing force per unit length and at 250 N/m (J/m^2), complete detachment (100% detachment) was observed. Hence we can conclude that the adhesion energy of Ti films on Si(100) surfaces ranges between 150 and 250 J/m^2.

The ion bombardment clearly enhances the adhesion, but the adhesion enhancement produced by Ar ion bombardment is inferior to the pretreatment by dilute HF acid etching. Figure 5 shows that the adhesion enhancement produced by etching is more effective than that by the ion bombardment. In this case, the adhesion energy exceeded 300 J/m^2 and the film could not be detached. The structures and compositions of both chemically etched and ion-bombarded interfaces were investigated by TEM and EDS. Figure 6 shows a Ti film deposited on a chemically etched Si(100) surface. An amorphous Ti–Si layer was formed on the Si(100) surface, and the adhesion of the Ti film was so strong that we could not detach the film from the substrate. Figure 7 shows the interface of a Ti film on an Ar ion-bombarded Si(100) surface. A few layers of amorphous Si were observed at the interface. Detachment by the peel method occurred at the a-TiSi/a-Si interface.

Next we consider the effect of the heat treatment. It is well known that heat treatment usually enhances adhesion, although exceptions have been reported. For example, the adhesion of Cu and Ag films deposited on glass substrates heated between room temperature and 200°C did not show appreciable enhancement [13, 14]. Figure 8 shows the results on the deposition temperature dependence of the adhesion energy of Au films on glass substrates. In this case, although the data scatter is large, enhancement of the adhesion was observed after 400°C. In many cases, heat treatment is effective for adhesion enhancement. However, in the case of Ti on Si, the temperature dependence is not simple. As shown in Fig. 9, heating at 623 K rather decreases the adhesion. Further heating at higher temperatures seems to be effective for the adhesion enhancement.

In order to investigate the temperature effect of the adhesion, we examined again the Ti/Si interfaces by AES, TEM, and EDS after the heat treatment. It is very interesting and important to note that Ar atoms were observed at the interface. Figure 10 shows the AES results on both the film and the substrate sides after the detachment. It was found that Si was clearly separated from the amorphous Ti–Si compound. In addition, Ar atoms were found to remain only on the Si surface and no traces of Ar atoms were observed on the Ti side. From observation of the interface by AES, the density of Ar atoms on the Si side does not seem to be changed by the heat treatment. But

A. Kinbara and I. Kondo

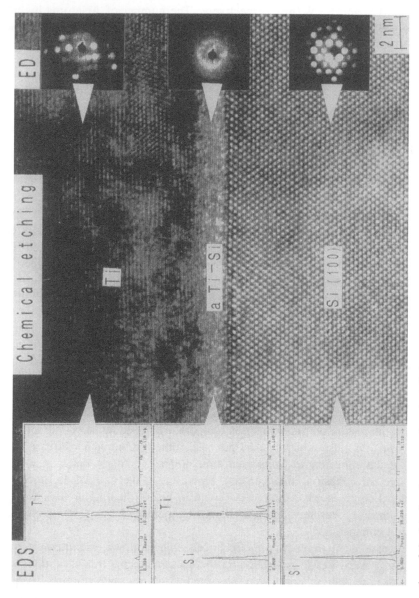

Figure 6. Observation of the interface between a deposited Ti film and a chemically etched Si(100) surface by TEM, electron diffraction (ED), and energy dispersive spectroscopy (EDS). TEM shows that a disordered layer is formed on the single-crystal Si substrate; the layer was confirmed to be amorphous by ED. EDS shows that the constituents of the layer are Ti and Si.

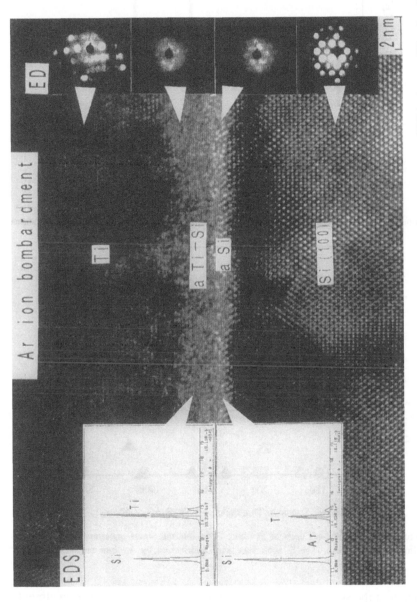

Figure 7. Observation of the interface between a deposited Ti film and an Ar ion-bombarded Si(100) surface by TEM, electron diffraction (ED), and energy dispersive spectroscopy (EDS). TEM shows that a disordered layer is formed on the single-crystal Si substrate; the layer was confirmed to be amorphous by ED. EDS shows that the layer is divided into two layers: one is a-Si and the other is a a-TiSi. In the a-Si layer, a trace of Ar was observed.

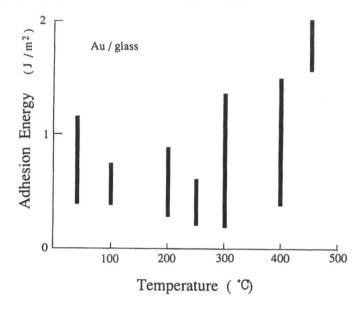

Figure 8. Deposition temperature dependence of the adhesion energy of Au films on glass substrates.

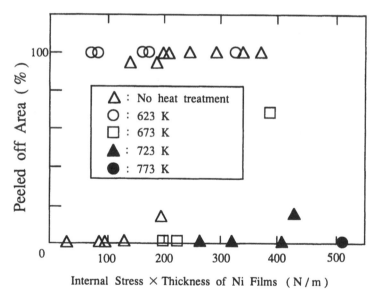

Figure 9. Percentage of peeled-off area of Ti films. The internal stress generated in Ni films is applied to Ti films to detach them. The Si substrates were bombarded by Ar ions and heated at different temperatures.

TEM showed that the structure changed at the interface. When the heat treatment was carried out, the a-Si layer seemed to be crystallized and the structure of the interface became wavy as shown in Fig. 11. EDS showed that Ar atoms precipitated at the top of the protrusions of the wave. The role of Ar is not clear at present but the disappearance of the amorphous layer and the precipitation of Ar atoms at particular

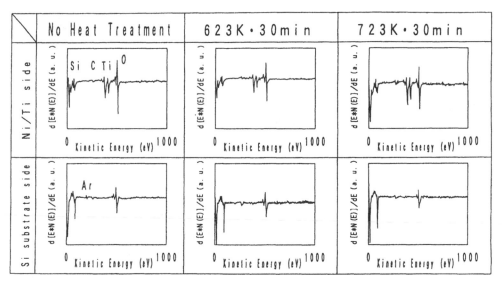

Figure 10. Auger electron spectra of a detached Ti film and a Si substrate. The Ti film (a Ni film is deposited on the Ti film) was peeled off from the substrate by Scotch tape. The surfaces of the substrate side and the tape side after peeling were investigated by AES. The heat treatment temperature after the sample preparation was taken as a variable.

places at the interface seem to be correlated with the adhesion. At the same time, the wave formation and the precipitation of Ar atoms may be correlated with the data scatter and the partial detachment of the films.

4. DISCUSSION

As mentioned before, adhesion has been evaluated by various methods but the relation among these methods is not clear.

In order to compare the results obtained by the scratch method with those obtained by the pull method, a relation has been given by Benjamin and Weaver [14] but its oversimplification has been shown to be inadequate. If we use an appropriate value for Brinell's hardness of the substrate, say 9 GPa, and apply Benjamin and Weaver's relation to our results on silver films on glass substrates shown in Fig. 3, the calculation shows that the minimum and maximum values of the scratch force of 1.8 and 4.3 mN correspond to adhesion (pull) strengths of 1.8×10^8 and 2.8×10^8 Pa, respectively. In other words, the conversion from force to stress using their relation gives values about one order of magnitude larger than those obtained by the pull method. The main reason for this large difference is clearly the oversimplification included in their relation. In addition, we have already suggested that the detached areas in scratching and pulling are very different from each other and this may cause the difference [12]. The detachment of the film initiates at the weakest point. As the rivet in the pull method covers a considerably large area, the area under the rivet can include many areas of weak adhesion spots, and thus the values measured by the pull methods are apt to be lower than the scratch test values. However, even in scratching,

A. Kinbara and I. Kondo

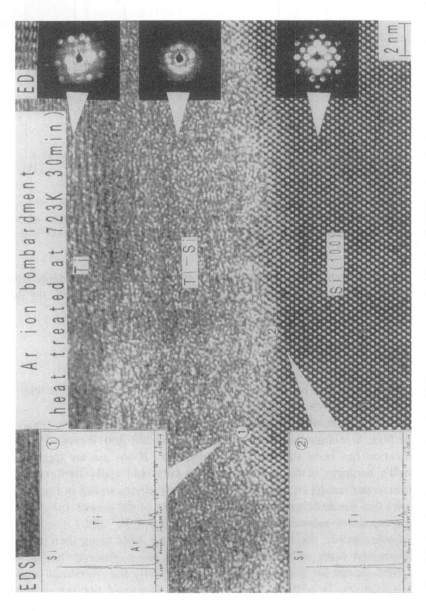

Figure 11. Observation of the interface between a deposited Ti film and an Ar ion-bombarded Si(100) surface after heat treatment at 723 K for 30 min by TEM, electron diffraction (ED), and energy dispersive spectroscopy (EDS). TEM shows that a disordered layer is formed between the single-crystal Si substrate and the Ti film; the constituents of the layer were confirmed to be Ti and Si. EDS shows traces of Ar in the protrusions ① at the interface but no traces were observed at the bottom ②.

partial detachment has been observed [12]. So, the role of the area detached in the adhesion measurement process still remains an unknown factor in the evaluation of the adhesion.

The mechanism of the scratching has not been fully understood so far. This process is complicated and consists of at least two processes. One is a momentum transfer or an impact process from the stylus to protrusions on the film surface. This process is important and it depends on the state of the surface roughness, size and mass of the stylus, etc. and in reality, we cannot take all of these factors into account. The other process is the elastic deformation process or indentation. A vertical load applied to the stylus presses and deforms the film and the substrate materials. The maximum stress appears right under the stylus if the tip of the stylus is a hemisphere. The stress F_v, is given by [13]

$$F_v = 0.578\left[\left(E_s/R\right)^2 W\right]^{1/3}, \tag{3}$$

where E_s is the Young modulus of the substrate, R is the radius of the stylus, and W is the load applied to the stylus. From this relation, the stored energy density, u, or the energy on a unit area under the stylus, U, is calculated as follows:

$$U = ud = F_v^2 d(1 - v)/E_f, \tag{4}$$

where d is the thickness of the film, v is Poisson's ratio of the film, and E_f is the Young modulus of the film material.

If this energy is consumed to detach the film from the substrate, we can estimate the adhesion energy [15] defined as the energy required to detach a unit area of the film from the substrate in a quasi-static process. Relation (4) was applied to the scratch test results shown in Fig. 3, and this energy was around 1 J/m^2. This is consistent with the energy obtained by the peel method.

Next, we consider the relation between the adhesion strength and the interfacial energy. As the force is expressed in terms of a derivative of the potential energy, for a rough estimation the adhesion force is considered to be the maximum slope of the curve of the adhesion energy vs. the separation between the film and the substrate. If the separation is typically 1 nm, then the adhesion energy corresponding to an adhesion stress of 1×10^7 Pa is approximately 1×10^7 Pa $\times 10^{-9}$ m $= 10^{-2}$ J/m$^2 = 10$ mJ/m^2. This is about one or two orders of magnitude smaller than the direct measurement by the peel method.

Ar ion bombardment has been shown to be effective for the improvement of adhesion. We have shown the effect on Ti films on a Si substrate and it is also applicable to plastic substrates. Figure 12 shows an example of the Ar ion bombardment effect of PTFE substrate. Au films were vacuum-deposited on PTFE. The irradiation time was taken as a parameter between 0 and 180 s. Again we observed partial detachment, but from the figure the effect of the bombardment is very clear. The surface of PTFE is activated by the ion bombardment and the surface activation seems to induce enhancement of the adhesion. A characteristic feature is that the initiation of

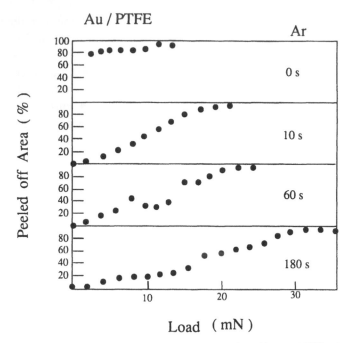

Figure 12. Percentage of peeled-off area by the scratch method of Au films on PTFE substrate bombarded by Ar ions prior to film deposition. The applied load was varied and the exposure time to Ar ions was taken as a variable.

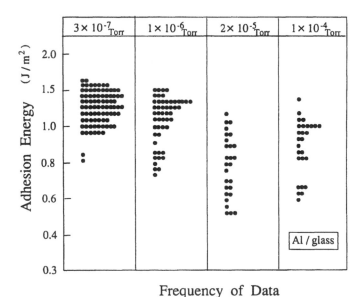

Figure 13. Adhesion energy of Al films vacuum-deposited onto glass substrates at different residual gas pressures. Because of the large scatter in the experimental data, many experiments were performed. One dot corresponds to one experiment. So, the horizontal axis corresponds to the frequency of a certain value of the adhesion energy.

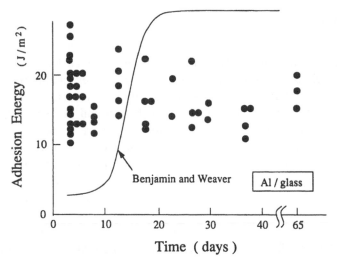

Figure 14. Time dependence of the adhesion energy of Al films on glass substrates, stored at ambient temperature and pressure. The results obtained by Benjamin and Weaver are also shown as a solid curve.

the detachment occurs at almost zero load, and complete detachment is dictated by the ion bombardment time.

It should be pointed out, however, that Ar ion bombardment has three effects: surface cleaning, the formation of an amorphous layer on the surface, and the inclusion of Ar atoms. The amorphous layer is closely connected with adhesion failure.

One of the effects of the ion bombardment is surface cleaning. Adsorption of gas atoms or, in general, contamination of the surface decreases the surface free energy and stabilizes the surface state. This means that the adhesion is decreased by the adsorption of gas atoms. So, one would expect the adhesion to be affected by the residual gas pressure during the film preparation. Figure 13 shows that the adhesion energy of Al films on glass substrates. The residual gas pressure during deposition ranged between 1×10^{-4} and 3×10^{-7} Torr. The scatter is large but on average we can observe that high vacuum deposition enhances the adhesion energy.

Finally, we should comment on the time dependence effect on adhesion. Benjamin and Weaver [14] showed that the adhesion changed with time. Most of the metal films on glass substrates showed an increase of the adhesion. The authors suggested diffusion of oxygen or water molecules from the top of the film surface to the bottom and ascribed this increase to oxide layer formation at the interface between the films and the substrates. We have performed an experiment on the time change of the adhesion energy of aluminum films. The results are shown in Fig. 14 together with Benjamin and Weaver's results. Our results do not show a significant time dependence. The thickness of our films is approximately 0.5 μm, which is probably higher than Benjamin and Weaver's thickness, although they did not indicate the thickness of their films. Hence, it is suggested that diffusion is small and the interface structure keeps its initial state.

5. CONCLUSION

The adhesion of metal films to several kinds of solid substrates was measured by scratch, peel, and pull methods. The quantities obtained by these methods have different dimensions but all the methods were useful in evaluating the adhesion. The experimental data obtained by these methods all showed scatter. The distribution of the data was approximated by Weibull's distribution and this suggests that the film detachment in the adhesion tests is similar to the fracture process of solid materials.

For the enhancement of the adhesion, the effect of Ar ion bombardment of the substrates was studied. This enhanced the adhesion but heat treatment followed by ion bombardment was more effective. Observations of the interface by TEM, ED, EDS, and AES showed that ion bombardment produced an amorphous layer on the substrate and that Ar atoms were included in this layer. The flatness of the interface was disordered by the heat treatment and the included Ar atoms precipitated at protrusions appearing in the interface. The disorder seems to be related to the enhancement of the adhesion.

Acknowledgements

We would like to express our sincere gratitude to Professor S. Baba, Dr T. Yoneyama, and Dr A. Kobayashi for their useful discussions.

REFERENCES

1. S. Krongelb, in: *Adhesion Measurement of Thin Films, Thick Films and Bulk Coatings*, K. L. Mittal (Ed.), p. 107. American Society for Testing and Materials, Philadelphia (1978).
2. J. L. Vossen, in ref. 1, p. 122.
3. K. L. Mittal, *J. Adhesion Sci. Technol.* **1**, 247 (1987).
4. K. L. Mittal, in ref. 1, p. 5.
5. H. E. Hintermann, *Wear* **100**, 381 (1984).
6. S. Baba, A. Kikuchi and A. Kinbara, *J. Vac. Sci. Technol.* **23**, 3015 (1986).
7. J. W. Obreimoff, *Proc. R. Soc. London, Ser. A* **127**, 290 (1930).
8. I. B. Yoon, *Jpn. J. Appl. Phys.* Suppl. 2, Pt. 1, 849 (1974).
9. I. Kondo, T. Yoneyama, K. Kondo, I. Takenaka and A. Kinbara, *J. Vac. Sci. Technol.* **A10**, 3166 (1992).
10. A. Kinbara and S. Baba, *Proc. Joint ASME/JSME Adv. Electronic Packaging*, W. T. Chen and H. Abe (Eds), p. 1. The American Society of Mechanical Engineers, New York (1991).
11. A. Kikuchi, S. Baba and A. Kinbara, *Thin Solid Films* **124**, 343 (1985).
12. A. Kinbara and S. Baba, *Thin Solid Films* **163**, 67 (1986).
13. A. Kikuchi, S. Baba and A. Kinbara, *J. Vac. Soc. Jpn* **25**, 258 (1982).
14. P. Benjamin and C. Weaver, *Proc. R. Soc. London, Ser. A* **254**, 163 (1960).
15. A. Kinbara, S. Baba and A. Kikuchi, *Thin Solid Films* **171**, 93 (1989).

Adhesion Measurement of Films and Coatings, pp. 87–94
K. L. Mittal (Ed.)
© VSP 1995.

Fracture mechanics tests for measuring the adhesion of magnetron-sputtered TiN coatings

D. MÜLLER, Y. R. CHO, S. BERG and E. FROMM*

*Max-Planck-Institut für Metallforschung, Institut für Werkstoffwissenschaft,
Seestrasse 92, D-W 7000 Stuttgart 10, Germany*

Revised version received 22 February 1993

Abstract— Experiments for determining the adhesion strength between TiN coatings and high-speed steel substrates have been performed using the three-point bend test and a modified shear test. Sample preparation is easier for the modified shear test; however, the interpretation of the results is more complicated. The three-point bend test yields fracture mechanics data such as the interface fracture energy G_c and the fracture toughness K_c. The most critical problem is to prepare a sample with a notch which initiates crack propagation at the coating/substrate interface. Various sublayers with weak adhesion have been tested as a notch, such as carbon, oxides, and metals, as well as thin mechanical slits. The results obtained so far demonstrate that the adhesion strength depends on the substrate cleaning treatment and on the contamination of the sputter gas with oxygen. Typical values of the fracture toughness K_c for non-sputter-cleaned substrates are in the range of $1-3$ MN m$^{-3/2}$. The advantages and disadvantages of the various methods used are analysed and critically discussed with respect to sample preparation, reproducibility, and data evaluation.

Keywords: Fracture mechanics tests; adhesion strength; thin film; coatings; TiN; fracture toughness.

1. INTRODUCTION

Quantitative determination of the adhesion strength of hard coatings on a metal substrate is a difficult task and a satisfactory solution of this problem has not yet been attained. For practical application, the adhesion properties of a coating system are usually characterized by the critical load measured in a scratch test or by micro-indenter data [1-4]. The values obtained with both methods are a mixture of the adhesion strength and contributions from the plastic and/or elastic properties of the substrate and the coating materials [5, 6]. The load at the coating/substrate interface is more in the form of compressive stress than tensile or shear stress, and thus, the interpretation of results is complex and not well developed with respect to the determination of adhesion strength data [3].

---

*To whom correspondence should be addressed.

Fracture mechanics test offer a more promising approach to this problem [7, 8]. These provide data, such as the fracture energy G_c, which are well-defined quantities for brittle materials. Extensive studies on the adhesion properties of metal/ceramic systems have been performed successfully by three- and four-point bend tests. Suga *et al.* [9, 10] have shown that G_c is closely related to the adhesion strength. This, of course, is also true for coating systems. Samples used for the three-point bend test have a notch which initiates the crack propagation at the position desired. The notch length and the radius of the notch tip affect the data measured. The notch length is easier to control than microcracks or other defects at the notch tip which give rise to a relatively larger scatter in K_c or G_c values, especially in the case of brittle materials [11]. The formation of a suitable notch exactly at the coating/substrate interface and the connection of different parts of the sample by a strong adhesive are major problems in sample preparation. Because of the scatter in the fracture mechanics data, a set of identical samples must be tested and the quantitative data evaluated with the aid of Weibull statistics [11].

Consequently, many samples coated within a single deposition run should be prepared under well-defined vacuum and sputtering conditions. This study explores various experimental techniques for sample fabrication.

For the preparation of samples under pure or well-defined impurity conditions, a UHV system has been built with various shutters and masks for special sputter cleaning and deposition treatments of the samples [12]. Two types of defects have been tested as a notch. Films of carbon, oxide films, and metal films of iron, aluminium, and copper with weak adhesion have been deposited on the high-speed steel (HSS) substrate before deposition of the TiN coating. Another kind of notch is a cut through the coating perpendicular to the surface after deposition and gluing only one side of the coating to the counterpart. The standard notch of a saw cut at the position of the interface does not work since it is much wider than the coating thickness (about 1 μm). First a modified shear test was employed. Samples are easier to prepare for this technique, but no quantitative theory is available at present for the interpretation of the 'fracture force' or maximum load measured. Therefore, interest was focused more on the development of methods based on the three-point bend test.

Fracture mechanics tests were performed for several series of samples treated under different surface cleaning processes before film deposition and with various amounts of impurity gases in the sputter gas atmospheres. The results are compared with data of the scratch test.

2. EXPERIMENTAL

2.1. Shear test

A HSS sheet 1.5 × 10 × 100 mm was used as the substrate for the shear test (see Fig. 1a).

One end was lapped, mechanically polished, and chemically cleaned. An oxide film produced by annealing the sample for 24 h at 300 °C was used as a notch. Before depositing the TiN film, part of this oxide film was removed mechanically or by sputter cleaning. A shear test sample was produced by gluing a similar HSS sheet on top of the TiN film (adhesive used: Gupalon 20 mono; supplier: Perma Bond, D-6730 Neustadt, Germany). Carbon layers serving as a notch did not work reliably since carbon can contaminate the uncoated sample surface during sputter cleaning, and at higher oxygen concentrations adhesion of the TiN coating to the carbon film fails, probably because of CO formation.

The samples were broken in a tensile testing machine. The coating was completely removed from the substrate and the 'fracture load', F_c, normalized to the non-notched area of the coating was measured. The advantage of the shear stress over the scratch test or the bend test is that only a shear stress is acting at the interface and cracks must start at the interface. Cohesive failure of the coating, which is frequently observed in the scratch tests, is prevented. The 'fracture load' obviously depends on the pretreatment of the samples [12]. Since no useful theory is available for the fracture load, this is rather a relative measure comparable to the critical load in the scratch test. It depends mainly on the interface bond strength, and thus on the presence of

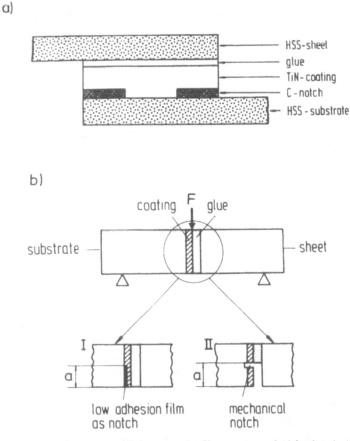

Figure 1. Samples for the fracture mechanics tests. (a) Shear test sample (glued region). (b) Bend test sample: (I) low adhesion film as a notch; (II) a cut perpendicular to the coating as a mechanical notch.

microcracks and other defects at the notch tip. Consequently, the scatter in the data
is larger than with scratch test data, where defects at the interface play a minor role
compared with the bulk properties of the coating.

2.2. Three-point bend test

HSS samples, $5 \times 10 \times 20$ mm, were coated on the top side (5×10 mm) and a
similar HSS bar was bonded as a counterpart on the coating. Two kinds of notches,
about 5 mm long, were used with these samples (see Fig. 1b). In addition to the
oxide films, evaporated metal films, 100 nm thick, were also tested. Copper films
showed the best results when compared with Al, Fe, and oxide films. The adhesion
of the copper film to the HSS substrate is weak but the bonding to the TiN coating
is strong enough to remove it from the substrate. The other notch type tested was a
diamond saw cut vertical to the coating/substrate interface (see Fig. 1b, II). Here only
the upper half of the film is fixed to the counterpart. Overgluing of the notch must be
carefully avoided in this design. The notch length and quality of sample preparation
are controlled by the compliance of the samples.

From the maximum load, F_c, measured in the bend test, the fracture toughness at
the interface, K_c, can be calculated:

$$K_c = F_c \frac{L}{bB^{3/2}} Y\left(\frac{a}{b}\right),$$

with $Y(a/b)$ being a function of the ratio of the notch length, a, to the sample height, b.
B is the thickness of the samples and L is the distance between the supports [9]. This
equation is an approximation derived from linear elastic fracture mechanics for brittle
materials in the limit of a very thin joint. The interface fracture energy, G_c, is defined
as

$$G_c = \frac{K_c^2}{E^*},$$

where E^* is the Young's modulus of the interface. Suga and co-workers showed that
the quantity G_c is proportional to the bond strength at the interface even if its value
is much smaller than G_c due to the contributions of other parameters [10].

2.3. Preparation of clean and contaminated coating systems

The effects of chemical cleaning of the substrates and the oxygen partial pressure
during magnetron sputtering on the fracture mechanics data were determined. Three
series of HSS samples were compared after cleaning in ultrasonic baths of alcohol
and acetone for 5, 10, or 15 min. The samples were not sputter-cleaned before TiN
deposition since the adhesion was too high to be measured with our tests. Another
four series of samples were coated at O_2 partial pressures of 10^{-6}, 10^{-4}, 10^{-3}, and
10^{-2} Pa in the Ar–N_2 sputter gas atmosphere (10^{-1} Pa Ar, 1.5×10^{-2} Pa N_2).

3. RESULTS AND DISCUSSION

Figure 2 shows the fracture toughness K_c as a function of the substrate cleaning time in an acetone–alcohol bath. The two types of samples with different notches, mechanical cut and poor adhesion copper film, show absolute K_c values which differ by a factor of about 2. The large scatter in individual data for samples with the same coating preparation obeys the Weibull distribution function. This is demonstrated in Fig. 3 by two data sets, one for the samples with a copper notch and 15 min cleaning time, the other for samples with a mechanical notch and 5 min cleaning time. The data scatter is probably caused by defects at the notch tip. The steeper slope of the curve for the samples with a copper notch (Weibull modulus $m = 13.6$) indicates that the stress condition is more uniform at notch tip than it is for samples with a

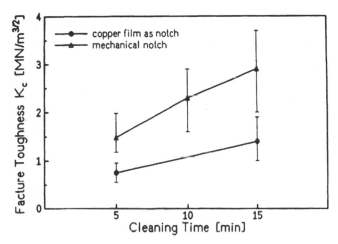

Figure 2. Interface fracture toughness K_c as a function of the cleaning time in an acetone–alcohol ultrasonic bath, measured with two notch types: a copper film and a mechanical cut.

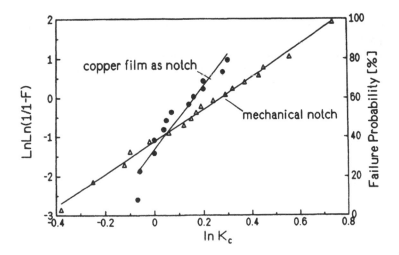

Figure 3. Weibull diagram for bend test samples with a mechanical or copper film notch.

mechanical notch ($M = 4.8$). Thus, crack initiation seems to be easier at the end of a poorly adhering film than at the edge of a saw cut at the mechanical notch.

Figure 4 shows the effect of oxygen contamination in the sputter gas atmosphere on the bend tests results on samples with a mechanical notch. O_2 partial pressures higher than 10^{-4} Pa reduce both the fracture toughness K_c and the fracture load F_c remarkably. *In situ* Auger electron spectroscopy experiments in a UHV system showed that at O_2 pressures higher than 10^{-4} Pa, increasing amounts of oxygen were incorporated into the TiN coating [13].

In Table 1, the F_c values for the shear test samples, the K_c values for the bend test samples, and the L_c values for the scratch test samples are shown for deposition runs with and without sputter cleaning. An increased adhesion strength is found by all

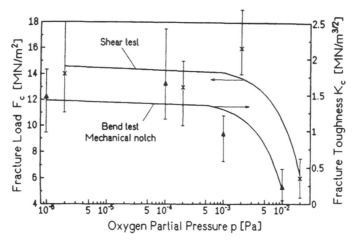

Figure 4. Influence of oxygen contamination on the adhesion strength measured with the shear test and the bend test.

Table 1.

Typical scratch test, shear test, and bend test data on samples with and without sputter cleaning before coating deposition

Test method	Substrate treatment		
	No sputter cleaning	5 min sputter cleaning	10 min sputter cleaning
Scratch test L_c (N)	7	30	70
Shear test F_c (MN m^{-2})	3	6	9
Bend test, mechanical notch K_c (MN m$^{-3/2}$)	1	> 6[a]	
Bend test, copper notch K_c (MN m$^{-3/2}$)	0.75	3	[b]

[a]Glue failure.
[b]Coating failure.

three test methods in sputter-cleaned samples. Typical data for K_c are 1 $MN\,m^{-3/2}$ for non-sputter cleaned samples and > 6 $MN\,m^{-3/2}$ for sputter-cleaned samples.

The maximum load measured in bend and shear tests is limited by the strength of the adhesives used for gluing and the cohesive strength of the coating. If the bond strength at the coating/substrate interface is higher than one of these two values, the samples separate at the coating/adhesive interface or the crack deviates from the coating/substrate interface, crosses the coating, and continues its way in the adhesive. The maximum values that could measured for the interface fracture energy G_c are in the range 200–250 $J\,m^{-2}$ (K_c < 8 $MN\,m^{-3/2}$), i.e. lower than the values for very good commercial hard coatings. Therefore, adhesives with a higher strength should be employed, or the size and dimensions of the samples should be changed in order to shift this upper limit to higher values.

4. CONCLUSIONS

It has been demonstrated that fracture mechanics tests can be performed with hard coatings on HSS substrates if the samples are prepared by appropriate techniques. A weakly bonded metal or carbon film or the oxide scale on one side of a coated surface can act as a notch in a shear test as well as in a bend test. The best results for the system TiN/HSS were obtained with 100 nm thick copper films. Another technique for the initiation of crack propagation along the coating/substrate interface is a saw cut perpendicular to the coating. The samples separate completely along the coating/substrate interface in the bond strength is not higher than the adhesive or cohesive strength of the coating. These two facts preclude the use of fracture mechanics tests for the investigation of very strongly bonded hard coatings on metals.

The results of three-point bend tests can be interpreted by the theories developed for metal/ceramic joints, and Weibull statistics can be used to define quantitatively K_c and G_c values in the framework of fracture mechanics. Metallographic analyses of surfaces of broken samples provide additional information on the crack propagation route and defects at the interface. Comparison of fracture mechanics data with the critical load of the scratch test, or microhardness data may reveal which quantities in these commonly used test methods are related to the adhesion strength and which to the mechanical properties of the coating ant the substrate.

REFERENCES

1. K. L. Mittal, *J. Adhesion Sci. Technol.* **1**, 247 (1987).
2. K. L. Mittal (Ed.), *Adhesion Measurement of Thin Films, Thick Films and Bulk Coatings.* STP No. 640, p. 7. American Society for Testing and Materials, Philadelphia (1978).
3. A. J. Perry, *Surface Eng.* **2**, 183 (1986).
4. Y. Tsukamoto, H. Kuroda, A. Sato and H. Yamaguchi, *Thin Solid Films* **213**, 220 (1992).
5. P. A. Steinmann, Y. Tardy and H. E. Hintermann, *Thin Solid Films* **154**, 333 (1987).
6. J. H. Je, E. Gyarmati and A. Naoumidis, *Thin Solid Films* **136**, 57 (1986).
7. W. D. Bascom, P. F. Becher, J. T. Bitner and J. S. Murday, in ref. 2, p. 63.
8. M. J. Filiaggi and R. M. Pilliav, *J. Mater. Sci.* **26**, 5383 (1991).

9. T. Suga and G. Elssner, *Z. Werkstofftech.* **16**, 75 (1985).
10. G. Elssner, T. Suga and M. Turwitt, *J. Phys. (Paris) Collog. C4* **46**, 597 (1985).
11. A. de S. Jayatilaka, *Fracture of Engineering Brittle Materials.* Applied Science Publishers, London (1979).
12. S. Berg, S. W. Kim, V. Grajewski and E. Fromm, *Mater. Sci. Eng.* **A139**, 345 (1991).
13. S. Berg, N. Eguchi, V. Grajewski, S. W. Kim and E. Fromm, *Surface Coat. Technol.* **49**, 336 (1991).

Adhesion Measurement of Films and Coatings, pp. 95–102
K. L. Mittal (Ed.)
© VSP 1995.

The effect of residual stresses on adhesion measurements

M. D. THOULESS[1],* and H. M. JENSEN[2]
[1]*IBM Research Division, T. J. Watson Research Center, Yorktown Heights, NY 10598, USA*
[2]*Department of Solid Mechanics, The Technical University of Denmark, DK-2800 Lyngby, Denmark*

Revised version received 2 September 1993

Abstract—The adhesion of films and coatings is often measured by determining the load required to separate them from their substrate. If there are residual stresses that are relaxed upon delamination, then an additional contribution to the energy-release rate will affect the measurements. These residual stresses may also cause a shift in the mode-mixedness of the interface crack which, in turn, can affect the interfacial toughness. To ensure an accurate interpretation of adhesion measurements, therefore, the effects of these stresses must be considered. These effects are discussed with particular reference to two commonly used test geometries: the blister test and the peel test.

Keywords: Residual stress; adhesion; peel test; blister test; fracture mechanics.

1. INTRODUCTION

Delamination of a thin film or coating occurs when the energy-release rate, or crack-driving force, $\mathcal{G}$, equals the fracture resistance of the interface, Γ. If delamination is accompanied by relaxation of any residual stresses that may be incorporated in the system, then the relaxation provides a contribution to the energy-release rate. Indeed, the residual stresses can be of a sufficient magnitude to induce delamination without any additional external load, and may be used to provide a fairly convenient means of measuring adhesion. A number of geometries with which residual stresses can be used to determine values of interfacial toughness are illustrated in Fig. 1: (a) delamination from an interfacial flaw situated at a free edge of the film, (b) delamination from a straight cut through the thickness of the film [1], (c) delamination from a circular cut in the film [2, 3], and (d) buckling-driven delamination [4–10]. The advantage of using residual stresses to measure adhesion is that the stress change providing the driving force for delamination is elastic, even if the original stress in the film exceeds the yield stress. Some of the errors associated with large-scale non-linear deformations which can be caused by externally applied loads are therefore avoided. In contrast, for example, almost all the work done in peeling a reasonably adherent metal film from a substrate is dissipated in plastic deformation of the film [11–13]. However, it should be emphasized that, in general, even adhesion measurements using residual stresses will not produce unambiguous geometry-independent results for the interfacial toughness except when the film is very hard, or the interface is very brittle. This is because the high stresses

---

*To whom correspondence should be addressed.

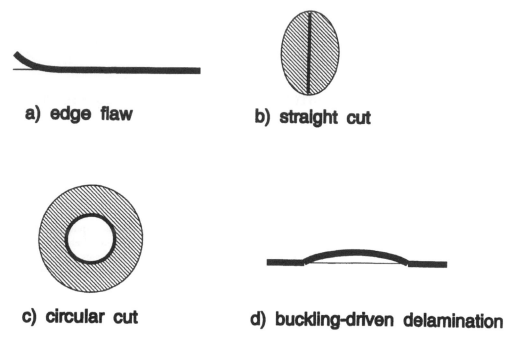

a) edge flaw   **b) straight cut**

c) circular cut   **d) buckling-driven delamination**

Figure 1. Schematic illustrations of several geometries with which residual stresses can be used to make adhesion measurements: (a) delamination from an interfacial flaw situated at a free edge of the film, (b) delamination from a straight cut through the thickness of the film, (c) delamination from a circular cut in the film, and (d) buckling-driven delamination.

associated with a crack tip will cause local non-linear deformations. If the zone in which these non-linear stresses occur is large compared with the film thickness, then the small-scale yielding condition required for rigorous measurements of the toughness will be violated. This can be of considerable significance when measuring the adhesion of thin films. For example, in the peel test, satisfaction of the small-scale yielding condition provides comparable restrictions on the relative values of the yield stress and toughness as the requirement that the peel force produce no plastic bending [14].

One of the features of interfacial fracture that has become appreciated in recent years is the importance of the symmetry of the crack-tip stress field. In particular, three modes of deformation are possible at the tip of a crack: modes I, II, and III, which correspond to normal, in-plane shear, and anti-plane shear, respectively. In axisymmetric and plane-strain geometries, which are appropriate for the blister and peel tests, there is no anti-plane shear and it is convenient to express the ratio of the other two modes by a phase angle, ψ, defined as

$$\psi = \tan^{-1} K_{II}/K_I, \tag{1}$$

where K_I and K_{II} are the mode-I and mode-II stress-intensity factors [15]. If the phase angle is $0°$, pure mode-I conditions exist at the crack tip; if the phase angle is $90°$, pure mode-II conditions exist. Experimental observations suggest that the toughness of an interface is frequently not a constant but depends on the phase angle. It should be emphasized that there is no *a priori* reason to assume that the fracture resistance is independent of the mode of loading, since the size and shape of any dissipative region

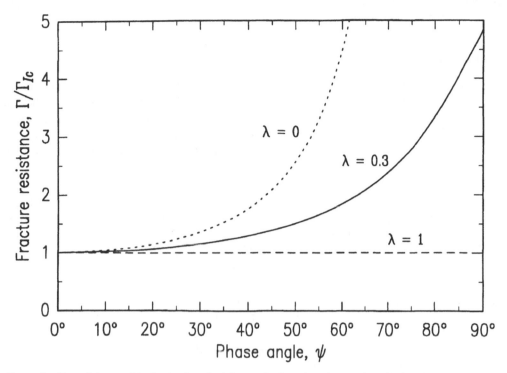

Figure 2. Plot of the empirical mixed-mode failure criterion given in equation (2) that illustrates how an interface toughness may vary with the phase angle.

near the crack tip (which is responsible for most of the toughness) will depend on the details of the elastic stress field. There appears to be no unique relationship between the failure criterion and the phase angle; it must be determined experimentally as part of the procedure of characterizing the interface of interest. One purely empirical expression that links the toughness of an interface to the phase angle, and appears to capture at least some of the essential elements of observed mixed-mode failure behavior, is

$$\Gamma(\psi) = \Gamma_{Ic}\left[1 + \tan^2(1 - \lambda)\psi\right], \tag{2}$$

where Γ_{Ic} is the toughness under pure mode-I conditions and λ is a fitting parameter. This expression is plotted in Fig. 2 for three values of λ. In the limit of $\lambda = 1$, this expression reduces to an energy-balance criterion in which the toughness is independent of the mode of loading. The opposite limit of $\lambda = 0$ is one in which only the mode-I component of the crack-tip stress field is responsible for propagating the crack; the mode-II component of the energy-release rate is dissipated in friction or plastic deformation. In practice, it appears that cracks can grow under pure mode-II conditions, and that λ lies somewhere between these two limits.

The relaxation of residual stresses, therefore, can influence the measurement of adhesion in two ways. Not only will it contribute to the energy-release rate, but, by possibly affecting the stress state at the crack tip, it may also change the value of the appropriate interfacial toughness. These effects are illustrated in the following sections by considering the specific examples of two commonly used adhesion-measurement techniques: the blister test and the peel test. For simplicity, the analyses assume that the elastic properties of the film and substrate are identical, and that the film is sufficiently hard so as to ensure small-scale yielding conditions.

2. BLISTER TEST

A blister test was performed on a model system to illustrate how the fracture resistance of an interface can depend on the phase angle. These experiments, described in detail in [16], were done using a thin sheet of mica bonded to an aluminum substrate by a brittle thermoplastic adhesive. Delamination along the mica/adhesive interface was driven by a point load applied through the substrate and against the back surface of the mica. A thermal-expansion mismatch between the mica and the substrate induced a compressive stress within the mica. The effect of such a residual stress on the phase angle and the energy-release rate of an interface crack in a blister test has been calculated in [16]. For example, an approximate plot showing how the energy-release rate varies with the residual stress is given in Fig. 3. In this plot the film thickness is h, Young's modulus and Poisson's ratio of the film are E and ν, the radius of the delamination is R, and the residual stress is σ_0. The effect of a tensile stress is to make the film more resistant to deformation so that, for a given applied load and geometry, the energy-release rate is reduced; conversely, the energy-release rate is increased by a compressive stress. Using the analysis of [16], the experimental data were converted to provide a plot of the fracture resistance against the phase angle for the mica/adhesive interface (Fig. 4). Superimposed on this figure is the empirical mixed-mode failure criterion of equation (2), where λ has been assigned a value of 0.3 and Γ_{Ic}

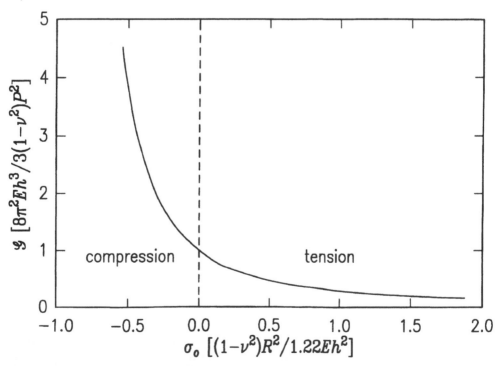

Figure 3. Approximate plot of the energy-release rate against the residual stress for the point-loaded blister test [16].

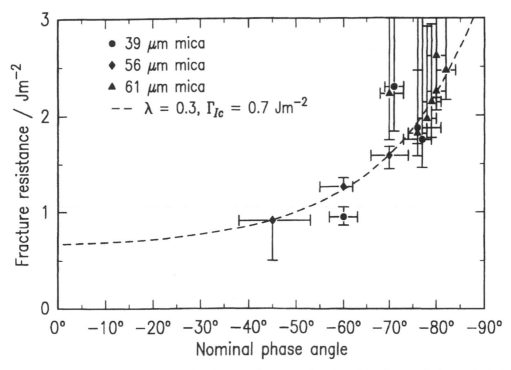

Figure 4. Experimental results for the interface toughness as a function of the phase angle for a mica/adhesive interface. Superimposed on this figure is a fit to the mixed-mode failure criterion given in equation (2) [16].

has been assigned a value of 0.7 J/m^{-2}. These values are in excellent agreement with the results obtained for the same interface using a different experimental configuration involving plane-strain buckling [9]. It should be appreciated that incorporation of the residual stress effects made a great difference to the calculations of the interface toughness. Interpretation of the data using an analysis that assumes the absence of residual stresses [17] produced results that were unreasonably low and not in agreement with other measurements.

Another effect of the compressive stress in this geometry is that it can cause perturbations along the circumference of the region of delamination. The mechanics of this behavior have been described in [8] and [16]. The resulting loss of axisymmetry caused some ambiguity in determining the appropriate radius of delamination and was responsible for some of the uncertainty in the values of toughness shown in Fig. 4.

3. PEEL TEST

If a linear-elastic film of thickness h, with no residual stress, is peeled from a substrate by a force P (per unit width) inclined at an angle θ to the substrate, the energy-release rate is

$$\mathcal{G} = \frac{P^2(1 - \nu^2)}{2Eh} + P(1 - \cos \theta).$$

(3)

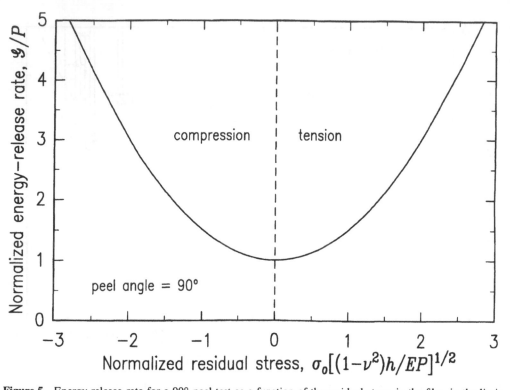

Figure 5. Energy-release rate for a 90° peel test as a function of the residual stress in the film, in the limit of small peel strains.

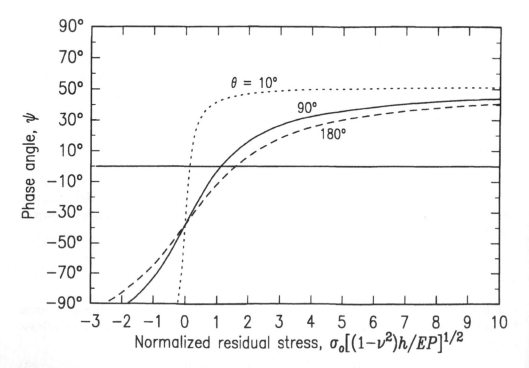

Figure 6. Variation of the phase angle with the residual stress for different peel angles in a peel test [14].

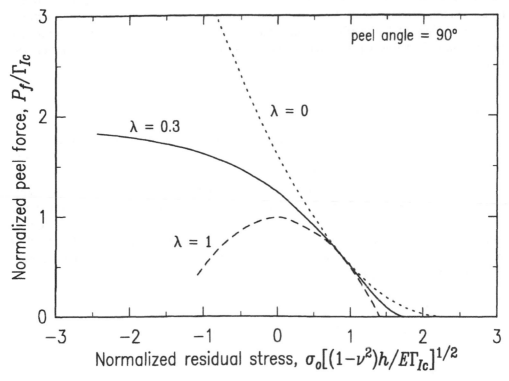

Figure 7. The predicted dependence of the force (per unit width) required to peel a residually stressed film from a substrate for different values of λ.

If there is a residual stress σ_0 in the film, the energy-release rate is [14]*

$$\mathcal{G} = \frac{P^2(1 - \nu^2)}{2Eh} + P(1 - \cos\theta) - \frac{P(1 - \nu^2)\sigma_0 \cos\theta}{E} + \frac{(1 - \nu^2)h\sigma_0^2}{2E}. \qquad (4)$$

This equation is plotted in Fig. 5 for the special case of $\theta = 90°$ and the small peel-strain limit ($P/Eh \to 0$) when it reduces to the limit derived by Kendall [18]. It can be noted that, under these conditions, compressive and tensile stresses have the same influence on the energy-release rate: a residual stress of either sign increases the energy-release rate.

If the residual stress had no effect on the phase angle, then the presence of any residual stress, whether compressive or tensile, would reduce the force, P_f, required to peel a film from a substrate. However, as shown in Fig. 6, the phase angle is very sensitive to the presence of a residual stress [14] and can vary from about 51° to −90°. (The sign convention chosen is such that a negative phase angle indicates a tendency for the interface crack to kink into the film.) If the results of Fig. 6 are combined with equation (4) and the mixed-mode failure criterion of equation (2) (which for this purpose has been assumed to be symmetrical about $\psi = 0°$), it is possible to compute how P_f varies with the residual stress. This is illustrated in Fig. 7 for three values of the fitting

---

*This equation can be derived by considering the energy changes involved as the crack increases by a unit area with respect to the undeformed film. The factor $\cos\theta$ in the third term disappears if the calculation is done for a unit increase in the substrate surface area. Equation (4) is consistent with the stress-intensity factors derived from knowing the bending moment and axial load acting at the crack tip [14, 15].

parameter λ. When $\lambda = 1$ (corresponding to a failure criterion that is independent of the phase angle), the maximum force for peeling occurs when there is no residual stress. For other values of λ, the effects of the phase angle are dominant, and a tensile residual stress results in a lower peel force, while a compressive stress results in a higher peel force.

4. CONCLUSIONS

The major point emphasized in this paper is that if residual stresses are relaxed while adhesion measurements are being made, then both the energy-release rate and the symmetry of the crack-tip stress field will be affected. The additional contribution to the crack-driving force provided by the relaxation of residual stresses must be incorporated when deducing the value of an interfacial toughness from experimental measurements. Furthermore, since a complete characterization of the fracture of an interface requires knowledge of the phase angle at which the toughness has been measured, the effect of the residual stresses on the phase angle must also be considered so that the applicability of the results to other geometries can be deduced.

Acknowledgement

We acknowledge the help given by Y. Wu in reviewing the manuscript.

REFERENCES

1. H. M. Jensen, J. W. Hutchinson and K.-S. Kim, *Int. J. Solids Struct.* **26**, 1099–1114 (1990).
2. M. D. Thouless, *Acta Metall.* **36**, 3131–3135 (1988).
3. R. J. Farris and C. L. Bauer, *J. Adhesion* **26**, 293 (1988).
4. H. Chai, C. D. Babcock and W. G. Knauss, *Int. J. Solids Struct.* **17**, 1069–1083 (1981).
5. G. Gille, in: *Current Topics in Materials Science*, E. Kaldis (Ed.), Vol. 12, pp. 420–472. North Holland, Amsterdam (1985).
6. A. G. Evans and J. W. Hutchinson, *Int. J. Solids Struct.* **20**, 455–466 (1984).
7. H. Chai, *Int. J. Fract.* **46**, 237–256 (1990).
8. J. W. Hutchinson, M. D. Thouless and E. G. Liniger, *Acta Metall. Mater.* **40**, 295–308 (1992).
9. M. D. Thouless, J. W. Hutchinson and E. G. Liniger, *Acta Metall. Mater.* **40**, 2639–2649 (1992).
10. H. M. Jensen, *Acta Metall. Mater.* **41**, 601–607 (1993).
11. K. S. Kim and N. Aravas, *Int. J. Solids Struct.* **24**, 417–435 (1988).
12. K. S. Kim and J. Kim, *J. Eng. Mater. Technol.* **110**, 266–273 (1988).
13. J. Kim, K. S. Kim and Y. H. Kim, *J. Adhesion Sci. Technol.* **3**, 175–187 (1989).
14. M. D. Thouless and H. M. Jensen, *J. Adhesion* **38**, 185–197 (1992).
15. J. W. Hutchinson and Z. Suo, in: *Advances in Applied Mechanics*, J. W. Hutchinson and T. Y. Wu (Eds), Vol. 29, pp. 63–191. Academic Press, New York (1992).
16. H. M. Jensen and M. D. Thouless, *Int. J. Solids Struct.* **30**, 779–795 (1993).
17. H. M. Jensen, *Eng. Fract. Mech.* **40**, 475–486 (1991).
18. K. Kendall, *J. Phys. D* **6**, 1782–1787 (1973).

Adhesion Measurement of Films and Coatings, pp. 103–114
K. L. Mittal (Ed.)
© VSP 1995.

Adhesion of diamond-like carbon films on polymers: an assessment of the validity of the scratch test technique applied to flexible substrates

B. OLLIVIER and A. MATTHEWS*

Research Centre in Surface Engineering, University of Hull, Hull HU6 7RX, UK

Revised version received 16 November 1993

Abstract—The validity of the scratch test as a method of assessing the adhesion of diamond-like carbon (DLC) on polymers has been studied. Sheets of 12 μm thick polyethylene terephthalate (PET) and 100 μm thick polypropylene (PP) were adhesively bonded to glass slides in order to perform the scratch tests. The critical load is defined as the load at which tensile cracks occur in the coating homogeneously throughout the scratch. It is shown that the type and the thickness of the adhesive used have an influence on the critical load value. However, the calculated values of the interfacial shear strength do not depend on the adhesive thickness, and qualitative results in agreement with the literature have thus been obtained. The influence of a nitrogen plasma pretreatment on the adhesion of DLC films on PET and PP has been determined by both scratch test and tensile test techniques. The results follow the same trend and show that the scratch test technique is a good tool for semi-quantitative comparisons.

Keywords: Scratch test; polymer (PP, PET); adhesion; DLC; thin films; plasma pretreatment.

1. INTRODUCTION

Thin films deposited on polymers are widely used in various industrial applications. One of the most important factors is that the coating should not delaminate from the substrate during long-term usage, and thus good adhesion is required. However, adhesion measurement of coatings on flexible polymers is generally achieved by fixing the polymer onto a rigid substrate using an adhesive, and by performing a test valid for rigid substrates. The use of adhesive has several limiting factors:

(i) it must have a strength exceeding the adhesion of the coating;

(ii) the adhesive must not penetrate and chemically affect the interface; and

(iii) it must not introduce additional stress during its setting.

The adhesion of coatings on flexible polymers is generally assessed means of a peel-off test. This method requires the use of two adhesives and is also limited to low-adhesion coatings (the coating must be completely removed from the substrate). A more suitable test consists of the tensile test, but little work has been done on this and the mechanics of the test are not well understood. The laser spallation test seems to be easily applicable to polymers without using an adhesive, but this technique is very sensitive and may be

---

*To whom correspondence should be addressed.

expensive. The scratch test method is generally considered as a standard test for rigid substrates. We have studied the application of the scratch test to flexible polymers, and more specifically the influence of the adhesive layer. The results show that the critical load alone is not a good indication of adhesion, but semi-quantitative results can be obtained if the composite hardness of the sample is taken into consideration.

2. THEORY

The scratch test technique consists of drawing a rounded diamond tip (typically 200 μm in radius) across the thin film surface. A load is gradually applied to the tip until the film is debonded from the substrate. Heavens was the first person to use this test to assess the adhesion of aluminium films on glass using an intermediate chromium layer [1]. Benjamin and Weaver [2] analysed the test in detail and related the critical value to the interfacial shear strength between the film and the substrate. Their studies showed that plastic deformation of the substrate occurred during the scratch test and that this deformation induced a shearing force at the film–substrate interface in front of the indenter. When a critical load was reached, this shearing force was equal to the interfacial shear strength and thus complete removal of the film from the substrate occurred. However, Butler *et al.* [3] observed that the process of scratch formation was very complex and concluded that no absolute values of adhesion could be deduced. Further studies also showed the limitation of the simple model described by Benjamin and Weaver [4–7]. Since then, different approaches have been developed to describe the scratch test using a dynamic model [8] or based upon energy theories [9, 10]. The scratch test remained, however, a standard adhesion test and the critical load is generally taken as a measure of adhesion, the main difficulty in the scratch test being the critical load determination. Recently it has been defined as the load for which either increased acoustic emission is generated [11, 12] or a change in friction is detected [13]. As films fail in different modes according to both the coating and the substrate properties, different criteria were assigned, such as lower and upper critical loads [14, 15] and threshold adhesion failure [16], to describe the observed effects.

The theory that we based our calculations on to assess the interfacial strength of DLC films on PET substrates is similar to the one-dimensional model described by Benjamin and Weaver [2]. Figure 1 shows the forces acting on the diamond tip during the scratch test. W is the load applied on the tip of radius R, F is the shearing force per unit area due to the deformation of the surface, and a is the radius of the circle of

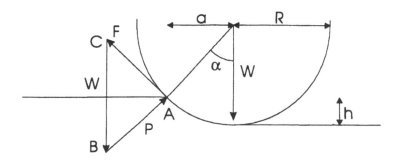

Figure 1. Forces acting on the diamond tip during the scratch test (after [2]).

contact and thus the half-width of the track left after the test. P is similar to a uniform hydrostatic pressure acting normal to the surface of indentation and can therefore be considered as the 'resistance to penetration' or 'hardness' of the substrate material. The forces represented by the triangle ABC can thus be represented as

$$F = P \tan \alpha = \frac{aP}{\sqrt{R^2 - a^2}}. \tag{1}$$

This expression is valid if there is a clear track formation. In our study, the critical load is not determined upon this criterion, but on crack formation; thus, the calculated value for the interfacial shear strength is lower than the actual value. Although this equation has been shown by many researchers to give imprecise adhesion values, we believe its use is appropriate in cases where only semi-quantitative comparisons are needed.

Under the assumption that the substrate material is plastically deformed, the hardness P can be related to the applied load and the radius of the contact circle as shown in Fig. 2:

$$P = \frac{W}{\pi a^2}. \tag{2}$$

This expression is slightly different from the one put forward by Benjamin and Weaver, as they assumed that the half-width a was much smaller than the stylus radius R. However, as our substrates are very soft (thin polymer sheets adhesively bonded to glass slides), the half-width of the tracks occurring during the scratch test is typically 70 μm (one-third of the radius of the diamond tip) and thus parameter a cannot be assumed to be negligible. The interfacial shear strength τ_0 and the critical load L_c can then be related by

$$\tau_0 = \frac{L_c}{\pi a \sqrt{R^2 - a^2}}. \tag{3}$$

Equations (1) and (3) permit the calculation of the interfacial shear strength in two ways; they also provide a means of checking the assumption of plastic deformation during the

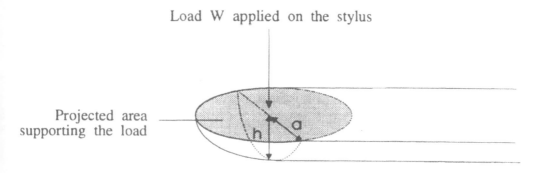

Figure 2. Relation between the effective hardness or penetration resistance and the applied load.

test. Experiments thus consist of drawing the stylus across the surface of the film until the criterion defining the critical load is observed. Then by measuring the half-width of the track created by applying the critical load, the interfacial shear strength of the sample can be calculated from equation (3). To measure the composite hardness of the sample consisting of the film, the polymer sheet, the adhesive, and the glass slide, an indentation is produced using the rounded diamond indenter under a certain load and the composite hardness is calculated from equation (2). Inserting this value into equation (1) allows the calculation of the interfacial shear strength τ_0.

The main problem with thin polymer sheets adhesively bonded to glass is that the nature of the adhesive and its thickness will affect the effective 'hardness' of the sample. Furthermore, as samples do not use the same adhesive thickness, their hardnesses vary. It has been shown, for brittle films on soft substrates, that the critical load value depends on the hardness of the sample [17]. Semi-quantitative comparisons cannot be achieved if only the critical load is considered as a measurement of adhesion; a calculation of τ_0 taking the hardness in account is therefore required.

3. EXPERIMENTAL

DLC films were deposited on polymer substrates in an RF CVD process, using acetylene as the plasma gas. The experimental conditions are summarized in Table 1.

The films produced were brown, semi transparent, and hard. Every deposition was carried out under similar conditions and the film thickness was measured using a Talysurf and was equal to 110 ± 5 nm.

The polymer sheets used were 12 μm thick polyethylene terephthalate (PET) and 100 μm thick polypropylene (PP). PP strips (5 mm wide, 25 mm long) were used for the tensile test. Both polymer sheets were adhesively bonded to glass slides prior to deposition in order to perform a scratch test, using either epoxy resin or Loctite 406. Polymers were bonded to glass slides by rolling the plastic sheet onto the glass incorporating an adhesive sandwich. As epoxy resin presents a higher viscosity than Loctite adhesive, the thickness uniformity of the Loctite adhesive was better than that of the epoxy resin: the thickness of the epoxy resin typically varied between 2 and 10 μm for a sample, whilst the Loctite adhesive thickness was more uniform, being between 2 and 3 μm. The surface roughness of the polymers was not significantly affected by this operation.

Table 1.
Experimental conditions for the deposition of DLC films

	Etching stage	DLC deposition
Plasma gas	Nitrogen	Acetylene
Pressure (Torr)	10	4.5
RF power (W)	200	100
DC offset voltage (V)	-275	-230
Temperature ($^\circ$C)	35	45
Deposition time (min)	0–7.5	10
Deposition rate (nm/min)	—	11

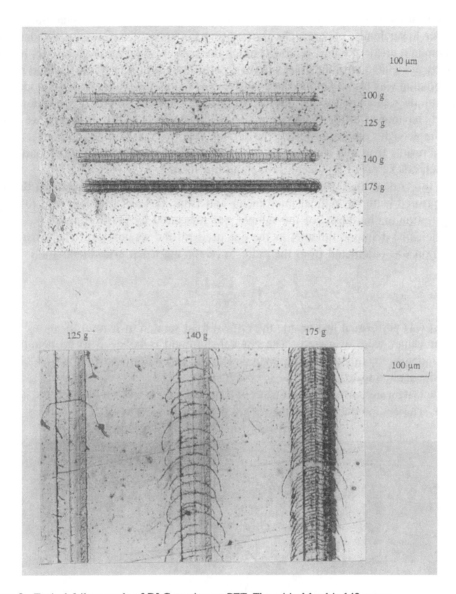

Figure 3. Typical failure mode of DLC coating on PET. The critical load is 140 grams.

The scratch test was performed with an adapted microhardness tester (Leitz Miniload2). A stylus with a Rockwell C type diamond was employed, the tip being ground to a radius of 200 μm. While a load was applied, the sample was moved at a constant speed of 0.1 mm/s. The load was then incrementally increased for a new scratch on an adjacent area. Although several authors have used the scratch test to assess the adhesion of metallic films on plastics, no precise criterion was defined to determine the critical load [18–23]. We defined the critical load as the load for which cracks occur in a homogeneous way throughout the film, as shown in Fig. 3 [24]. It was observed during the test that an increase in load over this threshold induced partial spallation of the film, but did not leave a clear channel. A similar mechanism has been observed by Je *et al.* with TiN films on soft substrates [25]. They defined the critical load as the load at which an increase in acoustic emission was detected. Although some cracks could occur for lower loads, photographs showed that the acoustic emission was related to a homogeneous array of cohesive cracks along the scratch. Spallation of the TiN film was generally associated with the cracks in the case of thick films (1500 nm), but no removal of the coating was found behind the cohesive cracks for thinner coatings (700 nm). The failure mode was shown to be due to the coating buckling, which is considered as a typical behaviour of brittle material on ductile substrates [17, 26]. Although we did not study the acoustic emission during the scratch test, we can compare our observations with the results from the study of Je *et al.* [25] and thus define the critical load as that for which cracks occur regularly throughout the scratch.

Once the critical load was reached during the scratch test, the half-width of the track was measured using an optical micrometer and from Talysurf profiles. The resistance to penetration or 'hardness' of the sample was measured by performing an indentation using the same diamond indenter as the one used for the scratch test. The resistance to penetration was calculated from the radius a of the mark left after indentation:

$$P = \frac{\text{Load}}{\pi a^2}. \qquad (4)$$

This test was performed just beside the critical load scratch in several locations and the different values were averaged. The conical diamond is believed to be better than a Knoop indenter for this test as the geometry in the indentation process is then closer to that in the scratch test.

Tensile tests were performed by means of a simple tensile tester manufactured in-house (Fig. 4). The equipment and the theory are described elsewhere [27].

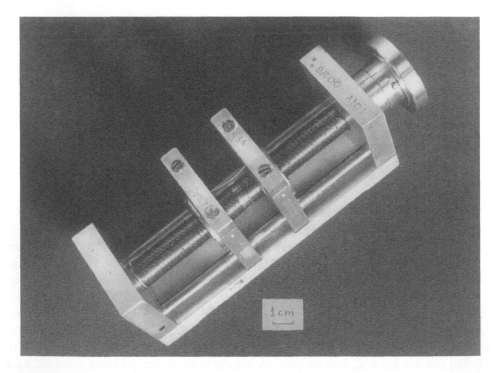

Figure 4. Tensile tester used for adhesion evaluation.

4. RESULTS AND DISCUSSION

4.1. Influence of the resistance to penetration of the surface on the critical load

Different thicknesses of epoxy resin were used to affix 12 μm thick PET sheets on glass. DLC films were deposited on several of these samples (Fig. 5) under identical experimental conditions and the adhesion of the DLC coating was measured using the scratch test method on several locations of each sample. By removing the PET sheet and creating a step through the epoxy resin, the adhesive thickness was then estimated from Talysurf profiles. The results are given in Figs 6 and 7.

Although these coatings theoretically present a similar adhesion to PET, it can be seen that the value of the critical load depends on the thickness of the epoxy resin underlayer. The thicker the epoxy resin, the higher the critical load. As the sample composite hardness is inversely proportional to the adhesive interlayer, the critical load is dependent on the sample composite hardness (Fig. 7). Figure 8 shows the effect of the adhesive interlayer thickness on the resistance to penetration of the sample (consisting of the DLC coating, the PET sheet, the epoxy resin interlayer, and the glass slide). The sample is more resistant to penetration if the adhesive layer is thin. As the epoxy resin is softer than glass, the thinner the epoxy layer, the harder the surface. Because stresses and deformation cannot expand through a thin epoxy layer as much as through a thick one, the surface becomes more resistant to penetration.

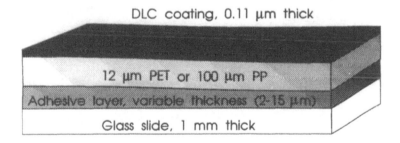

Figure 5. Typical sample used for the scratch test.

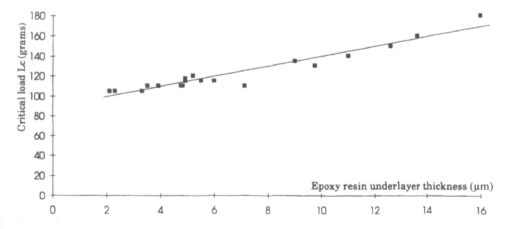

Figure 6. Influence of the adhesive interlayer thickness on the critical load L_c (DLC films on PET).

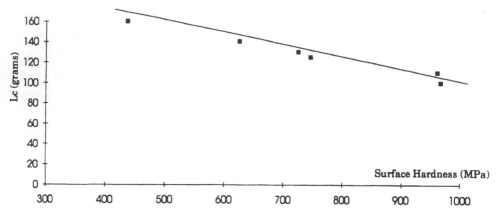

Figure 7. Influence of the sample composite hardness on the critical load, the composite hardness being inversely proportional to the adhesive interlayer thickness.

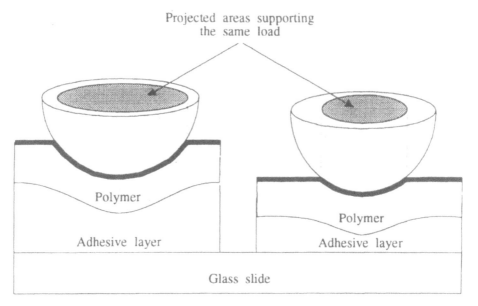

Figure 8. Adhesive layer thickness and resistance to penetration of the surface. As the glass is harder than the adhesive, it will limit deformations. The surface is more resistant to penetration for a thin adhesive interlayer because stresses cannot expand as much through a thin layer as through a thick one. Therefore the projected area supporting the load is greater if the adhesive is thick and the pressure on the coating is lower. As a certain pressure is required to induce shear stresses large enough to remove the coating, the critical load for the 'hard' surface will be lower than that for the softer surface.

The deformation of samples with a thick adhesive interlayer is larger than that of a low thickness epoxy resin sandwich specimen; therefore, the contact area between the stylus and the polymer substrate is larger for soft surfaces (thick adhesive layer). When the same load is vertically applied, the pressure at the area of contact is thus low if the surface is soft, and higher for a harder surface. If a certain pressure is required at the area of contact between the stylus and the film to induce a shear stress P large enough to remove the coating, this pressure will be reached with a lower load for the harder surface than for the softer substrate; thus, the critical load for the hard substrate will be lower than that of the soft surface, as shown schematically in Fig. 8. This also means that the critical load is not always truly representative of the adhesion of the coating on polymers adhesively bonded to glass, and therefore it is required that the adhesion of such samples be assessed taking into account both the critical load and the resistance to penetration of the surface.

4.2. Validity of the method

In order to assess the validity of the scratch test, we performed scratch tests on DLC films on PP and PET substrates. The polymers were stuck onto glass using Loctite 406 as the adhesive, and were pretreated prior to DLC deposition using a nitrogen plasma for different etching times. A nitrogen plasma has been shown to be an efficient treatment to increase the adhesion of metallic films on PET [28], and in a previous study [27] we showed that the adhesion of DLC films on PET was increased using a nitrogen plasma pretreatment, but the increase was not as important as in the case of metallic films. Nitrogen plasma pretreatments have been shown to modify the energy of PP surfaces and also to increase dramatically the adhesion between aluminium films and PP [29, 30]. Comparison were made between tensile test 'adhesion' values and scratch tests performed on identical samples, but it must be kept in mind that these quantitative values are not absolute adhesion values, but only a way to establish semi-quantitative comparisons, which can be related to the concept of 'practical adhesion' defined by Mittal [31].

Figures 9a and 9b show the results obtained using different nitrogen etching times prior to DLC deposition. Variations of the critical load are given in Fig. 9a and the interfacial shear strength values calculated from equations (1) and (3) are displayed in Fig. 9b. As discussed previously, it can be seen that if the critical load were the only parameter taken into account for adhesion, the nitrogen plasma pretreatment could be concluded as having no effect on adhesion enhancement. However, the trend shown by the calculated values, taking into account both the critical load and the resistance to penetration, is similar to the results given in the literature [29, 30]. The results obtained from the tensile test are qualitatively similar to those calculated using equation (3), but quite different quantitatively.

The difference between the calculated values from the two equations may be due to several factors: it has been noticed that for the same load, the half-width of the scratch left by the stylus in motion is larger than the radius of the contact circle due to a static indentation. The theory described previously is based on the assumption that deformations occur in a plastic mode, and so these two values should theoretically be equal under the full plastic mode. During the 'hardness' test, the reading of the radius of the contact circle may be affected by elastic recovery.

During the scratch test, the area supporting the load is geometrically smaller than in the static indentation, and because of the motion, a higher resolved force is applied on this surface due to combined stresses. Under the same load, the half-width of the track left by the stylus in motion is thus greater than the radius of the static indentation. Therefore the resistance to penetration is 'underestimated' compared with the real deformation induced by the stylus in motion, so that the values calculated from equation (3) are greater than those calculated using equation (1).

The effect of the nitrogen plasma pretreatment on the interfacial shear strength of DLC films on 12 μm thick PET substrates is shown in Fig. 10. The shear strength

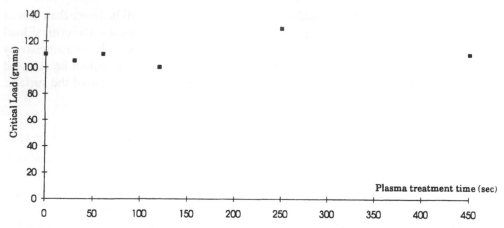

Figure 9a. Influence of nitrogen plasma pretreatment on the critical load required to remove DLC films on PP substrates.

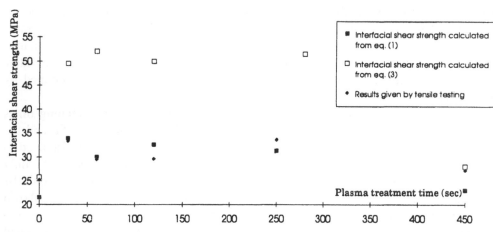

Figure 9b. Influence of nitrogen plasma pretreatment on the interfacial shear strength of DLC films on PP substrates. Calculations are made from data of Fig. 9a.

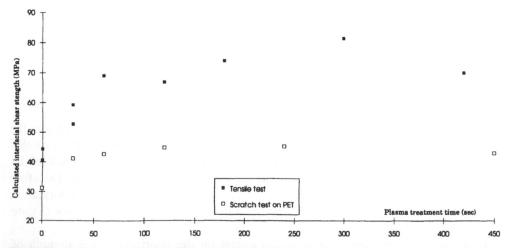

Figure 10. Influence of nitrogen plasma pretreatment on the interfacial shear strength of DLC films on PET substrates. Shear strengths have been assessed by both scratch test and tensile test.

values assessed by the scratch test were calculated from equation (1). They can be compared with the results obtained on similar samples using the tensile test: the results show the same trend qualitatively (increase in 'adhesion'), but no valid comparison can be made from a quantitative point of view. This conclusion shows the limit of the scratch test, as it can be used only for semi-quantitative comparisons on flexible substrates (PP or PET).

5. CONCLUSION

A simple model for the scratch test applied to flexible substrates adhesively bonded to glass has been derived from a simple theory for rigid substrates. The critical load cannot be considered as a precise indication of the adhesion for such samples as it depends on the thickness of the adhesive layer. However, by measuring the resistance to penetration of the sample surface and by calculating the interfacial shear strength between the film and the polymer, it is possible to obtain semi-quantitative comparisons. The results provided by the scratch test technique about the influence of nitrogen plasma pretreatment on the adhesion of DLC films on PP and PET are in qualitative agreement with those obtained using the tensile test and also with those available in the literature. But no conclusions can be drawn about the validity of the scratch test to give absolute interfacial shear strength values, since the sample loading is complex, the theory is approximate (derived from a simple model applied to rigid substrates), and the criterion for the critical load determination is not the one required in the case of the rigid substrate theory. However, the scratch test appears to be a good tool for semi-quantitative comparisons if both the critical load and the surface 'hardness' of the sample are taken into account for the calculation of interfacial shear strength values.

REFERENCES

1. O. S. Heavens, *J. Phys. Radium* **11**, 355 (1950).
2. P. Benjamin and C. Weaver, *Proc. R. Soc. London, Ser. A* **254**, 163 (1960).
3. D. W. Butler, C. T. H. Stoddart and P. R. Stuart, *J. Phys. D, Appl. Phys.* **3**, 887 (1970).
4. K. L. Chopra, *Thin Film Phenomena*, p. 313. McGraw Hill, New York (1969).
5. P. Benjamin and C. Weaver, *Proc. R. Soc. London, Ser. A* **274**, 267 (1963).
6. J. L. Mukherjee, L. C. Wu, J. E. Greene and M. Pestes, *J. Vac. Sci. Technol.* **12**, 850 (1975).
7. K. L. Mittal, *Electrocomp. Sci. Technol.* **3**, 21 (1976).
8. M. T. Laugier, *Thin Solid Films* **76**, 289 (1981).
9. M. T. Laugier, *Thin Solid Films* **117**, 243 (1984).
10. S. J. Bull and D. S. Rickerby, in: *Advanced Surface Coatings*, D. S. Rickerby and A. Matthews (Eds), p. 333. Blackie Publisher, Glasgow (1991).
11. A. J. Perry, *Thin Solid Films* **78**, 77 (1981).
12. P. A. Steinmann, Y. Tardy and E. Hintermann, *Thin Solid Films* **154**, 333 (1987).
13. S. J. Bull, D. S. Rickerby, A. Matthews, A. Leyland, A. R. Pace and J. Valli, *Surface Coatings Technol.* **36**, 503 (1988).
14. H. Ronkainen, K. Holmberg, K. Fancey, A. Matthews, B. Matthes and A. Broszeit, *Surface Coatings Technol.* **43/44**, 888 (1990).
15. H. Ronkainen, S. Varjus, K. Holmberg, K. S. Fancey, A. R. Pace, A. Matthews, B. Matthes and E. Broszeit, in: *Mechanics of Coatings*, D. Dowson, C. M. Taylor and M. Godet (Eds). Leeds-Lyon 16 Tribology Series, p. 17. Elsevier, Amsterdam (1990).
16. J. Oroshnik and W. K. Croll, in: *Adhesion Measurement of Thin Films, Thick Films and Bulk Coatings*, K. L. Mittal (Ed.), p. 158. ASTM, Philadelphia (1978).
17. S. J. Bull, *Surface Coatings Technol.* **50**, 25 (1991).

18. C. Weaver, *Adhesion, Fundamentals and Practice*, p. 46. McLaren, London (1969).
19. C. Weaver, *J. Vac. Sci. Technol.* **12**, 18 (1975).
20. L. F. Goldstein and T. J. Bertone, *J. Vac. Sci. Technol.* **12**, 1423 (1975).
21. P. Phuku, P. Bertrand and Y. De Puydt, *Thin Solid Films* **200**, 263 (1991).
22. O. Knotek and F. Loffler, in: *Metallized Plastics 2: Fundamental and Applied Aspects*, K. L. Mittal (Ed.), p. 141. Plenum Press, New York (1991).
23. A. A. Galuska, *J. Vac. Sci. Technol.* **A10**, 381 (1992).
24. A. Dehbi-Alaoui, B. Ollivier and A. Matthews, *Vide Couches Minces, Suppl.* **261**, 221 (1992).
25. J. H. Je, E. Gyarmati and A. Naoumidis, *Thin Solid Films* **136**, 57 (1986).
26. P. J. Burnett and D. S. Rickerby, *Thin Solid Films* **154**, 403 (1987).
27. B. Ollivier and A. Matthews, *J. Adhesion Sci. Technol.* In press
28. Y. De Puydt, Y. Novis, M. Chtaib, P. Bertrand and P. Lutgen, *Vide Couches Minces, Suppl.* **243**, 157 (1988).
29. L. Garby, B. Chabert, D. Sage, J. P. Soulier and L. Porte, *Vide Couches Minces, Suppl.* **243**, 153 (1988).
30. F. Arefi, V. Andre, F. Tchoubinel, J. Amouroux and M. Goldman, *Vide Couches Minces, Suppl.* **243**, 167 (1988).
31. K. L. Mittal (Ed.), in: *Adhesion Measurement of Thin Films, Thick Films and Bulk Coatings*, pp. 5–17. ASTM, Philadelphia (1978).

Adhesion Measurement of Films and Coatings, pp. 115–126
K. L. Mittal (Ed.)
© VSP 1995.

Adhesion improvement of RF-sputtered alumina coatings as determined by the scratch test

F. RAMOS* and M. T. VIEIRA

Mechanical Engineering Department, University of Coimbra, Largo D. Dinis, 3000 Coimbra, Portugal

Revised version received 18 February 1993

Abstract—One of the most important problems to overcome in the production of coatings, a major technology in surface modification, is the maximization of coating to substrate adhesion. In spite of its limitations, the scratch technique is most used in adhesion characterization studies. In this study, the adhesion of alumina coatings deposited by RF magnetron sputtering on AISI M2 high speed steel (HSS) was evaluated by the scratch test. An attempt to improve the adhesion of the alumina coatings was investigated by the deposition of an intermediate layer of Ti or TiN. The influence of the substrate surface roughness, the intermediate layer thickness, and the sputter etch cleaning on the coating adhesion was studied.

Keywords: Sputtering; alumina coatings; adhesion; scratch test; cohesive failure; adhesion failure.

1. INTRODUCTION

Alumina, as a bulk material, exhibits high oxidation resistance, high chemical stability, high hardness, and good wear resistance, which makes it desirable for use in applications where these properties are indispensable for a successful performance in service. However, in a large number of applications, the use of this material is strongly but its low toughness and poor machinability. This has led to the production of alumina coatings where the 'good' properties can be exploited whilst the 'poor ones' can be avoided by the use of tougher substrate materials to be coated, for example steel. Using different deposition techniques and/or different deposition conditions, it is possible to adapt the alumina-coated component to a large spectrum of applications. Nevertheless, the production of these coatings, as in the case of other coatings, has faced a serious problem, namely poor adhesion or at least not sufficiently good adhesion to show a satisfactory behaviour in service conditions. Thus, the use of an intermediate layer which is chemically more reactive than Al_2O_3 could potentially improve the adhesion between the alumina coating and the substrate.

---

*To whom correspondence should be addressed.

The selection of an intermediate layer can be a difficult task since the requirements for the coating/substrate composite are often very complex and several compromises must be made. It is known that in chemical vapour deposition (CVD) techniques this problem has been overcome by using an intermediate coating, usually TiC, Ti(C,N) or TiN, as a bonding layer between the performance coating and the substrate [1–4]. In fact, the majority of CVD-coated cemented cutting tools are produced on this basis, which has allowed the achievement of better performing tools. Other successful results, where the life times of tools are increased several times, have been obtained for TiN-coated high speed steel (HSS) tools using physical vapour deposition (PVD) processes [5–8]. These results are possible only if good adhesion between the two materials can be achieved.

Finally, it is known that reactive metals which form stable oxides usually show good adhesion to other oxides [9]. Taking into account the fact that titanium is a metal which easily reacts with oxygen, producing very stable oxides, it is expected that it will perform well as a bonding layer between the Al_2O_3 coating and the HSS substrates. In fact, chemical reactions can occur between Ti and the oxygen of both the alumina coating and the native HSS surface oxides.

The aim of this work was to improve the adhesion of RF-sputtered Al_2O_3 to HSS substrates, using a pure Ti or TiN coating as an intermediate bonding layer. The influence of differently deposited Ti and TiN coatings, the thickness of the intermediate bonding layer, the surface finish of the substrates, and the ion bombardment cleaning was considered in order to achieve the best Al_2O_3/HSS adhesion. The adhesion evaluation of the different Al_2O_3/Ti/HSS and Al_2O_3/TiN/HSS composites was made using the scratch test [10–12].

2. MATERIALS AND METHODS

2.1. Materials

All the coatings were deposited on heat-treated AISI M2 high speed steel (HSS). The surfaces of the samples were prepared with different roughnesses using 1000 and 4000 mesh grit paper and 3 μm diamond paste, leading to average roughnesses R_a of 0.2, 0.07, and 0.05 μm (average roughness R_a defined according to ISO 4287/1 or DIN 4768/1).

The coating materials were deposited from Ti and α-Al_2O_3 sintered targets. The process gases were argon (99.9997%) and nitrogen (99.9995%); the latter was used as a reactive gas for the growth of the TiN intermediate layers.

2.2. Methods

2.2.1. Deposition of the coatings. All the depositions were carried out in an Edwards ESM 100 sputtering apparatus with a two-cathode configuration. Some alterations were made to the apparatus in order to allow it to operate with two different and independent power supplies (RF and DC). Except for these modifications, the standard

ESM 100 sputtering equipment was also fitted with an RF substrate biasing unit which was used to etch the substrates by ion bombardment before deposition and to bias the coating during its growth.

Alumina coatings were produced using a deposition power, P_{dep}, of 6.4 W/cm^2; a deposition pressure, p_{dep}, of 0.5 Pa; and a substrate bias, V_s, of 0 V. Figure 1 shows a standard alumina coating deposited under these conditions.

A systematic study was carried out in order to determine the deposition conditions for the intermediate bonding layer which gave the best adhesion between the functional coating and the HSS substrate.

As already mentioned, the selected intermediate bonding layers were Ti and TiN. Titanium was deposited on the HSS substrates at various deposition pressures and substrate biases in order to study the influence of these parameters on the Ti coating properties. The deposition conditions used in the production of Ti coatings are listed in Table 1. For the TiN coatings, only one deposition condition was studied based on the best adhesion results obtained for TiN coatings produced in the same deposition equipment based on other current research studies. This condition is also shown in Table 1.

Ti and TiN coatings were characterized mainly with respect to morphology and adhesion. The coatings for which a good bonding behaviour could be expected were selected for the second part of this work.

Substrates with different surface finishes were prepared, and intermediate layers with different thicknesses were deposited on the HSS substrates. Before deposition,

Figure 1. Al$_2$O$_3$ coating. Deposition conditions: $P_{dep} = 6.4$ W/cm^2; $p_{dep} = 0.5$ Pa; $V_s = 0$ V.

Table 1.
Ti and TiN deposition conditions

Coating	Deposition powder (W/cm^2)	Deposition pressure (Pa)	Substrate bias (V)	N$_2$/Ar flux ratio
RF Ti	6.4	0.1–1	0–200	—
DC Ti	4.5	0.1–0.5	0–200	—
DC TiN	6.4	0.5	0	1/6

the HSS substrates were also etched by ion bombardment for different times: 0, 20, and 60 min. The role of these parameters on the adhesion of the composite was evaluated.

2.2.2. Characterization of the coatings. Ti, TiN, Ti + Al$_2$O$_3$, and TiN + Al$_2$O$_3$ coated samples were inspected using both optical and scanning electron microscopy (SEM). A Nikon HFX-II optical microscope and a JEOL JSM-T330 scanning electron microscope were used. The Al$_2$O$_3$ samples were gold-sputtered to overcome their electrical insulating characteristics and thus make SEM observations possible.

Among the large number of methods investigated for coating adhesion evaluation, the scratch test is widely used to evaluate the adhesion of thin hard coatings on softer substrates [13]. Although the scratch test cannot provide a real adhesion measurement, it is commonly used to compare film quality with respect to the strength of the interface between similar coating/substrate samples, thus giving the coatings 'practical adhesion' [14]. In general, two different critical loads, L_{c1} and L_{c2} can be defined according to the failure events observed in the sample. The critical load L_{c1} is determined by the first cohesive failure (conformal cracking), while the critical load L_{c2} corresponds to the first sign of a lack of adherence between the film and the substrate (spalling).

The scratches were performed in a CSEM-Revetest (Neuchatel, Switzerland) in a continuous normal applied load mode, in the range 0–200 N and under the following standard conditions: diamond tip radius $R = 0.2$ mm; scratching speed $dx/dt = 10$ mm/min; loading rate $dL/dt = 100$ N/min. Four scratches were made on each sample, and average L_c values are presented. The morphology of the scratch tracks was observed by optical and scanning microscopy and the type and length of coating failure were determined. The use of energy dispersive X-ray spectroscopy (EDXS) permitted the locus-of-failure indentification.

3. RESULTS

3.1. Ti coatings

An increase in the deposition pressure in the case of RF-sputtered Ti coatings transforms a featureless morphology ($p_{dep} = 0.1$ Pa) into a columnar one ($p_{dep} = 0.5$ Pa; Figs 2a and 2b). The opposite effect occurs when the negative substrate bias is increased (Fig. 2c). These results are in accordance with the literature [15, 16], where it is recognized that the decrease in deposition pressure and the increase in negative substrate bias lead to more compact coating morphologies, due to increasing adatom mobility and to the ion shot peening effect on the growing film.

Bearing in mind these results, similar deposition conditions were selected for the deposition of Ti films by DC sputtering. The morphology of DC Ti films is different from that of RF films, exhibiting a homogeneous type T morphology [17] for a deposition pressure $p_{dep} = 0.5$ Pa and a substrate bias $V_s = -200$ V (Fig. 2d).

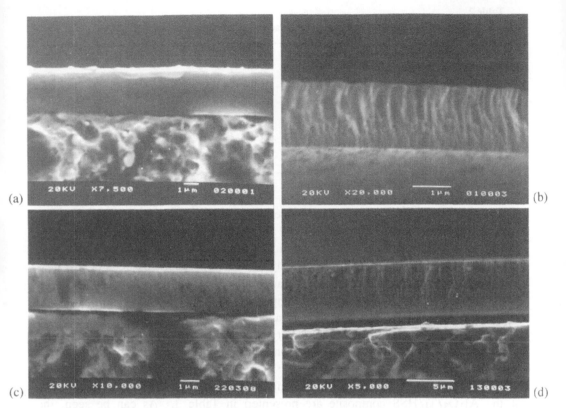

Figure 2. Ti coatings deposited under different deposition conditions; influence of the deposition pressure and negative bias on the coating morphology. (a) RF sputtering — $p_{dep} = 0.1$ Pa, $V_s = 0$ V; (b) RF sputtering — $p_{dep} = 0.5$ Pa, $V_s = 0$ V; (c) RF sputtering — $p_{dep} = 1.0$ Pa, $V_s = -200$ V; (d) DC sputtering — $p_{dep} = 0.5$ Pa, $V_s = -200$ V.

Some authors believe that thin films with fibrous and columnar morphologies support external stresses better than dense and compact coatings; a kind of 'pseudoelasticity' of the columnar coatings caused by the relatively poor cohesion between the single grains will be responsible for this behaviour [18].

The scratch test could not be applied to Ti coatings owing to their ductility. All the scratches carried out on these coatings showed an extrusion effect, even for small loads. Thus, the selection of the deposition conditions to precoat HSS substrates was based solely on morphological aspects.

In this work, two Ti deposition conditions were selected in order to obtain coatings with two different homogeneous morphologies using both RF and DC sputterings:

Condition A (RF sputtering): $P_{dep} = 6.4$ W/cm^2; $p_{dep} = 0.1$ Pa; $V_s = 0$ V

Condition B (DC sputtering): $P_{dep} = 4.5$ W/cm^2; $p_{dep} = 0.5$ Pa; $V_s = -200$ V.

For the final choice of condition A or B to be used in the deposition of the intermediate layer, HSS substrates ($R_a = 0.05$ μm, etching time = 20 min, and $V_s = -800$ V) were coated under conditions A and B with a 0.1 μm thick Ti film. The thickness of this film was chosen taking into account the results obtained by Bull *et al.* [19].

Table 2.

Influence of the Ti deposition conditions on the adhesion of the system $Al_2O_3/Ti/HSS$

Composite	Ti film thickness (μm)	Total thickness (μm)	Critical load (N)	
			L_{c1}	L_{c2}
Al_2O_3/HSS	—	1.8	—	5–10
$Al_2O_3/Ti/HSS$ (condition A)	0.1	1.8	22 ± 1	25 ± 2
$Al_2O_3/Ti/HSS$ (condition B)	0.1	1.8	25 ± 1	30 ± 2

Afterwards, the Ti precoated samples were immediately coated with Al_2O_3. The results of the scratch test performed on these samples are presented in Table 2, where it can be clearly seen that the critical load values obtained are significantly higher than those measured on Al_2O_3 deposited directly on the HSS substrates.

In spite of the increase in the critical load values observed with the use of a Ti intermediate layer, we also attempted to improve adhesion further by optimizing the substrate roughness, the Ti film thickness, and the substrate sputter etch cleaning. For evaluating the influence of these parameters, the Ti intermediate layer was deposited under condition B. Although both conditions show similar L_{c1} and L_{c2} values, higher values are obtained for condition B.

The range of parameters studied and their influence on the adhesion characteristics of the $Al_2O_3/Ti/HSS$ composite are presented in Table 3. As can be seen, the variation in the parameters can lead to a change in the locus of failure and to an appreciable increase in the L_{c1} and L_{c2} values, when compared with the critical loads for the Al_2O_3/HSS composite.

For the same thickness of the intermediate layer, the increase in substrate roughness inhibits the failure by conformal cracking, and only adhesion failure is observed (Fig. 3). This phenomenon can be explained by the variation of the stress field (in di-

Table 3.

Influence of the Ti intermediate layer thickness, the substrate roughness (R_a), and the sputter etching time on the adhesion of the system $Al_2O_3/Ti/HSS$

Intermediate layer thickness (μm)	$R_a = 0.05\ \mu$m		$R_a = 0.07\ \mu$m		$R_a = 0.20\ \mu$m	
	L_{c1}	L_{c2}	L_{c1}	L_{c2}	L_{c1}	L_{c2}
0.1	20 ± 2	30 ± 6	—	43 ± 5	—	29 ± 10
0.3	30 ± 3	31 ± 4	—	31 ± 4	—	19 ± 10

Sputter etch cleaning time (min)	$R_a = 0.05\ \mu$m		$R_a = 0.07\ \mu$m		$R_a = 0.20\ \mu$m	
	L_{c1}	L_{c2}	L_{c1}	L_{c2}	L_{c1}	L_{c2}
0	39 ± 3	49 ± 10	35 ± 6	45 ± 10	—	23 ± 3
20	20 ± 2	30 ± 6	—	43 ± 5	—	29 ± 10
60	—	28 ± 3	25 ± 2	36 ± 2	—	22 ± 4

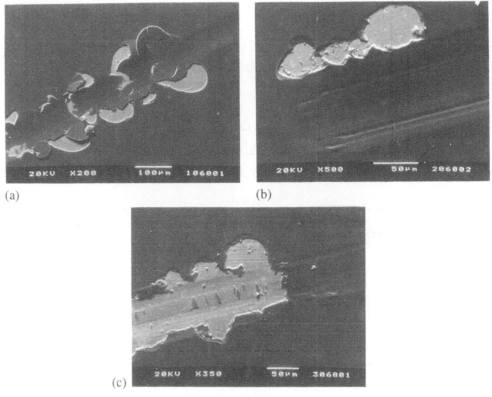

(a)   (b)

(c)

Figure 3. Adhesion failure of the Al_2O_3/Ti/HSS composite (sputter etch cleaning time = 20 min, Ti thickness = 0.15 μm). (a) $R_a = 0.05$ μm; (b) $R_a = 0.07$ μm; (c) $R_a = 0.20$ μm.

rection and/or magnitude) induced by surface roughness. Also, the effective bonding area between the coating and the substrate increases for the same scratched area.

EDXS analysis reveals that adhesion failure occurs only at the Ti/HSS interface and not at the Al_2O_3/Ti interface, since titanium was never detected in the detached zones observed (Fig. 4).

It should be stressed that for the highest roughness, the L_{c2} values decrease, particularly when the Ti film thickness is greater than 0.1 μm, which allows us to conclude that this parameter has a negative effect on the adhesion properties of the Al_2O_3/Ti/HSS assembly.

The decrease of adhesion with high surface roughness can be attributed to an increase in the surface friction coefficient, which is known to reduce the critical load of brittle materials fracture [20, 21].

Regarding the influence of the etch cleaning time, it can be seen that the highest L_{c2} value was obtained when the substrates were only ultrasonically cleaned. Generally, the scratch test critical load decreases with increasing sputter etching time. In fact, between the ultrasonic cleaning and final pump-down of the deposition chamber, the HSS surfaces can be easily oxidized (at a pressure of 10^{-3} Pa, it takes only 1 s to form an oxide monolayer [22]). The presence of oxygen on the HSS surfaces

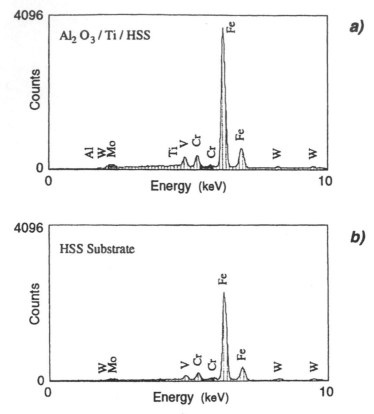

Figure 4. EDXS analysis. (a) Al_2O_3/Ti/HSS ($R_a = 0.07$ μm); detached zone at the end of the scratch; (b) HSS substrate.

probably promotes the chemical reaction between the Ti coating and the oxidized surface [23].

The decrease of the L_{c2} values with increasing sputter etching time, particularly for an etching time of 60 min, can be attributed to the decrease of the HSS substrates' chemical reactivity, owing to the smaller amount of oxides on the HSS surfaces. The oxidized surfaces exposed to the plasma are sputtered away, leaving a more oxide-free surface for Ti deposition.

As previously mentioned, the thickness of the intermediate layer can play an important role in film failure. It was observed in this study that Ti thicknesses greater than 0.1 μm tended to reduce the L_{c2} values of the Al_2O_3/Ti/HSS composite. This trend can be a consequence of the differences between the mechanical properties of Ti and Al_2O_3, which have an increasing contribution to the final behaviour of the composite with increasing Ti layer thickness. In an extreme case, a very thick Ti intermediate layer could act as a softer substrate than HSS. For this reason, intermediate layers should be no thicker than what is necessary to promote the gettering effect and thus minimize other undesirable influences on the assembly. Moreover, the thicker the intermediate layer, the greater the tendency for the occurrence of failure inside this layer [19].

3.2. DC TiN coating

The role of TiN coatings as an intermediate bonding layer was studied with respect to the influence of its thickness and of the substrate roughness.

TiN coatings were deposited on HSS using the deposition conditions presented in Table 1. The film obtained showed a typical columnar morphology and adhesion failure at a critical load $L_{c2} = 50$ N. These properties, particularly the high value of the critical load for adhesion failure, permit us to expect a suitable behaviour of the TiN coating when used as an intermediate bonding layer, i.e. with a lower thickness.

The influence of the TiN layer thickness and the HSS surface roughness on the adhesion of the Al_2O_3/HSS composite is summarized in Table 4.

Table 4.
Influence of the TiN intermediate layer thickness and the substrate roughness (R_a) on the adhesion of the system Al_2O_3/TiN/HSS

Intermediate layer thickness (μm)	$R_a = 0.05\ \mu$m		$R_a = 0.07\ \mu$m		$R_a = 0.20\ \mu$m	
	L_{c1}	L_{c2}	L_{c1}	I_{c2}	L_{c1}	L_{c2}
0.1	25	42	30	55	21	40
0.3	22	56	18	62	22	42

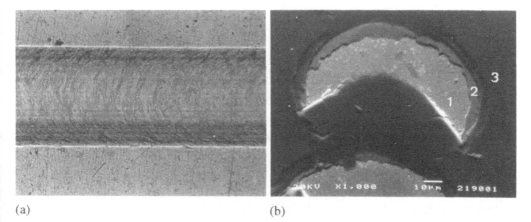

(a)   (b)

Figure 5. Adhesion failure of the Al_2O_3/TiN/HSS composite (TiN thickness = 0.3 μm). (a) Cohesive failure (conformal cracking); (b) adhesion failure (spalling): zone 1 — HSS substrate; zone 2 — TiN intermediate layer; zone 3 — Al_2O_3 coating.

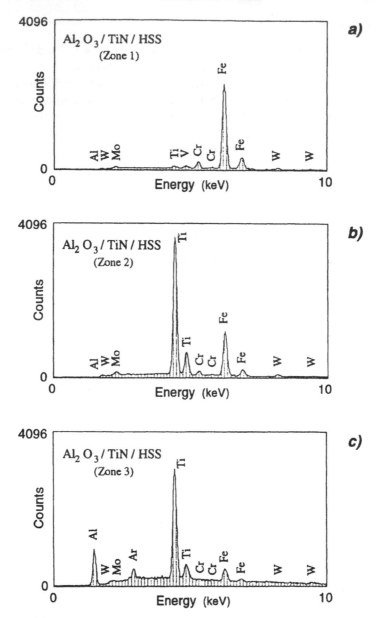

Figure 6. EDXS analysis of the adhesion failure of the Al_2O_3/TiN/HSS composite. (a) On the HSS substrate (zone 1); (b) on the intermediate TiN coating (zone 2); (c) on the Al_2O_3/TiN/HSS composite (zone 3).

The results obtained reveal that with a TiN intermediate layer, irrespective of the substrate roughness, cohesive failure in the AL_2O_3/TiN/HSS system is always observed, and this is followed by adhesion failure of the coating at much higher loads. This effect is quite different from that found for the Ti intermediate layer, where the increase of substrate roughness inhibits the appearance of cohesive failure.

Cohesive failure (conformal cracking) and adhesion failure (spalling) of the Al_2O_3/TiN/HSS system are shown in Figs 5a and 5b, respectively. When adhesion failure is observed, it occurs at the TiN/HSS interface, as can be concluded from the EDXS spectra presented in Fig. 6.

The use of a TiN intermediate layer instead of a Ti layer increases the L_{c2} critical load of the coating assembly. This effect is a natural consequence of the better mechanical strength of the TiN layer, which becomes more evident when thicker intermediate layers are compared. On the other hand, studies concerning TiN coatings show that thicker layers give higher values of the critical load [24].

4. CONCLUSIONS

The presence of titanium and titanium nitride as bonding intermediate layers can significantly improve the adhesion of Al_2O_3 coatings to M2 high speed steel.

Ti intermediate layers of approximately 0.1 μm thickness appear to maximize the Al_2O_3/HSS adhesion, particularly when the average surface roughness of the substrates is kept below 0.07 μm. These conditions not only increase the resistance to spalling (adhesion failure), but also avoid, in some cases, the cohesive failure of the Al_2O_3 coatings.

The increase of the titanium nitride intermediate layer thickness increases the scratch test critical load L_{c2} values *vis-à-vis* those obtained for the Al_2O_3/HSS composite. Nevertheless, when TiN films are used, first cohesive failure is always observed. The L_{c1} values observed for Al_2O_3/TiN/HSS are similar to those obtained for Al_2O_3/Ti/HSS when cohesive failure occurs in this system.

The presence of a Ti intermediate bonding layer can improve by approximately five times the adhesion of RF-sputtered alumina coatings to M2 high speed steel, which consequently contributes to a better performance of the Al_2O_3/HSS composite.

REFERENCES

1. J. N. Lindstrom and R. Johannesson, *J. Electrochem. Soc.* **123**, 555–559 (1976).
2. P. A. Dearnley, *Surface Coatings Technol.* **29**, 157–177 (1986).
3. I. Lehermitte-Sebire, R. Colmet and R. Naslain, *Thin Solid Films* **138**, 221–233 (1986).
4. S. Vourinen and L. Karlsson, *Thin Solid Films* **214**, 132–143 (1992).
5. W. D. Sproul and R. Rothstein, *Thin Solid Films* **126**, 257–263 (1985).
6. H. Randhawa, *J. Vac. Sci. Technol.* **A4**, 2755–2758 (1986).
7. A. Matthews, *Chart. Mech. Eng.* **32**, 31–34 (1985).
8. J. E. Sundgren, *Thin Solid Films* **128**, 21–44 (1985).
9. G. Katz, *Thin Solid Films* **33**, 99 (1976).
10. A. J. Perry, *Thin Solid Films* **107**, 167–180 (1983).
11. A. J. Perry, *Surface Eng.* **2**, 183–190 (1986).
12. T. A. Steinmann, Y. Tardy and E. Hintermann, *Thin Solid Films* **154**, 333–349 (1987).
13. K. L. Mittal, *J. Adhesion Sci. Technol.* **1**, 247 (1987).
14. K. L. Mittal (Ed.), *Adhesion Measurement of Thin Films, Thick Films and Bulk Coatings*, pp. 5–17. ASTM, Philadelphia (1978).
15. J. A. Thornton, *J. Vac. Sci. Technol.* **A4**, 3059 (1986).
16. C. W. Chen and C. S. Alford, *J. Vac. Sci. Technol.* **A6**, 128–133 (1988).

17. J. A. Thornton and A. S. Penfold, in: *Thin Film Processes*, J. L. Vossen (Ed.), pp. 75–133. Academic Press, New York (1978).
18. J. T. Roth, E. Brozeit and K. H. Kloos, in: *Plasma Surface Engineering*, E. Brozeit (Ed.), Vol. 2, pp. 837–844. DGM Informationsgesellschaft. Verlag, Oberursel (1989).
19. S. J. Bull, P. R. Chalker, C. F. Ayres and D. S. Rickerby, *Mater. Sci. Eng.* **A139**, 71–78 (1991).
20. J. M. Challen and P. B. Oxley, *Wear* **53**, 229 (1979).
21. D. R. Cilroy and W. Hirst, *Br. J. Appl. Phys.* **2**, 1784 (1969).
22. B. N. Chapman, *J. Vac. Sci. Technol.* **11**, 106–113 (1974).
23. J. A. Thornton, in: *Deposition Technologies for Films and Coatings*, R. F. Bunshah (Ed.), p. 243. Noyes Publications, New Jersey (1982).
24. F. H. M. Sanders, *Mater. Sci. Eng.* **A139**, 85–90 (1991).

Adhesion Measurement of Films and Coatings, pp. 127–142
K. L. Mittal (Ed.)
© VSP 1995.

Scratch indentation, a simple adhesion test method for thin films on polymeric supports

GEORGE D. VAUGHN,[1] BRUCE G. FRUSHOUR[1,*] and WILLIAM C. DALE[2]

[1]*Monsanto Company, The Chemical Group, Advanced Performance Materials Division, St. Louis, MO 63167, USA*
[2]*Monsanto Company, The Chemical Group, Physical and Analytical Sciences Center, Indian Orchard, MA 01151, USA*

Revised version received 17 November 1993

Abstract—We have utilized scratch indentation as a test method for evaluating the strength and adhesion of polymeric substrates coated with metal and diamond-like carbon (DLC) films. Metallized films were prepared by coating poly(ethylene terephthalate) with copper, using a proprietary electroless deposition process. The copper adhesion was evaluated using a standard tape-pull test, a special inverse peel test in addition to scratch indentation, and a good correlation was observed among the methods. The most critical experimental parameter in the scratch indentation method was the stylus radius. Good differentiation of adhesion among samples was much more evident using an 800 μm tip radius as opposed to 200 μm. We concluded that a simple reliable measurement of practical adhesion could be based on commercially available scratch indentation equipment.

The DLC coatings are hard and brittle compared to the ductile copper. The scratch indentation tests were quite sensitive to the strength of the DLC coatings. Adhesion was not directly measured but was inferred from optical analysis of the scratch trace. One of the instruments contained an acoustic emission detection channel and noise burst were recorded that could be directly associated with the cracking of the DLC coating when tip styli of 200 and 800 μm were used. Another instrument of quite different design utilized a stylus tip radius of 1 μm and could be operated in displacement control. Distinct drops in the normal force curve were observed when the stylus broke through the DLC coating. The critical load, L_c, was measured as a function of the coating thickness and could be fitted to a classical physical model based on the breaking strength of a thin disk.

Keywords: Scratch test; scratch indentation test; electroless deposit; peel test; diamond-like films; adhesion; metal films.

1. INTRODUCTION

For the past several years, we have been involved in the development of 'performance surfaces' based on a functional surface material, often supported on a thicker polymer layer. Several classes of surfaces have been investigated, where the desired function may be mechanical, electrical, chemical, optical, or some combination. The surface layer itself may have some complicated internal or interfacial structure. Each of these structures presents real or potential problems in adhesion which critically affect performance. A universal adhesion test has been elusive, and each functional surface must

---

*To whom correspondence should be addressed.

be considered as a separate case. We report here the results and successful assessment of adhesion using small-scale scratch indentation testing in two particular cases: metal and diamond-like carbon (DLC).

We currently utilize a proprietary electroless deposition process to metallize roll film and synthetic fabrics [1]. The metal coatings are of the order of 125 μm thick. The DLC coatings are applied to plastic substrates, such as polycarbonate, that are typically 3.2 mm (1/8 in.) thick [2]. This paper addresses the need for a relatively rapid and simple test for adhesion that can be applied to both the development and the manufacturing of these products. The test must eventually be applicable to a wide variety of planar configurations, starting with simple geometrical shapes and extending to intricate printed patterns. After a careful examination of adhesion test methods, we decided to evaluate scratch indentation, and, in the case of the metallized materials, compare this method with a variation of the conventional peel test method used in the plating industry.

The application of scratch indentation as a rapid and reliable method for evaluating the adhesion of thin films has been described in review articles by Mittal [3] and Wu [4]. In this method a stylus is pulled across the surface of a coated substrate to induce coating failure by delamination, cracking, or some other failure event. Two testing modes are common: constant load and increasing load. In constant load testing, the load on the stylus is kept constant as the stylus is pulled across the surface. This is repeated with a higher load until failure point or critical load (L_c) is obtained. In the increasing load mode, the load is ramped up constantly during a single scratch until the coating fails and the load at failure is taken as L_c. While the design of the scratch testers does vary, most include, in addition to normal force, some combination of lateral force, acoustic emission, scratch depth, and microscopic observation for determining L_c. All of the experiments in this study used the increasing force mode.

The experimental difficulties in scratch indentation and instrumental complexity depend strongly on the radius of the stylus tip. Relatively simple mechanical designs are sufficient for the 50–1000 μm range instruments. However, a new generation of instruments featuring radii of the order 1 μm are becoming commercially available. Experiments in this range are much more demanding, but offer the opportunity of true micromechanical measurements of film material properties since the stylus tip radius is now of the order of the film thickness. Both types of instruments were used in this study and are described in the Experimental section.

Our first objective was to determine whether scratch indentation could discriminate among metallized polymeric materials which exhibited obvious differences in adhesion, based on simple tape tests and some qualitative observations. In essence, we are developing a test of practical adhesion [5] because the measurement will yield a number that hopefully reflects the inherent or basic strength of the interfacial bond but is then modified by extraneous effects introduced by the test procedure, film thickness, and other sample parameters. Preliminary results with random samples of metallized polymer films demonstrated that the scratch test did generate reliable values of practical adhesion. We next set out to systematically vary some of the basic parameters of the metallized film, i.e. the metal thickness, the plastic film substrate, the substrate thickness, and the method of deposition, to determine the generality of using L_c as a measure of practical adhesion.

A second and more exploratory goal of this work was a feasibility study to determine whether scratch indentation would provide fundamental information about the strength and adhesion of DLC coatings on polymeric substrates. To this end, we were able to examine the effect of DLC film thickness on L_c utilizing two quite different scratch indentation instruments with stylus tip radii of 1 and 800 μm. The dependence of L_c on the film thickness obtained using the smaller radius was consistent with a classical physical model based on the breaking strength of a thin disk.

2. EXPERIMENTAL

2.1. Preparation of metallized films

Most of the work was done using copper-metallized poly(ethylene terephthalate) (PET) films. The unmetallized commercial PET substrate film was supplied as biaxially oriented, either 100 or 125 μm thick, and was obtained from two vendors. The films were coated on one side by the manufacturer with either a slip coating or a proprietary coating that increased the adhesion of the metal. Much of our investigation concerned two samples of copper on PET that exhibited large differences in practical adhesion as measured by a simple tape test. These materials were prepared using an electroless deposition method described later. Some measurements were also made on copper-coated polyimide film metallized in a similar fashion.

The copper thickness for the electroless copper films was determined using transmission electron microscope images of the metallized film cross-section. The metal thicknesses for the sputtered samples were measured with a DEKTAK 3030 profilometer, manufactured by the Sloan Technology Division of Veeco Instruments.

2.1.1. (a) Catalyzation process. A catalytic metal solution [1] was prepared from 0.08 g of Pd (OAc)$_2$, 10 g of acetone, and 1.5 g of water. A water-soluble polymer solution was prepared from 0.2 g of poly(vinyl alcohol) (MW 125 000, 88 mol% hydrolyzed), 0.1 g of Triton® X-100 (25% aqueous solution, surfactant), 37.9 g of water, and 0.073 g of triethylamine. The catalyst solution was slowly added to the polymer solution and 50 g water and 0.025 g magnesium sulfate were added. This was coated on PET using a 25 μm wet-film applicator. The samples were activated at 160°C for 9 min.

(b) Electroless plating. A MaCuDep 54 bath (MacDermid) was used, but it was diluted 27% with water. Samples were plated for 2.5 min at 35°C.

2.1.2. Sputtering. We also prepared copper-coated PET films by a sputtering process using a Materials Research Corporation (MRC 903) sputtering device. This enabled us to vary the thickness and to directly compare films made by the sputtering and electroless processes.

2.2. Preparation of hard coatings

The diamond-like carbon (DLC) coatings were prepared by Daniels [2] using a radio frequency planar diode process to generate a plasma from a hydrocarbon feed gas, and then depositing the DLC coating from this plasma the 3.2 mm thick polycarbonate substrate. The DLC coating thickness was varied from 1.4 to 3.3 μm by adjusting the feed gas pressure, hydrocarbon content, and other process parameters.

2.3. Tape-pull testing

We performed this test by adhering 25 mm wide squares of metallized film onto a rigid substrate using Scotch brand tape that had adhesive on both sides. The copper side was facing up. Next, the surface was scored very lightly to cut only through the copper layer, thus creating 100 individual 2.5 mm wide squares. To this surface, one end of a strip of Scotch brand number 610 single-coated tape was pressed on firmly and then pulled away in the vertical direction using a smooth 1 s duration pull. Each individual square of the scored copper coating was examined under a 10× power microscope to determine if any copper had been removed, and the percentage of squares in which no copper was removed was taken as a practical measure of the adhesion strength.

The electroless plated samples designated **EMD-LOW** and **EMD-HIGH** (Table 1) were labeled as such because of the tape-pull results. The percentage of copper remaining on the PET surface after the pull was 22 and 93%, respectively. The adhesion strength measured by the tape-pull was high for the sputtered material (**SP-80**, Table 1), i.e. no copper was pulled away from 95% of the test area.

Table 1.

Comparison of the CSEM Micro Scratch Tester data with the inverted peel test data for copper-coated PET

Sample	200 μm tip		800 μm tip		Inverse 90° peel (g/cm)
	3 N/min (N)	10 N/min (N)	10 N/min (N)	30 N/min (N)	
SP-80[a] (80 nm)[b]	0.5	0.8, 0.7	> 10	8.0, 10.1, 9.8	NA
SP-120[a] (120 nm)	1.3, 1.3	1.5	> 10	11.3, 11.1	NA
EMD-LOW[c] (100 nm)	0.48, 0.39	0.69, 0.77	1.48, 1.45, 1.41	NA	8.2 (9)[d]
EMD-HIGH[c] (100 nm)	0.58	1.71, 2.11	4.50, 5.43, 3.78	NA	14 (5)[d]

[a]SP = sputtered.
[b]Metal thickness.
[c]EMD = electroless metal deposit; **LOW** = failed tape test; **HIGH** = passed tape test.
[d]Number of replicates.

90° Peel Configuration for Cu/PET

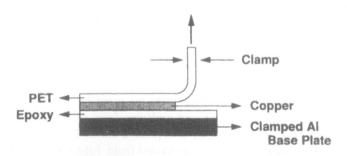

Figure 1. The inverted peel test configuration.

2.4. Peel testing

The most common adhesion test in the plastics plating industry is the 90° peel test (ASTM B 533-92). This test requires that the metal be ductile and thick enough to be bent back and pulled from the plastic substrate at a 90° angle. One typically begins with an electroless layer and then builds a much thicker layer using an electrolytic process. Final thicknesses are typically 35 μm, which for copper is equivalent to 300 g/m^2, and approximately 2 h of electrolytic plating are required. However, the peel test used in this study was modified to allow measurement of adhesion directly between the thin electroless metal layer and the plastic film. Because the plastic film substrate is flexible and the metal film is very fragile, we used an inverted peel test [6]. The metal surface was fixed to an aluminum base plate using epoxy adhesive, and the flexible film was pulled from the metal using a standard 90° peel test jig. A diagram of the inverted peel test is shown in Fig. 1. A tab for grabbing the PET film with the tensile machine grips (peel initiation point) was prepared by simply scraping off the metal for a total distance of 2.5 cm from the end of the test piece. An epoxy adhesive (Devcon® 5-Minute Epoxy) was applied to the aluminum base plate and the metallized film was pressed onto the surface, with a tab of the substrate separated from the aluminum base by tape (peel initiation point). The samples were allowed to cure for 24 h at ambient temperature. The peel test was performed on an Instron Tester with the base plate clamped to a frame designed to provide horizontal motion coordinated to the clamp motion such that the line of propagating peel remained directly below the vertically moving clamp holding the free tab. A crosshead speed of 2.5 cm/min was used throughout, and peel forces were measured on a 20 pound (89 N) capacity load cell. The force output showed an initial

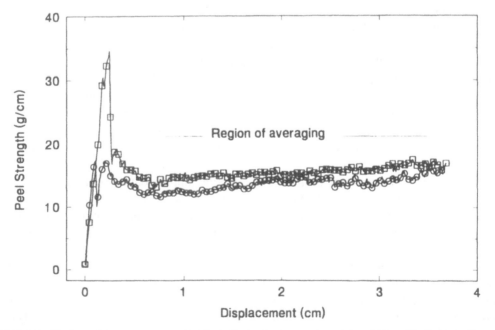

Figure 2. Peel strength curves (representative) obtained using an inverted peel test. The initial spike is ignored and the peel strength is obtained by averaging the data over the region indicated (0.16–0.55 cm).

(highly variable) spike, followed by a relatively constant peel force. This peel propagation force was averaged over a large part of the length and used to calculate the peel strength. Typical peel force traces are shown in Fig. 2. The peel test is only useful if the adhesion between the metal coating and aluminum base provided by the epoxy exceeds the adhesion between the metal and the substrate film. In some systems with very high metal-to-substrate adhesion, the epoxy bond will fail and the test becomes invalid.

2.5. Scratch indentation testing

The majority of the work was done using a Micro Scratch Tester manufactured by the Centre Suisse d'Electronique et de Microtechnique S.A. (CSEM), Neuchâtel, Switzerland, and operated in the increasing normal load mode. The scratching speed was kept constant at 10 mm/min for all experiments, and loading rates of 3, 5, 10, and 30 N/min were used. The two styli were diamond indenters ground to a 120° cone (Rockwell C profile), with a spherical apex of either 200 or 800 μm. Some measurements were also taken on a developmental instrument located in the Center for Interfacial Engineering (CIE) at the University of Minnesota [7]. This instrument, designated later as a CIE scratch tester, is based on the design of T. W. Wu [4] at IBM. It uses a stylus with a tip radius of 1 μm and depth-ramping (displacement) control, usually at 15.0 nm/s for metallized surfaces and 25.0 nm/s for the DLC hard coatings. Other tip sizes are available.

The scratch test method is basically a comparison test [8]. The critical load, L_c, determined by this method is widely regarded as being representative of coating adhesion behavior [3–5, 8]. However, extracting a pure measure of adhesion from L_c is difficult because L_c is also dependent on the testing parameters (loading rate, stylus tip radius) and sample parameters (coating thickness and substrate hardness). Three different detection modes for determining the L_c values of coated samples are built into the Micro Scratch Tester: optical microscopy, acoustic emission, and tangential force measurements. For the ductile copper coatings used in this study, only the microscopic detection mode was sensitive to the coating failure at L_c. After the sample is scratched, the instrument returns to the starting point, replaces the stylus with an optical microscope, and enables the operator to view the entire length of the scratch. The point along the scratch at which delamination first occurs is determined by the operator and the corresponding scratch distance and normal force are read out by the instrument. This normal force is L_c.

Acoustic emission was not useful for the ductile copper coating; only a continuous background noise was observed. With brittle materials, such as the hard organic coatings used here, we were able to observe discrete acoustic events that could be associated with the formation of cracks in the coating surface. This will be discussed later in Section 3.2.

Microscopic detection of L_c was also required when testing copper-coated films with the CIE instrument. The scratches were typically less than 100 μm long, so a scanning electron microscope was needed to determine the point of failure. However, with the DLC coatings, distinct decreases in the normal force could be detected as the stylus penetrated the coating at L_c. This is the expected behavior for failure of the coating while operating the CIE instrument in displacement control.

3. RESULTS

3.1. Copper on PET film

Optical photographs of the scratch traces of the copper-coated PET films designated as having low adhesion (**EMD-LOW**) and high adhesion (**EMD-HIGH**) in Table 1 are compared in Fig. 3. As expected, the copper coatings deformed in a ductile manner when tested with the Micro Scratch Tester. A much higher normal load was re-

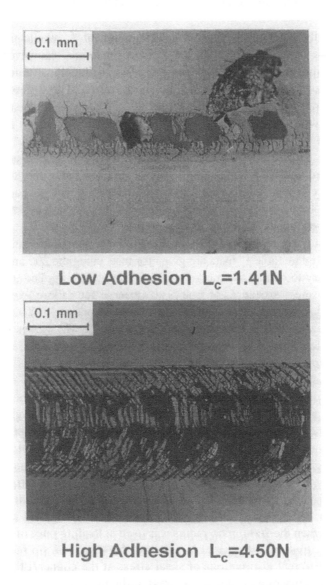

Figure 3. Scratch traces of copper-coated PET showing low and high adhesion. The section of the traces near L_c is shown. The loading rate and the scratching rate were the same for both samples, i.e. 10 N/min and 10 mm/min. The width of the scratch at L_c is larger for the sample with high adhesion simply because failure occurs further along the scratch. The beginning of the scratches is on the left side.

quired to cause significant delamination of copper for **EMD-HIGH**. We determined a value for the critical load, L_c, by examining the scratch trace using the attached microscope (see Section 2.5) and estimating the point along the scratch where approximately 20% of the copper was missing from an area element that spans the trace and is of the order of 0.1 mm wide. The values of L_c obtained in this manner for the metallized films tested using several loading rates and two stylus tip radii are summarized in Table 1. Both the normal force and the acoustic emission traces were smooth curves that exhibited no breaks, peaks, or other discontinuities that one could associate with failure or delamination. We will see later, however, that these two detection modes do respond to failure events for the DLC materials, where the coatings are brittle.

We were initially skeptical that visual determination of L_c would provide the basis for a reproducible assessment of adhesion in these materials. However, we have now tested the same samples in three testing sessions (two identical instruments but different operators) and have obtained essentially equivalent results. The difference between L_c for the low and high adhesion films has averaged 3.2 N for the three sets of data when using the stylus with the 800 μm tip radius (and the data in Table 1 for **EMD-LOW** and **EMD-HIGH** are the third set).

The peel force data for **EMD-LOW** and **EMD-HIGH** in Fig. 2 are also included in Table 1 and are consistent with the L_c values obtained using the 800 μm tip. This inverse peel force test is not practical for routine testing, but it does create failure surfaces that can be analyzed by surface characterization techniques that then provide insight regarding the locus of failure and debonding mechanisms.

The L_c values for copper-coated films made using the electroless and sputtered processes are compared in Table 1. Data are given for tests using the 200 and 800 μm radius tips. It is necessary to choose the maximum load that will be applied to the film during the test, and this must exceed L_c if one is to observe the failure event. The scratch length and test time were kept at 10 mm and 1 min, respectively. This means that the loading rate increases with the maximum applied load. The L_c values for all materials in Table 1 were less than 3 N when the smaller tip radius was used, and experiments were run at maximum loads of 3 and 10 N. With the larger tip radius, L_c values up to 11.1 N were observed, and it was necessary to use maximum loads of 10 and 30 N. As one might expect, we see the effects of both the tip radius and the loading rate on the value of L_c, and these are factors that must be considered in developing a method for the measurement of practical adhesion. However, a complete explanation of these effects is beyond the scope of this paper and will be considered in future work.

L_c increases significantly with the tip radius for a given material and we see greater differentiation between materials as well. For example, the 3.2 N difference in L_c between **EMD-LOW** and **EMD-HIGH** was observed using the 800 μm tip at a loading rate of 10 N/min. The difference in adhesion between **EMD-HIGH** and **EMD-LOW** was less evident when the 200 μm tip radius was used at loading rates of 3 and 10 N/min. This issue of sensitivity to differences in adhesion for these two tip radii is probably a consequence of different distributions of shear stress at the copper/PET interface. The larger tip, we think, places a greater shear stress at the interface and induces less cutting into the copper film for a given load. This result suggests that a proper choice of the stylus tip radius is essential if scratch indentation is to be used as a measure of practical adhesion.

The L_c values for the sputtered materials are also sensitive to the tip radius, but in addition we can see the effect of the thickness. L_c increases with thickness under all conditions. The values of L_c for the sputtered materials are much larger in comparison with the electroless metal deposition material when using the 800 μm tip radius. Unfortunately, the comparison cannot be made at the same loading rate. When the sputtered materials were tested at a loading rate of 10 N/min, the maximum load of 10 N was reached prior to L_c. One can conclude that L_c is greater than 10 N versus a range of 1.4–5.4 N (Table 1) for the materials prepared by electroless deposition. However, when the loading rate is increased to 30 N/min, the L_c values for the sputtered materials fall in the range from 8 to 11 N. We did not test the electroless deposited films at the higher loading rate. It is interesting to note that the tape-pull test could not differentiate between the **SP-80** and **EMD-HIGH** materials. They both showed excellent copper adhesion (see Section 2.3). This result suggests that the scratch indentation test will discern differences in copper adhesion between materials that might otherwise be considered equivalent based on the tape test.

A mild loading rate effect is apparent in the tests done with the 200 μm tip radius, where L_c increases slightly with the rate. The origin of this effect is not clear at this point, but it could be a consequence of a viscoelastic response of the polymeric substrate to the loading rate. Some deformation of the substrate will most likely occur during the test, particularly as the tip plows through the copper coating and approaches the interface. It is well known that the modulus of polymers will increase with the deformation rate [9], and such an increase might reduce the substrate deformation during the test and hence reduce the deformation of the copper coating. This would, in turn, reduce the contact area between the tip and the coating so that a higher load would be required to cause the coating to fail.

The CIE instrument gives information similar to that obtained by the Micro Scratch Tester on the copper-coated films, but on a much smaller scale since the tip radius is now 1 μm. L_c was determined by first scratching the sample and then inspecting the scratch using an SEM. Figure 4 shows traces for copper-coated polyimide films before and after annealing at 70°C for 3 h. The normal force curves for these scratch tests are compared in Fig. 5. The scratch distance at which failure occurred and the corresponding L_c value are indicated for each curve. Delamination of the copper coating occurs from 20–25 μm in the unannealed sample versus 45–50 μm in the annealed case, indicating greater adhesion of the copper after annealing. No discernible features in the normal force curves stand out as indicating the onset of delamination. The slope is slightly greater for the annealed versus the control (118 N/m of scratch length versus 99 N/m). This might be expected if the annealing made the polyimide substrate stiffer.

3.2. Diamond-like carbon

We now describe the scratch test measurements with the DLC coatings. The results reported here were obtained with the Micro Scratch Tester, using a stylus tip radius of 200 μm, and the CIE instrument, where the radius is 1 μm. A magnified portion of a typical scratch test trace and the acoustic emission curve obtained using the Micro Scratch Tester are compared in Fig. 6. The first distinct crack, oriented convex with respect to the scratching direction, was observed at a distance of 3.1 mm from the beginning of the scratch, and, correspondingly, the first peak in the acoustic emission

136   G. D. *Vaughn* et al.

Copper on Polyimide

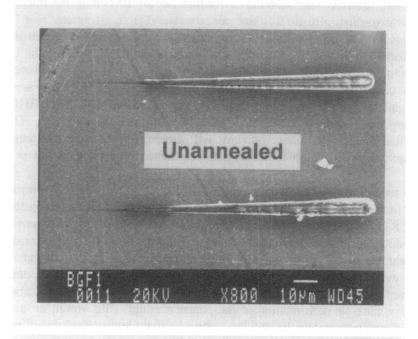

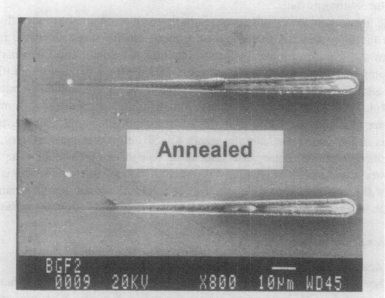

Figure 4. SEM photographs of scratches created by the CIE scratch tester. The samples are copper-coated polyimide (unannealed and annealed at 70°C for 3 h). The point of delamination is shown. The beginning of the scratches is on the left side.

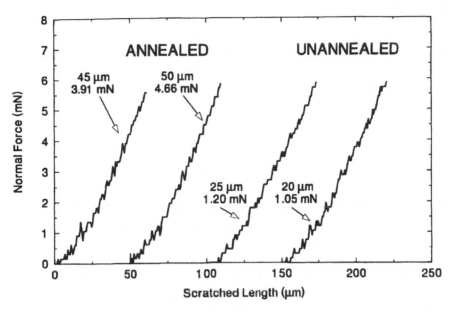

Figure 5. Scratch traces from copper-coated polyimide using the CIE instrument with a 1 μm stylus tip radius. Duplicate scratches are shown for the annealed and unannealed samples. The scratch length and normal force where delamination was observed are shown for each scratch. Actual traces are shown in Fig. 4. The individual traces on the abscissa are offset by 50 μm.

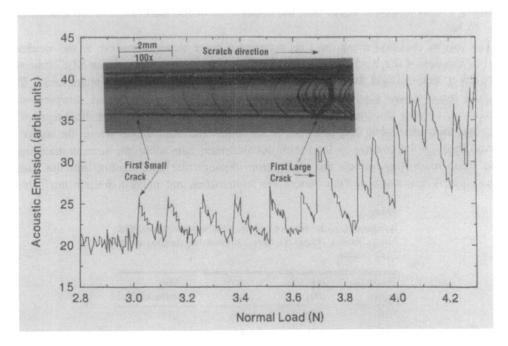

Figure 6. Acoustic emission curve generated during scratch testing of DLC-coated polycarbonate using the CSEM Micro Scratch Tester with a 200 μm stylus tip radius. A photograph of the scratch trace is included. The scratch direction is from left to right, and the isolated small tensile cracks line up with their corresponding acoustic emission peaks on the emission curve. The first large bending crack is also shown. The loading rate and scratching speed were 10 N/min and 10 mm/min, respectively; therefore the distance traversed by the stylus was 1 mm/N.

curve is seen at a load of 3.04 N, which is also 3.04 mm along the scratch, given the loading rate of 10 N/min and scratch rate of 10 mm/min. It seems reasonable to associate the acoustic emission peak with the formation of this crack. A series of these convex cracks and their associated acoustic emission peaks follow, all at approximately equally spaced intervals, until a load of approximately 3.8 N is reached and then both the pattern of cracks and the acoustic emission peaks become much more complex. Large cracks are now observed (only the first one is shown in Fig. 6) and they are oriented in the opposite direction with respect to the scratching direction.

This pattern of cracks fits well with a generalized description by Burnett and Rickerby of the cracking pattern seen in hard wear-resistant coatings [10]. The first cracks can be attributed to the tensile cracking mode and are formed as a result of the tensile frictional stresses present behind the trailing edge of the stylus. The larger cracks result from the bending of the coating beneath the stylus. Both types of cracks leave their distinct imprint on the acoustic emission trace.

The loads at which the small and large cracks first appear are undoubtedly a measure of the coating strength and integrity. These loads are given in Table 2 (plotted in Fig. 7) for three different coating thicknesses. One can see that the required loads increase with thickness. Also shown is the relative change in the acoustic emission baseline near the appearance of the first large crack, which increases as the coating becomes thicker. Limited testing was done using the 800 μm tip radius stylus, and the results were qualitatively identical to those obtained with the 200 μm radius, but the large bending cracks tended to occur near the upper limit of the load range of the instrument, i.e. 30 N.

The results obtained using the CIE instrument were clearly sensitive to the mechanical properties of the hard coating. Unlike the softer copper films, these DLC coatings showed a well-defined load drop at L_c after the coating had been penetrated. The normal force versus scratch length is plotted in Fig. 8 for a series of samples with increasing coating thickness. The critical load increases with the film thickness. Recall that this instrument operates in a penetration-controlled mode where the stylus is driven through the coating at a constant displacement rate while the force is measured. The sharp load drop reflects the penetration through the hard coating into the softer substrate. A few static indents (where the stylus does not make a scratch but instead

Table 2.

Load and acoustic emission values, obtained using the CSEM Micro Scratch Tester, for three different thicknesses of the DLC coating

Thickness (μm)	L(sc)[a] (N)	L(lc)[b] (N)	AE[c] (relative)
1.20	1.67	2.58	3.5
2.13	2.71	3.53	5.0
3.04	2.88	3.64	9.0

[a]Load at which first small crack appears.
[b]Load at which first large crack appears.
[c]Increase in baseline of AE trace where large cracks begin to occur. (All values are the average of two measurements.)

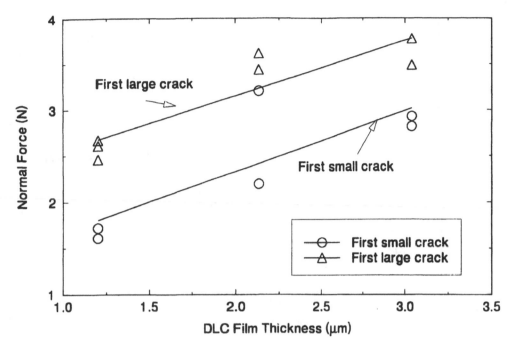

Figure 7. The normal force corresponding to the appearance of the first small and large cracks is plotted against the DLC film thickness for the DLC-coated polycarbonate. The data were obtained using the CSEM Micro Scratch Tester with a 200 μm radius tip.

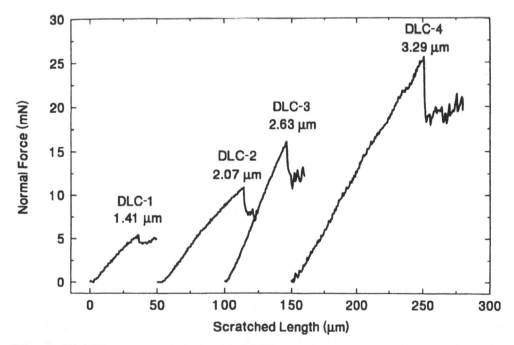

Figure 8. Normal force curves obtained using the CIE scratch tester (1 μm tip radius) for DLC-coated polycarbonate with different coating thicknesses. Each curve is offset by 50 μm on the abscissa.

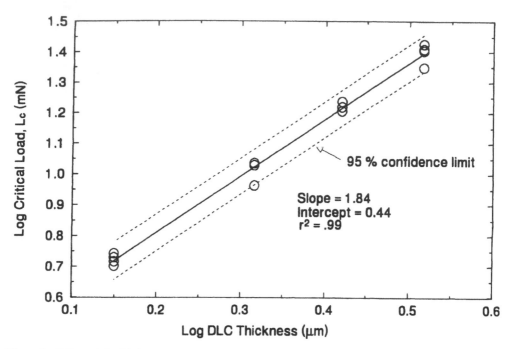

Figure 9. Effect of the DLC coating thickness on the critical normal force required to penetrate the coating when using the CIE scratch tester.

is driven vertically down through the film) indicated that the L_c values of the coatings decreased by only 10–25% from those of the scratch indentation test and were not strongly dependent on the scratch rate. Since the static indentation minimizes the amount and rate of energy dissipated under the indenter, one would expect a lower value of L_c in comparison with the value obtained during the scratch indentation test. Properly interpreted, such data could provide important information about the capacity of the system to dissipate energy, and an important distinction between viscoelastic (or plastic) systems and the more brittle elastic materials commonly evaluated by scratch testing.

The dependence of the critical load on the coating thickness is shown in Fig. 9. The 1 μm radius data show a nearly second-power dependence of L_c on the thickness. Such a mathematical relationship is consistent with a simple physical model, such as an edge-supported disk whose diameter scales linearly with its thickness and which fractures at a constant value of stress at the lower surface [11]. Interestingly, an exponent of exactly 2 lies outside the 99% confidence limits on the slope; refitting the data with an exponent forced to be 2 reduced r^2 (the goodness of fit) from 0.991 to 0.984. This may reflect deviations from the simple physical model due to factors such as the contribution of the softer polycarbonate substrate to the fracture mechanism.

The adhesion of the DLC coating to the substrate seems to be rather good since no large-scale delamination was observed at any of the three stylus tip radii (1, 200, and 800 μm). Spallation of the coating around the compression cracks of hard coatings in general can be an indication of adhesion failure, and this was not observed to a great degree. In future studies, we plan to include DLC samples representing a range of adhesion levels to address this key issue.

Table 3.

Comparison between L_c (CIE scratch tester) with the Taber haze and scratch rank values for four different thicknesses of the DLC coating

Sample	L_c^a	Thickness (μm)	Taber[b] haze	Steel wool[c] scratch rank
DLC-1	5.3 ± 0.2	1.41	3.53	5
DLC-2	10.4 ± 0.8	2.07	0.78	4
DLC-3	16.6 ± 0.6	2.63	0.33	3
DLC-4	24.9 ± 1.9	3.29	0.03	2

[a] Mean ± standard deviation, replicates = 3 or 4.
[b] ASTM 1044 D.
[c] See [12].

The scratch tests seem to be measuring primarily (almost entirely in the case of the 1 μm radius tip) the properties of the coating. An excellent correlation is found with the standard Taber abrasion test (ASTM 1044D) and the steel wool scratch test [12], which subject the coating to abrasive treatment. In the Taber test, grit-impregnated rubber wheels are held against the coating and the subsequent haze from the scratching is measured. An unscratched sample would have no haze and receive a rating of zero. The steel wool test is similar to the Taber test, but the ranking is based on the number of scratches produced by rubbing a steel wool pad across the coated surface at a controlled speed and load. An unscratched sample would have no haze and receive a rating of zero; thus, the higher ratings correlate with a lower scratch resistance. In Table 3, the CIE critical loads, thickness, Taber rating, and steel wool scratch rank are compared, and one can clearly see a good correlation between L_c and the Taber rating and scratch rank. This is further evidence that the CIE scratch test is primarily sensitive to the coating properties.

4. CONCLUSION

We have shown scratch indentation testing to be a very practical tool for characterizing plastic substrates that are coated with ductile metals or hard brittle DLC coatings. In the case of metallized films, we were able to identify test conditions that provided at least a reliable and practical measure of adhesion.

For our purposes, it has proven to be more useful than the inverse peel test and more discriminating than the tape-pull test. Additional work needs to be done to understand the effects of stylus tip radius, loading rate, metal thickness and substrate hardness and any other factors that combine with the basic adhesion to give the measured value of practical adhesion. It would be especially gratifying to account for the effects of variations in metal thickness.

Our observations with the DLC materials showed that scratch indentation testing provides information about the coating strength that correlates with critical end-use performance properties such as scratch resistance. With these materials, the adhesion could not be directly determined because the coatings became highly damaged by the stylus before any large-scale delamination was observed. One can infer, however, that the practical adhesion was very good from the way that the pieces of broken DLC coating remained adhered to the substrate.

Acknowledgements

We give special thanks to Dr Ludvik Martinu and Pierre Leroux at École Polytechnique in Montreal. Most of the Micro Scratch Tester data were taken on their instrument. We also thank Bill A. Burniski, Donald E. Williams, Oleg V. Karin, Charles I. Pechmann, Edward F. Tokas, Paul D. Garrett (Monsanto Company); Robert A. Mendelson (retired from Monsanto, now with Exxon Chemical); Roger Lambert (CSEM); W. W. Gerberich, Shankar Venkataraman, Ridha Berriche, and D. L. Kohlstedt (University of Minnesota Center for Interfacial Engineering).

REFERENCES

1. G. D. Vaughn, U.S. Patent 5 082 734 (1992).
2. B. K. Daniels, *Proc. Annu. Tech. Conf. Soc. Vac. Coaters* **35**, 260 (1992).
3. (a) K. L. Mittal, *J. Adhesion Sci. Technol.* **1**, 247 (1987).
   (b) K. L. Mittal, *Electrocomponent Sci. Technol.* **3**, 21 (1976).
4. T. W. Wu, *J. Mater. Res.* **6**, 407 (1991).
5. K. L. Mittal (Ed.), *Adhesion Measurements of Thin Films, Thick Films and Bulk Coatings*, STP No. 640. American Society for Testing and Materials Philadelphia, PA (1978).
6. K. Nakamae, S. Tanigawa and T. Matsumoto, in: *Metallized Plastics 1: Fundamental and Applied Aspects*, K. L. Mittal and J. R. Susko (Eds), pp. 235–245. Plenum Press, New York (1989).
7. S. Venkataraman, D. L. Kohlstedt and W.W. Gerberich, *J. Mater. Res.* **7**, 1126 (1992).
8. P. A. Steinmann, Y. Tardy and H. E. Hintermann, *Thin Solid Films* **154**, 333 (1987).
9. F. W. Billmeyer, Jr, *Textbook of Polymer Science*, 3rd edn, pp. 311–320. John Wiley, New York (1984).
10. P. J. Burnett and D. S. Rickerby, *Thin Solid Films* **154**, 404 (1987).
11. (a) C. B. Bucknall, *Toughened Plastics*, p. 288. Applied Science Publishers, London (1977).
    (b) S. Timoshenko and S. Woinowsky-Krieger, *Theory of Plates and Shells*, 2nd edn. McGraw-Hill, New York (1959).
12. Open Circuit Self-Contained Breathing Apparatus for Firefighters — Faceplate Lens Abrasion Test, National Fire Prevention Association 1981 Standard, Section 4.9 (1992).

Adhesion Measurement of Films and Coatings, pp. 143–160
K. L. Mittal (Ed.)
© VSP 1995.

On the evaluation of adhesion of coatings by automatic scratch testing

T. Z. KATTAMIS

Department of Metallurgy, Institute of Materials Science, University of Connecticut, Storrs, CT 06269-3136, USA

Revised version received 25 March 1993

Abstract—Automatic scratch testing is an expedient technique for comparatively evaluating the cohesive failure load and adhesion failure load of thin coatings on various substrates. In combination with SEM exmination of the scratch track, this technique has been used herein to detect and evaluate various effects on coating strength and adhesion. For soft Triballoy T-800 and Stellite SF-6 cobalt-base coatings on 4340 low alloy steel, adhesion was found to be strong and failure was found to be cohesive in the coating. In the presence of a plated chromium interlayer, pre-existing cracks lowered substantially the cohesive failure load, which was also lowered by an increase in the coating deposition pressure. The spacing of transverse cracks within the coating was found in all cases to decrease with increasing applied normal load. In soft aluminum coatings on depleted uranium (DU)–0.75% Ti alloy specimens, alloying aluminum with magnesium or zinc enhanced the coating strength and adhesion. In (Al–Mg) coatings on this substrate, a smoother surface led to a lower friction coefficient and a higher adhesion failure load. In hard, thin TiN coatings on 17-4 PH steel, a lower bias voltage applied to the substrate yielded higher cohesive and adhesion failure loads. In hydrogenated amorphous SiC thin coatings on 4340 steel, loss of hydrogen by annealing converted the residual compressive stresses into tensile stresses and lowered both the cohesive and the adhesion failure loads. Finally, automatic scratch testing proved helpful in determining delamination loads in multilayer TiN/Ti/TiN coatings on DU–0.75% Ti alloy.

Keywords: Cohesive failure load; adhesion failure load; friction coefficient; automatic scratch test; cobalt-based coatings; aluminum–magnesium coatings; TiN coatings; SiC coatings; depleted uranium substrate; coated 4340 steel.

1. INTRODUCTION

Scratch adhesion testing consists in displacing the coated specimen at constant velocity under a stylus which applies a stepwise [1] or countinuously [2] increasing normal load to the coating surface. With increasing load, the coating/substrate deformation generates stresses which, at a given load, result in permanent damage, such as chipping of the coating (cohesive failure) or flaking (adhesion failure). The smallest load leading to unacceptable damage is the 'critical' load. Various coating failure criteria have been used to define the critical load, such as initiation of cracking, chipping of the coating, spalling, initiation of flaking [3], or complete removal of the coating from the scratch track [1]. In essence in each case the criterion adopted for characterizing the failure of the coating should depend on the nature and extent of the damage that can possibly be tolerated for that particular application. Thus, the presence of a small number of microcracks may be acceptable in a protective coating which is anodic (sacrificial) with respect to the substrate [4]. In this case the critical load should not be identified with the cohesive failure load, L_C, which is the minimum load at which crack initiation occurs within the coating, but rather with the adhesion failure load, L_A, which is the minimum load at which the crack reaches the coating/substrate interface, causing detachment of the coating and exposure of the substrate. On the other hand, the presence of a single microcrack may be catastrophic when the protective coating is cathodic (noble) with respect to the substrate and the cohesive failure load should then be taken as the critical load.

The automatic scratch tester used herein, CSEM-Revetest (Centre Suisse d'Electronique et de Microtechnique, CSEM, Neuchâtel, Switzerland) is equipped with a Rockwell C diamond stylus cone of 120° and tip radius of 200 μm and a resonant acoustic emission (AE) detector. During testing, the AE signal intensity, the applied normal force F_n, the tangential (frictional or scratching) force F_t, and, in a more recent version of the apparatus, the lateral force F_l, which is applied laterally by the sample to the stylus, are recorded. In the present investigation, the AE signal intensity, and/or the frictional force F_t, and the friction (scratching) coefficient $\mu^* = F_t/F_n$ were plotted vs. the applied normal load, F_n.

Sekler et al. [5] considered the AE signal as a discontinuous stochastic event whose amplitude, related to the damage size, or repetition rate, related to the damage occurrence frequency, cannot be predicted. They correctly emphasized that the statistical nature of the failure probability requires that several scratches be made on a given sample, averaging the resulting critical load values. They also established that with TiN and TiC coatings on steel or cemented carbides F_n and F_l could not be used for determining the critical load, because they were not significantly affected by the damage. On the other hand, the tangential force F_t could be used [6] in conjunction with AE.

The scratch test does not measure directly the adhesion, but rather the load required for disruption of the coated specimen, either at the interface or in the interfacial region [7]. This load represents the 'practical adhesion' [7]. It has been used to compare the performance of different coating systems on similar substrates and/or of similar coating systems on different substrates [8–10]. AE vs. F_n diagrams for a given coating/substrate system may be used for comparing different manufacturing processes or sets of process variable values, for evaluating the effect of heat treatments on coating cohesion and adhesion, for quality control during production, and for following the degradation of protective coatings with time.

In a recent comprehensive study, Steinmann et al. [11] examined the various intrinsic parameters, which are related to the scratch test itself, and extrinsic parameters, which are related to the coating/substrate combination, and their effect on the critical load. The intrinsic parameters examined were the loading rate, scratching speed, indenter tip radius, and diamond tip wear. The extrinsic parameters studied were the substrate hardness, coating thickness, coating roughness, substrate roughness prior to coating, friction coefficient between the tip and coating, and friction force in the scratching direction. The most important conclusions and recommendations formulated [11] were as follows: (a) The critical load depends on the loading rate, dL/dt (N min^{-1}), and the scratching speed, dx/dt (mm min^{-1}), and hence on the dL/dx (N mm^{-1}) ratio, where L is the load (in N), x is the distance (in mm) and t is time (in s). It also depends on the stylus tip radius, R, according to a relationship which includes the coating hardness and thickness. Hence, it is recommended to keep the following operating parameters constant in order to minimize the number of those which influence the critical load: $dL/dx = 10$ N mm^{-1} and R (hemispherical tip radius) $= 200$ μm or conical angle $= 120°$; (b) the scratch test corresponds to the deformation of the coating–substrate assembly. Thus, the critical load increases with substrate hardness and coating thickness; (c) the sample arithmetic average surface roughness should be less than 0.3 μm in order to obtain a representative critical load; and (d) the critical load decreases as the friction coefficient increases.

Scratch testing is combined with examination of the scratch track morphology and the coating surface morphology in the vicinity of the scratch by optical and scanning electron microscopy (SEM). Detection of delamination and exposure of the substrate may be facilitated by electron microprobe analysis (EMPA), or energy dispersive spectroscopic analysis (EDS) coupled with SEM.

In this paper, examples are selected from previous and current work on the cohesion and adhesion of various soft and hard thin coatings on soft substrates illustrating the usefulness and potential of automatic scratch testing in comparatively evaluating the mechanical behavior of basically similarly coated systems. More specifically, the application of automatic scratch testing in evaluating the adhesion failure load in multilayer coatings and in detecting the effect of pre-existing microcracks on coating failure are discussed, as well as the effects of some process parameter values, alloying the coating, friction coefficient and residual stresses on coating strength and adhesion.

2. EFFECT OF PRE-EXISTING MICROCRACKS ON COATING FAILURE

For this investigation several disk-shaped (32 mm diameter × 7 mm thick) 4340 low alloy steel specimens, as well as chromium-plated 4340/Cr steel specimens, were coated by the cathodic arc plasma PVD process, using a coating deposition pressure of 3 or 25 mTorr and a bias of -100 V (Table 1). The steel specimens had previously been heat-treated and hardened by austenitization at 1103 K for 1 h followed by an oil quench and temper at 643 K for 2 h, followed by a final oil quench. The coatings consisted of (1) Triballoy T-800 [28.5% Mo–17.5% Cr–3.0% (Ni + Fe)–3.4% Si–0.05% C–balance Co] and (2) Stellite SF-6 (19% Cr–8% W–0.7% C–13% Ni–3% Si–1.7% B–balance Co). A certain number of specimens coated with either of these alloys were electrolytically preplated with a chromium layer about 18 μm thick prior to coating. The coated specimens were prepared at Professor Ramalingam's laboratory, University of Minnesota, using the 'Multi-Arc' process, where cathodic arc plasma PVD involved multiple arc evaporation of the coating material directly from solid targets [12]. From experience, for superior adhesion the vapor energy level was maintained high (50–150 eV) and the level of ionization very high (80–95%).

The thickness of various coatings and the volume fraction microporosity were evaluated in polished transverse sections. The average coating thickness is given in Table 2. The volume fraction microporosity was roughly the same in all T-800 and SF-6 coatings, varying between 0.029 and 0.032.

2.1. T-800-coated 4340 steel specimen processed at 3 mTorr

The AE signal intensity, the tangential or frictional force F_t, and the friction coefficient μ^* are plotted in Fig. 1a vs. normal load between 0 and 80 N. The surfaces of all T-800 and SF-6 coatings consist of spherical or flattened particles of a very wide size distribution from submicrometer to about 8 μm, as illustrated by the scanning electron (SE) micrographs of Figs 2 and 5. In all specimens, with load application the particles on the coating surface are plastically deformed and smeared in the vicinity of the stylus. As the load increases, the deformed particles gradually coalesce together, forming a continuous, plastically deformed track. Figure 2a shows an SE micrograph of a scratch at a location corresponding to a load of about 40 N.

Microscopic observations of the scratch revealed the presence of hairline cracks within the plastically deformed individual particles at a location corresponding to a load of about 6 N, which appears to be the cohesive failure load (Fig. 1a). However, the first major transverse microcrack corresponds to a load of about 15 N. The spacing between these transverse microcracks within the coating decreases; hence, the number of microcracks per unit length of scratch (microcrack density) increases with increasing applied load, as previously reported by Wu [13] for another system.

Based on previous work by Bull *et al.* [14] and Rickerby [15], the following simplistic explanation for the dependence of transverse crack spacing on applied normal load is proposed: Assuming uniaxial tension and ignoring any elastic effects in the substrate, the elestic strain energy per unit volume of coating is

$$U = \frac{1\sigma^2}{2E}, \tag{1}$$

where E is Young's modulus and σ is local stress, which depends on the elastic–plastic indentation stress, the tangential friction stress, and the internal stresses within the coating. The local stress σ is proportional to F_n; $\sigma = K F_n$, where K is a constant. Assuming a coating thickness t and parallel transverse cracks of similar geometry and of spacing λ, possibly inclined to the scratch direction and penetrating into the substrate, as microscopically observed in a transverse section, the volume over which the elastic strain is released will be λt and the released elastic energy will be

$$V = \frac{\lambda t}{2} \frac{\sigma^2}{E}. \tag{2}$$

This energy is equal to Ω, the energy required to form the next crack. Ω is equal to the energy of the generated crack surface augmented by the energy associated with

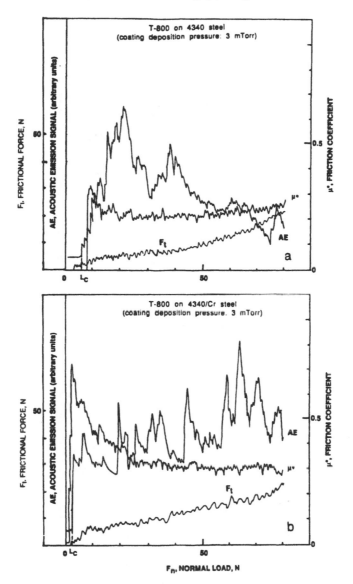

Figure 1. Acoustic emission signal intensity, friction coefficient, and frictional force vs. normal load. (a) T-800-coated 4340 steel specimen processed at 3 mTorr; (b) T-800-coated 4340/Cr-plated steel specimen processed at 3 mTorr.

crack tip plasticity. The energy Ω may be assumed to be approximately the same for all parallel cracks. Substituting Ω for V in equation (2) yields

$$\lambda = \frac{2E\Omega}{K^2 F_n^2 t}. \tag{3}$$

Thus, for a given coating thickness, λ is inversely proportional to the square of applied normal load.

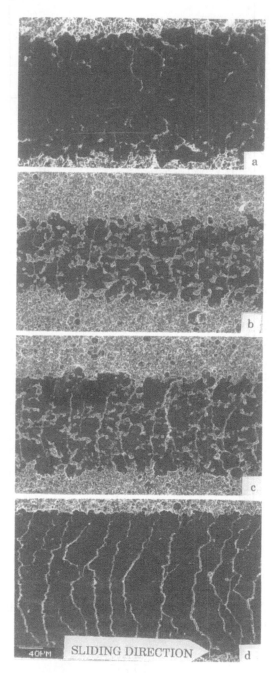

Figure 2. Scanning electron micrographs of a scratch on (a) a T-800-coated 4340 steel specimen processed at 3 mTorr, corresponding to a load of 40 N, and (b–d) a T-800-coated 4340/Cr-plated steel
specimen processed at 3 mTorr, corresponding to loads of 20, 31, and 52 N, respectively.

Table 1.

Average cohesive failure load L_C, adhesion failure (critical) load L_A, and friction coefficient μ^* between 0 and 80 N

Substrate	Coating	Coating deposition pressure (mTorr)	μ^*	L_C (N)	L_A (N)
4340	T-800	3	0.21	6.95	—
4340	T-800	25	0.23	4.87	—
4340/Cr	T-800	3	0.35	2.27	—
4340/Cr	T-800	25	0.31	1.95	—
4340	SF-6	3	0.31	8.24	—
4340	SF-6	25	0.30	5.90	—
4340/Cr	SF-6	3	0.27	2.20	—
4340/Cr	SF-6	25	0.30	1.80	—

In the AE curve of Fig. 1a, there are indications of a series of crack nucleation and growth events corresponding to various AE signal intensities that occur at different normal loads, without clearly defined adhesion load or adhesion failure events. Of these nucleation events, the most important occur at normal loads of 6, 15, 20, 32, 37, and 75 N. These results combined with microscopic observations and EMPA to identify exposure of the substrate clearly indicate that no delamination of the metallic coating occurs. Thus, adhesion between the metallic coating and the metallic substrate appears to be strong [12]. It therefore appears that each crack initiates and propagates across the coating and traverses directly into the underlying substrate. As stored elastic energy is released, the cracks propagate and stresses are relieved. With further advance of the diamond stylus, another build-up of stored elastic energy occurs within the coating and substrate until a new crack initiates within the coating and propagates across the coated specimen. This mechanism is repeated periodically all along the scratch.

For each scratch the average friction coefficient was calculated by area integration. Average value of L_C and μ^* for T-800 and SF-6 coatings are summarized in Table 1. They correspond to five or six scratches performed on each specimen.

2.2. T-800-coated 4340/Cr steel specimen processed at 3 mTorr

The AE signal intensity and the friction coefficient are plotted in Fig. 1b vs. normal load between 0 and 80 N. SE micrographs of the scratch are shown in Figs 2b–2d. As with the previous case, Fig. 1b indicates the presence of a series of crack growth events which correspond to various AE signal intensities and occur at different normal loads with no clear delamination load, indicating strong adhesion between the metallic coating and the metallic substrate. The load corresponding to the nucleation and propagation of the first microcrack which was observed is 2 N instead of $L_C = 8$ N for the non-plated 4340 steel substrate. One explanation for earlier crack formation in chromium plate involves the formation of unstable chromium hydrides, CrH to CrH_2, during electrodeposition; these decompose spontaneously to α-Cr and free hydrogen. With this decomposition there is a volume shrinkage exceeding 15%. Thus, the

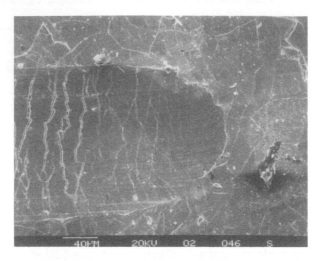

Figure 3. Scanning electron micrograph of a scratch on a TiN-coated 4340/Cr-plated steel specimen processed at 25 mTorr, corresponding to a load of about 80 N.

chrome plate restrained by the substrate is subjected to tensile forces, resulting in the formation of microcracks which are normal to the surface [12]. The mud-crack pattern of these microcracks is illustrated in Fig. 3, which exhibits a top view of a thin, transparent TiN coating on 4340 with a similar chromium interlayer. These pre-existing cracks, which have an average spacing of about 28 μm, may propagate easily within the coating and may be responsible for the smaller crack spacing observed in this specimen (about 26 μm) compared with a 38 μm spacing in the non-plated specimen. Figures 2b–2d illustrate the descrease in transverse crack spacing with increasing normal load.

2.3. SF-6-coated 4340 steel specimen processed at 3 mTorr

The AE signal intensity and the friction coefficient are plotted in Fig. 4a for loads between 0 and 80 N. The SE micrograph of Fig. 5a illustrates the geometry and internal morphology of the scratch. The absence of delamination and the repetitive nucleation of microcracks with increasing load are analogous to those observed with the two previous T-800 coatings on 4340 and 4340/Cr steels. The first major transverse microcrack appears at a load of about 8 N. The microcracks are of the conformal type, parallel to the leading edge of the stylus and are presumably caused by bending of the coating ahead of the stylus. Their spacing decreases with increasing load.

2.4. SF-6-coated 4340/Cr steel specimen processed at 3 mTorr

The AE signal intensity and the friction coefficient are plotted in Fig. 4b for loads between 0 and 80 N. The geometry and internal morphology of the scratch are illustrated by the SE micrographs of Figs 5b and 5c. Its spacing decreases with increasing

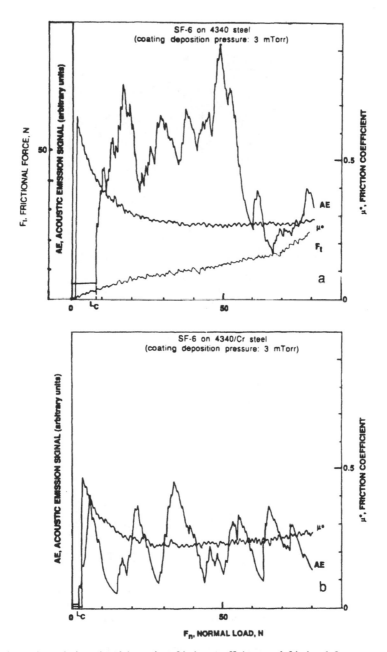

Figure 4. Acoustic emission signal intensity, friction coefficient, and frictional force vs. normal load. (a) SF-6-coated 4340 steel specimen processed at 3 mTorr; (b) SF-6-coated 4340/Cr-plated steel specimen processed at 3 mTorr.

load. Again, in this case there is a succession of crack initiation events and absence of delamination. As with the previous T-800 coatings on 4340 and 4340/Cr steels, microcrack spacing is smaller on the chromium-plated 4340 steel.

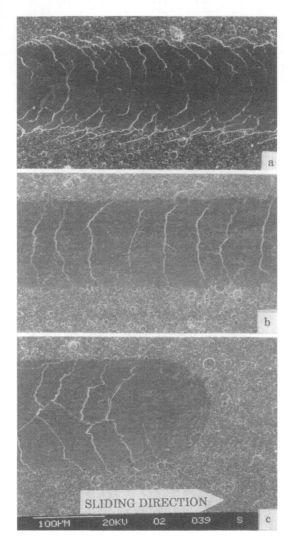

Figure 5. Scanning electron micrographs of a scratch on (a) an SF-6-coated 4340 steel specimen processed at 3 mTorr, corresponding to a normal load of about 40 N, and (b, c) an SF-6-coated 4340/Cr-plated steel specimen processed at 3 mTorr, corresponding to 50 and 75 N, respectively.

2.5. Comparison of cohesive failure loads of T-800 and SF-6 coatings

T-800 and SF-6 coatings deposited on 4340 steel at a pressure of 3 mTorr have approximately the same thickness (Table 2). An SEM examination of the coating surface morphology qualitatively indicated that the SF-6 coating is rougher than the T-800 and this may account in part for the higher friction coefficient of SF-6 (Table 1). The microhardness of T-800 is slightly higher than that of SF-6 (Table 2); hence,

Table 2.

Average coating thickness and microhardness (VHN)

Substrate	Coating	Thickness (μm)	Microhardness (VHN)
4340	SF-6[a]	5.12	857
4340	T-800[a]	5.07	1005
4340/Cr	SF-6[a]	4.98	735
4340/Cr	T-800[a]	5.02	823

[a]Coating deposition pressure: 3 mTorr.

its yield strength and cohesive failure load would also be expected to be higher. Table 1 shows instead that $L_C = 6.95$ N for T-800 and $L_C = 8.24$ N for SF-6. This discrepancy could not possibly be attributed to the difference in friction coefficient between the two coatings.

For both types of coating, the presence of a pre-cracked chromium interlayer leads to a decrease in the average cohesive failure load of the coating (Table 1). In this system of well-adhering layers and substrate, the pre-existing cracks need only propagate to cause cohesive failure of the coating. The required load for crack propagation is lower than that required for initiation; hence, in the presence of the pre-cracked chromium interlayer the average cohesive failure load is reduced.

3. EFFECT OF PROCESSING PARAMETER VALUES ON COATING COHESION AND ADHESION

3.1. Comparison of T-800- and SF-6-coated 4340 and 4340/Cr steel specimens processed at 25 mTorr with homologous specimens processed at 3 mTorr

All T-800-coated and SF-6-coated 4340 and 4340/Cr steel specimens processed at 25 mTorr exhibited a similar behavior during scratch testing with the same general form of the AE vs. F_n curve, as with the homologous systems coated at a pressure of 3 mTorr. However, in all cases, the corresponding average cohesive failure loads, L_C, are lower (Table 1), presumably because at a higher deposition pressure there is an increased risk of contaminating the coating surface with impurity atoms.

3.2. Comparison of TiN-coated 17-4 PH steel specimens processed at two different substrate bias voltages

Two TiN-coated 17-4 PH steel samples were prepared by cathodic arc plasma PVD with the same process parameter values except the substrate bias voltage, which was -35 V for specimen I and -120 V for specimen II. For both specimens the average coating thickness was about 2 μm, the volume fraction microporosity was about 0.021 and the friction coefficient μ^* at a given normal load was approximately the same (about 0.15 at 20 N). For specimen I, the measured average cohesive and adhesion failure loads were higher than those for specimen II (Table 3). This decrease in adhesion failure load with increasing substrate bias voltage has been previously

Table 3.

Average cohesive failure load L_C and adhesion failure load L_A between 0 and 70 N for TiN-coated
17-4 PH steel specimens I and II

Specimen	L_C (N)	L_A (N)
I (bias −35 V)	30.32	39.27
II (bias −110 V)	18.14	31.17

reported by Burnett and Rickerby [16] for TiN coatings, about 1.5 μm thick, on stainless stell. They had measured an increase in hardness and in internal stress with increasing bias voltage, which they attributed to a finer grain size and a higher growth rate, respectively.

4. EFFECT OF ALLOYING ON COATING STRENGTH AND ADHESION

Elemental aluminum, zinc, and magnesium, and alloyed (Al–Zn) and (Al–Mg) coatings were deposited on depleted uranium (DU)–0.75 wt% Ti disk-shaped specimens (25.40 mm in diameter and 6.35 mm thick). The specimens had previously been heat-treated at 1123 K, water-quenched at 0.46 K min^{-1} and aged at 623 K for 16 h to obtain the desired microstructure, which consisted of a major α-phase and a minor U_2Ti phase. The alloy had a martensitic structure and a grain size of about 400 μm. The disks were ground and polished with 600 grade grit silicon carbide paper to a surface roughness of about 0.25 μm and were degreased and ultrasonically cleaned in detergent, rinsed, and air-dried prior to placement into the deposition chamber. They were coated by Nuclear Metals, Concord, MA using the cathodic arc plasma (PVD) process in a Multi-Arc Vacuum System (Multi-Arc Vacuum Systems, St. Paul, MN). Details of the coating procedure have been reported elsewhere [4]. The alloyed (Al–Zn) and (Al–Mg) coatings were deposited using evaporation of elemental metallic rather than pre-alloyed cathodes.

The (Al–Mg) coating surface consisted of an agglomerate of spheroidal or flattened particles with a very wide size distribution from submicrometer to about 30 μm. Figure 6a illustrates the variation of AE, F_t, and μ^* with F_n between 0 and 70 N and shows $L_C = 22.6$ N, $L_A = 31.0$ N, and $\mu^* = 0.28$. Optical microscopy showed the first microcrack at 20.5 N and the first substrate exposure at 29.3 N. The (Al–Zn) coating was spongy with pores up to 1 μm in size, but smoother. All four types of coating were about 6 μm thick. Their average cohesive and adhesion failure loads, which were determined using six scratches per specimen, are reported in Table 4, as are Vickers microhardness measurements.

The four coatings have approximately the same friction coefficient (about 0.30). The measured volume fraction microporosity within the coatings varied from 0.025 to 0.027. There is a definite increase in L_C with alloying, most likely due to the increase in yield strength (through solid solution and precipitation hardening mechanisms), and hence in the stress required for crack initiation. A similar increase was observed in the adhesion failure load.

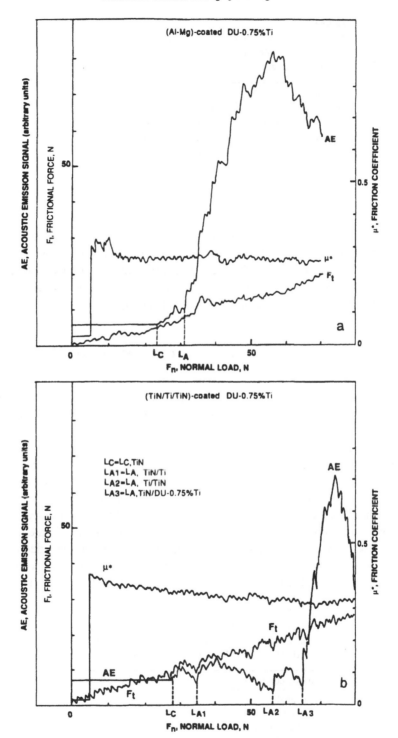

Figure 6. Acoustic emission signal intensity, friction coefficient, and frictional force vs. normal load. (a) (Al–Mg)-coated DU–0.75% Ti; (b) multilayer (TiN/Ti/TiN)-coated DU–0.75% Ti.

Table 4.

Average cohesive failure load L_C and adhesion failure load L_A
for Al, Zn, (Al–Zn) alloy, and (Al–Mg) alloy-coated DU–0.75%
Ti alloy speciments

Coating	L_C (N)	L_A (N)	Microhardness (VHN)
Al	15.7	23.4	15.46
Zn	15.2	25.3	32
(Al–Zn)[a]	24.5	32.6	46.2
(Al–Mg)[b]	23.1	31.8	63.9

[a] Aluminum-rich α-phase containing 1.23 wt% Zn.
[b] Aluminum-rich α-phase containing 2.12 wt% Mg.

Although no X-ray diffraction phase analysis was carried out, it appears from previous electrochemical polarization scans and observations after long exposure tests [17] that both (Al–Zn) and (Al–Mg) coatings are probably alloys, not mixtures of elemental phases. In fact, whereas aluminum is cathodic (noble) with respect to the substrate, (Al–Zn) and (Al–Mg) are anodic (sacrificial).

5. EFFECT OF THE FRICTION COEFFICIENT ON THE COATING STRENGTH AND ADHESION

In order to examine the possible effect of the surface friction (scratching) coefficient on the cohesive failure load L_C and critical (adhesion failure) load L_A in an (Al–Mg)-coated DU–0.75% Ti specimen processed as described above, the rough specimen surface was lightly polished with 1, 0.3, and 0.05 μm grade diamond paste, cleaned, and subjected to scratch testing after each polishing stage. The testing conditions were almost identical, with the exception of any coating thickness change resulting from the light polishing. The results are summarized in Table 5, which gives the measured average adhesion failure loads corresponding to the friction coefficients. There is a detectable decrease in the scratching coefficient, most likely because the ploughing component decreases with decreasing surface roughness. The average adhesion failure load gradually increases with decreasing scratching coefficient. The energy of crack initiation and propagation is

Table 5.

Average friction (scratching) coefficient μ^* between
0 and 70 N, and average adhesion failure load L_A for
an (Al–Mg)-coated DU–0.75% Ti specimen

Surface condition	μ^*	L_A (N)
As-coated	0.29	31.80
Polished (1 μm paste)	0.27	31.91
Polished (0.3 μm paste)	0.25	33.12
Polished (0.05 μm paste)	0.21	37.41

balanced by the release of stored elastic energy, which consists of stored energy around the indentation, stored energy due to internal stresses within the coating, and stored energy due to frictional stress. With decreasing friction coefficient, a higher normal load will be required to provide the strain energy required for crack initiation and propagation; hence, the cohesive and adhesion failure loads will increase. The increase in L_A with decreasing μ^* has been previously reported and modelled [14].

6. EFFECT OF RESIDUAL STRESSES ON THE COATING STRENGTH AND ADHESION FAILURE LOAD

A series of SiC-coated 4340 steel specimens were prepared by plasma-enhanced chemical vapor deposition (PECVD) with the following deposition conditions: 50 μm pressure; a gas mixture of 4.2 s cm^3 min^{-1} (centimeters cubed per minute, measured at standard temperature and pressure) SiH$_4$, 6.7 s cm^3 min^{-1} CH$_4$, and 10 s cm^3 min^{-1} Ar; an upper electrode at 100 W; and a lower electrode grounded. The coatings were amorphous and contained in the as-coated condition an excess amount of entrapped hydrogen gas, most likely in solid solution, which as previously reported [18, 19] establishes a biaxial residual compressive stress within the coating, which was found [18, 19] to be independent of the coating thickness and to decrease with increasing ion beam energy during coating. During subsequent annealing, hydrogen evolution leads to shrinkage of the coating with a gradual decrease in thickness and ultimately establishment of biaxial residual tensile stresses within the coating, which are also independent of the coating thickness and increase with increasing ion beam energy. Annealing treatments were carried out at 923 K for 10 and 20 min, prior to scratch adhesion testing. Average values of L_C and L_A corresponding to five scratches are reported in Table 6.

In the as-deposited condition, under compressive residual stresses the cohesive failure load is higher than in the annealed state in which compression is replaced by tension, most likely because for these specimens crack initiation is easier under tensile conditions. After 10 min annealing, most likely the compressive residual stresses were replaced by weak tensile stresses and this may explain the increase in adhesion failure load. After 20 min annealing, higher tensile stresses made delamination easier and the adhesion failure load decreased.

Table 6.

Average cohesive failure load L_C and adhesion failure (critical) load L_A for a PECVD-processed a : SiC : H coating on 4340 low alloy steel under various conditions

Coating condition	L_C (N)	L_A (N)	Thickness (μm)
As-deposited	3.81	4.18	0.95
Annealed (923 K, 10 min)	2.34	5.32	0.90
Annealed (923 K, 20 min)	1.81	3.23	1.23

Table 7.

Average cohesive failure load L_C, adhesion failure (critical) load L_A, and friction coefficient μ^* for multilayer TiN/Ti/TiN on DU–0.75% Ti alloy specimens

L_C $(= L_{C,TiN})$	27.18 N
L_{A1} $(= L_{A,TiN/Ti})$	32.55 N
L_{A2} $(= L_{A,Ti/TiN})$	52.17 N
L_{A3} $(= L_{A,TiN/DU-0.75Ti})$	62.73 N
μ^*	0.34

7. EVALUATION OF ADHESION FAILURE IN MULTILAYER COATINGS

7.1. Multilayer (TiN/Ti/TiN)-coated DU–0.75% Ti specimen

A multilayer TiN/Ti/TiN coating consisting of three layers with respective thicknesses of 3.21, 2.17, and 3.07 μm was processed as described above for the soft metallic coatings and is reported in detail elsewhere [20]. The upper TiN surface was relatively smooth with only some occasional microscratches, fine TiN droplets, and microcavities, most likely caused by extraction of surface particles during handling and cleaning.

Typical AE, F_t, and μ^* curves vs. F_n between 0 and 80 N are illustrated in Fig. 6b. For this particular scratch, the shape characteristics of AE and, to a lesser extent, F_t indicate that crack initition within the upper TiN layer occurs at $L_C = 27.5$ N. The crack reaches the interface between the upper TiN and the titanium layers, causing delamination at $L_{A1} = 34.5$ N. It subsequently propagates through the titanium layer, reaching the interface between Ti and the lower TiN layer and causing delamination at $L_{A2} = 56.7$ N. Finally, the crack reaches the interface between the lower TiN layer and the substrate, and causes delamination at the critical load $L_{A3} = 64.9$ N. Metallographic observations facilitated by the golden color of TiN, the silvery color of titanium, and the pinkish color of DU confirmed the above interpretation of the measurements. Average values using six scratches are given in Table 7.

8. CONCLUSIONS

Automatic scratch testing is an expedient technique for comparatively evaluating the cohesive and adhesion failure loads of thin coatings on various substrates.

(1) In soft Triballoy T-800 and stellite SF-6 cobalt-based coatings on 4340 low alloy steel,

    (a) adhesion is strong and failure is cohesive in the coating;

    (b) in the presence of a plated chromium interlayer, pre-existing cracks lower the cohesive failure load;

(c) for all coatings, a higher coating deposition pressure lowers the cohesive failure load; and

(d) the spacing of transverse cracks within the coating decreases with increasing applied normal load.

(2) In soft aluminum coatings on DU−075% Ti alloy specimens, alloying aluminum with magnesium or zinc enhances the coating strength and adhesion.

(3) In (Al−Mg) coatings on DU−0.75% Ti alloy specimens, a smoother surface leads to a lower friction coefficient and a higher adhesion failure load.

(4) In hard, thin TiN coatings on 17-4 PH steel, a lower bias voltage applied to the substrate yields higher cohesive and adhesion failure loads.

(5) In hydrogenated amorphous SiC thin coatings on 4340 steel, loss of hydrogen by annealing converts the residual compressive stress into tensile stress and lowers both the cohesive and the adhesion failure loads.

(6) Finally, automatic scratch testing can be very helpful in determining delamination loads in multilayer TiN/Ti/TiN coatings on DU−0.75% alloy.

Acknowledgements

Financial support of part of this work by the Advanced Technology Center for Precision Manufacturing (PMC) at the University of Connecticut is gratefully acknowledged. Part of the work was conducted at the US Army Research Laboratory, Watertown, MA 02172. I am grateful to Brent Chambers, MIT for preparing the silicon carbide-coated steel specimens.

REFERENCES

1. O. S. Heavens, *J. Phys. Rad.* **11**, 355 (1950).
2. H. E. Hintermann and P. Laeng, in: *Haftung als Basis für Stoff*, p. 87. Deutsche Gesellschaft für Metallkunde, Breman (1981).
3. A. J. Perry, *Thin Solid Films* **81**, 357 (1981).
4. T. Z. Kattamis, F. Chang and M. Levy, *Surface Coat. Technol.* **43/44**, 390–4901 (1990).
5. J. Sekler, P. A. Steinmann and H. E. Hintermann, *Surface Coat. Technol.* **36**, 519–529 (1988).
6. J. Valli and U. Makela, *Wear* **115**, 215 (1987).
7. K. L. Mittal (Ed.), in: *Adhesion Measurement of Thin Films, Thick Films and Bulk Coatings*, ASTM STP 640, pp. 5–17. American Society for Testing Materials (1978).
8. A. J. Perry, *Thin Solid Films* **107**, 167 (1983).
9. P. A. Steinmann and H. E. Hintermann, *J. Vac. Sci. Technol.* **A3**, 2394 (1985).
10. B. Hammer, A. J. Perry, P. Laeng and P. A. Steinmann, *Thin Solid Films* **96**, 45 (1982).
11. P. A. Steinmann, Y. Tardy and H. E. Hintermann, *Thin Solid Films* **154**, 333–349 (1987).
12. T. Z. Kattamis, K. L. Bhansali, M. Levy, R. Adler and S. Ramalingam, *Mater. Sci. Eng.* **A161**, 105–117 (1993).
13. T. W. Wu, *J. Mater. Res.* **6**, 407–426 (1991).
14. S. J. Bull, D. S. Rickerby, A. Matthews, A. Leyland, A. R. Pace and J. Valli, *Surface Coat. Technol.* **36**, 503–517 (1988).
15. D. S. Rickerby, *Surface Coat. Technol.* **36**, 541–557 (1988).
16. P. J. Burnett and D. S. Rickerby, *Thin Solid Films* **154**, 403–416 (1987).
17. F. Chang, M. Levy, B. Jackman and W. B. Nowak, *Surface Coat. Technol.* **39**, 721–731 (1989).

18. H. S. Landis, PhD Thesis, Department of Materials Science and Engineering, MIT, Cambridge, MA (1988).
19. A. S. Argon, V. Gupta, H. S. Landis and J. A. Cornie, *Mater. Sci. Eng.* **A107**, 41–47 (1989).
20. F. C. Chang, M. Levy, R. Huie, M. Kane and P. Buckley, T. Z. Kattamis and G. R. Lakshminarayan, *Surface Coat. Technol.* **49**, 87–96 (1991).

Adhesion Measurement of Films and Coatings, pp. 161–174
K. L. Mittal (Ed.)
© VSP 1995.

Continuous microscratch measurements of thin film adhesion strengths

SHANKAR K. VENKATARAMAN,[1] JOHN C. NELSON,[1] ALEX J. HSIEH,[2]
DAVID L. KOHLSTEDT[3] and WILLIAM W. GERBERICH[1,*]

[1] *Department of Chemical Engineering and Materials Science, University of Minnesota, Minneapolis, MN 55455, USA*
[2] *U.S. Army Materials Technology Laboratory, Watertown, MA 02172, USA*
[3] *Department of Geology and Geophysics, University of Minnesota, Minneapolis, MN 55455, USA*

Revised version received 17 June 1993

Abstract—The adhesion strengths of metal/ceramic, metal/polymer, and polymer/polymer interfaces have been characterized using the continuous microscratch technique. In these experiments, a conical diamond indenter was driven simultaneously into a thin film at a rate of 15 nm/s and across the film surface at a rate of 0.5 μm/s until a load drop or other discontinuity occurred, indicating film failure. The *critical load* at failure of the thin film was taken as a measure of the adhesion strength. For metal/ceramic systems such as Cr thin films on Al_2O_3, and for diamond-like-carbon (DLC) films on glass, clear load drops provided an accurate measure of the adhesion strengths. For metal/polymer systems such as Cu thin films on PET, a change in the loading pattern and periodic cracking events along the scratch track provided evidence of film delamination. For DLC films on polycarbonate substrates, the carbon thin film cracked before it delaminated. For bulk polymers such as polycarbonate and polystyrene/polypropylene, crack growth occurred by a stick-slip mechanism. Using a model developed in an earlier paper, the practical work of adhesion for the Cr/Al_2O_3 system was determined to be 0.09 J/m^2 and that for the DLC/polycarbonate system was 0.05 J/m^2. The fracture toughnesses of the polycarbonate and polystyrene were 0.81 and 0.2 MPa m$^{1/2}$, respectively. These numbers are in good agreement with those obtained by other methods for these systems.

Keywords: Adhesion; thin films; microscratch; toughness.

1. INTRODUCTION

In the last decade, the use of thin film materials has proliferated in a variety of micro-electronic, optical, and biomedical applications [1–3]. Thin films are widely used as insulators or conductors in the magnetic and electronic packaging industries and are often used as corrosion-, wear-, and abrasion-resistant materials. The durability and functionality of a thin film depend not only on the hardness and modulus of the film, but also on the adhesion between the film and the substrate that supports the film. Several techniques have been utilized to evaluate the adhesion properties of thin films, the most common being the peel test [4], the bulge test [5], laser spallation [6], the thin film tensile test [7], and the scratch test [8–11]. However, most of these tests are only qualitative and involve fundamental problems concerning measurable adhesion range and data accuracy. For years, the scratch adhesion test has provided a simple and rapid means of

---

*To whom correspondence should be addressed.

assessing the adherence of thin films to hard substrates, despite a basic lack of understanding of the mechanics of the testing technique.

In a scratch test, a diamond stylus is drawn over the sample (film) surface under a step-wise [8, 9] or continuously increasing normal force [10–14], until the film detaches from the substrate. The critical load at which the film is delaminated has been used as a qualitative measure of the adhesion strength of the film. For the same bi-material system, the critical load can be used to qualitatively predict differences in adhesion due to differences in film deposition techniques, annealing conditions, residual stresses, thickness, etc. A number of researchers have attempted to analyze the scratch process, because the scratch technique is relatively simple to use, yields reproducible results, and involves the testing of extremely small volumes of material. The models of Benjamin and Weaver [15, 16], Laugier [17], Bull et al. [18], Burnett and Rickerby [19, 20], and Venkataraman et al. [11, 12] have transformed the scratch test from one that only measured the critical load to failure into one that provides the more fundamental quantity of strain energy release rate. In this study, the adhesion strengths of metal/ceramic, metal/polymer, and polymer/polymer interfaces were characterized using a high-resolution microindenter in the microscratch mode. The analysis of Venkataraman et al., which has been successfully applied to metal thin films on ceramic substrates [11–14], was used to determine the strain energy release rate (practical work of adhesion) and interfacial fracture toughness of the various bimaterial systems. Preliminary investigation into the adhesion of diamond-like carbon (DLC) films to glass and polycarbonate substrates was also attempted. The scratch technique was also used to determine the fracture toughness of bulk polycarbonate.

2. EXPERIMENTAL

Chromium films, 0.2 μm in thickness, were deposited onto single crystal sapphire substrates of [11$\bar{2}$0] orientation using a DC magnetron sputtering unit. Prior to deposition, the substrates were cleaned with acetone and subjected to a short sputter etch. The films were deposited at rates of 0.15 nm/s with a substrate temperature of less than 80°C. Copper films 0.1 μm in thickness, were laid down on poly(ethylene terephthalate) (PET) by a Flectron™ metallization process developed by Monsanto Company. Two different kinds of Cu/PET samples were used, each having a different intermediate layer between the Cu and PET. The nature of the intermediate layer is proprietary. One-micrometer-thick polystyrene films were laid down on polypropylene substrates by spin casting. DLC films of 0.1 μm thickness were deposited onto glass and polycarbonate substrates by direct ion-beam deposition using a mixture of methane and argon gases. Finally, bulk, doped polycarbonate samples were also prepared for microscratch testing.

A continuous microintender [10, 11] was used in the microscratch mode to evaluate the adhesion strengths of these films. The load and depth resolution of the machine were 16 μN and 5 Å, respectively. Scratch tests were performed using a conical diamond indenter having a nominal tip radius of either 1 or 5 μm and an included angle of 90°. The indenter was translated simultaneously in the horizontal and vertical directions (across and into the sample), at rates of 0.5 μm/s and 15 nm/s, respectively. The normal load and depth of penetration of the indenter into the sample were continuously recorded during each scratch experiment. The scratch test was stopped when a load drop was detected indicating that the film had delaminated from the substrate or when evidence of cracking or fracture events in the film showed up as a series of small load drops in the scratch loading curves. After a series of scratches were made on a sample, it was necessary to examine the resulting scratch tracks with scanning electron and optical microscopes in order to locate the start of failure events such as cracks and fractures.

3. RESULTS

The results for all the experiments can be discussed under the following categories: (a) metal on ceramic; (b) DLC on glass and polymer; (c) metal on polymer; (d) polymer on polymer; and (e) bulk polycarbonate.

3.1. Metal on ceramic

Two representative load vs. scratched distance plots for the Cr/Al_2O_3 system are presented in Fig. 1a. To avoid overlap, the scratch loading curves have been offset from each other by 50 μm. In each case, the load increases as the indenter goes across the sample surface, and deeper into the sample. As the indenter pushes the film material ahead of it, a stress builds up in the film. When this stress exceeds the interfacial shear strength, the film debonds from the substrate. This point is registered as a load drop in the load–scratched distance curves. The first curve represents a scratch test that was not stopped just after the first load drop. In this curve, the first load change (drop) occurs at about 10 mN. After that point, the scratch curve shows periodic load increases and drops. In the second curve, the scratch test was stopped immediately after the first load drop. The load at which the first load drop occurs is highly reproducible.

The scanning electron microscopy (SEM) observations in Fig. 1b illustrate the features of scratch track 1 in Fig. 1a. Comparison of the SEM micrographs with the load–scratched distance plots reveals that a load drop occurs at the point at which the film first debonds from the substrate, the critical load to failure. In the case of scratch track 1, each load drop corresponds to a failure in the film. Hence some kind of start–stop crack growth (at the interface) mechanism is in operation. The critical loads, L_{cr}, were averaged over all the scratch experiments performed on this sample and are summarized in Table 1.

3.2. DLC on glass and polymer

The scratch loading curve of the DLC on glass, shown in Fig. 2a, is similar to that seen for metal on ceramic. In each test, a sharp load drop occurs at about 10.5 mN as the film delaminates from the glass substrate. The SEM micrograph in Fig. 2b shows the delaminated portion of the film.

For the DLC on polycarbonate, a typical scratch loading curve is shown in Fig. 3a. The SEM micrograph of the scratch track (Fig. 3b) shows cracking of the film at low loads followed by film failure at higher loads. Clear delamination of the film was not observed. It appeared that the polycarbonate, because of its higher

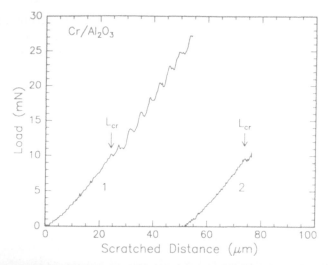

Figure 1. (a) Typical scratch loading curves for the Cr/Al_2O_3 system showing load drops at 10 mN. The two curves have been offset by 50 μm for clarity. Note the start–stop crack growth mechanism in curve 1. (b) Scanning electron micrograph of the scratch track 1, whose scratch loading curve is given in (a). Film delaminations correspond to the load drops seen in the scratch loading curve.

Table 1.
Practical work of adhesion (W_{ad}) and interfacial fracture toughness (K_i) for various bimaterial systems. All columns represent averages over four independent experiments. Here t is the film thickness and L_{cr} is the critical normal load

Sample	t (μm)	L_{cr} (mN)	G_i, W_{ad} (J/m²)	K_i (MPa$\sqrt{m}$)
Cr/Al_2O_3	0.2	10	0.09	0.16
DLC/glass	0.1	10.5	0.04	0.15

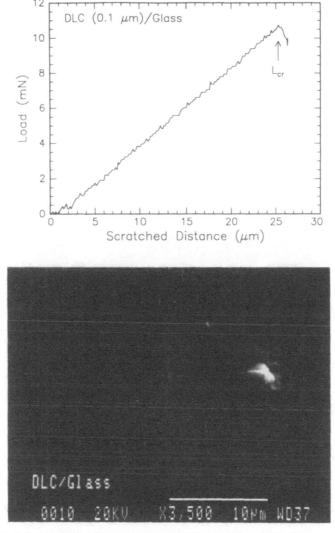

Figure 2. (a) Typical scratch loading curve for the diamond-like-carbon film on glass substrates showing a load drop at 10.5 mN. (b) Scanning electron micrograph of the scratch experiment whose scratch loading curve is shown in (a).

compliance and toughness compared to DLC, deformed to large elastic strains without fracturing but generated tensile cracks in the film.

3.3. Metal on polymer

Both types of Cu/PET samples were tested using the microscratch technique. A series of scratch tests were performed on each sample. Typical scratch loading curves for the two samples, shown in Fig. 4a, indicate first a change in the slope of the scratch loading curve and then a sudden load drop. An SEM micrograph of typical scratch tracks in the first Cu/PET sample is presented in Fig. 4b. The change in slope of the scratch loading curve occurs when the stress field below the

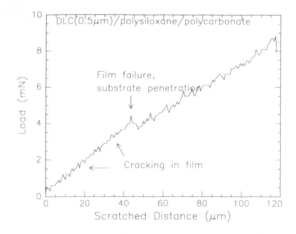

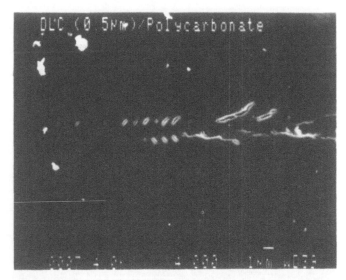

Figure 3. (a) Scratch loading curve for the diamond-like-carbon film on polycarbonate substrates showing continuous load discontinuities which correspond to film cracking. (b) Scanning electron micrograph showing the continuous cracking and delamination in the DLC film on polycarbonate.

indenter interacts with the intermediate layer and the substrate. The load drop occurs when the indenter penetrates through the film and intermediate layer into the substrate. The loads at which the slope change occurs for the two samples are 2.0 and 3.5 mN. Sharp drops in load occur at 5.25 and 7.5 mN, respectively, for the two samples. No area of delamination of the Cu film could be measured; delamination occurred on a local scale, as seen by the localized periodic buckling in Fig. 4b.

3.4. Polymer on polymer

A typical scratch loading curve and the corresponding SEM micrograph for the 1.0 μm thick polystyrene film on polypropylene are shown in Figs 5a and 5b. Stick–slip failure dominates in this system. The small loading discontinuities in

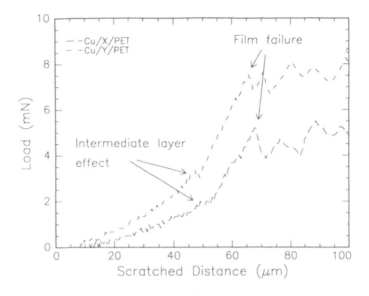

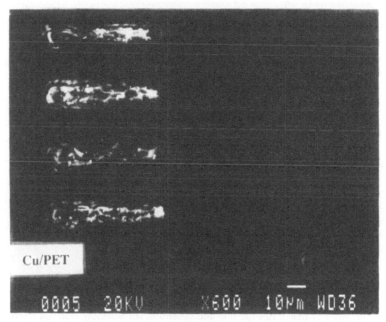

Figure 4. (a) Typical scratch loading curves for the two different Cu/PET samples showing differences in slope changes and load drops. (b) Typical scanning electron micrograph of scratch experiments performed on the first Cu/PET sample showing film deformation and failure.

the scratch loading curves coincide with the formation of the chevron-shaped gouges seen in the SEM micrograph. A schematic diagram of the indenter–polymer interactions producing these gouges is shown in Fig. 6. As the stress builds up in front of the moving indenter, a crack initiates at the bottom of the tip and propagates first into the polystyrene and later into the polypropylene. This crack propagates under a shear stress. Because of the stresses acting on the face of

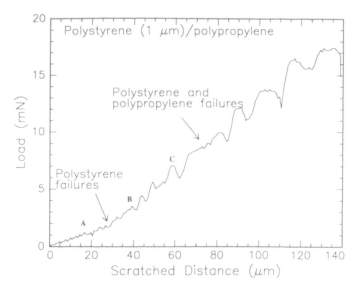

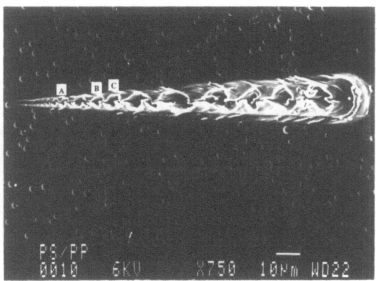

Figure 5. (a) Scratch loading curve for the 1 μm polystyrene film on polypropylene. Initially single load discontinuities are seen as the indenter penetrates the film. Later double loading discontinuities are seen as the indenter penetrates the film and substrate. (b) Scanning electron micrograph of the scratch experiment whose scratch loading curve is represented in (a). The loading discontinuities correspond to the chevon-shaped gouges seen in the picture.

the indenter tip, the indenter rides up over the built-up material. The polymer then elastically springs back beneath the tip and the process is repeated again and again. The gouges increase in size as the load becomes larger. Once the indenter penetrates through the film and into the substrate, it is in contact with two polymer surfaces. A double gouge is seen in the SEM micrograph; correspondingly, two discontinuities occur in the scratch loading curve. The area of a particular gouge is the fracture area due to the load at the corresponding discontinuity.

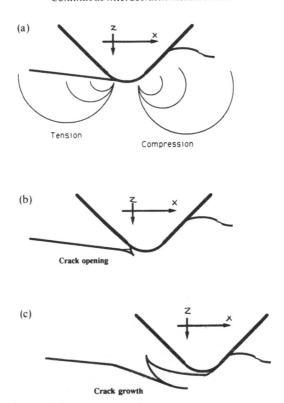

Figure 6. Schematic diagram of the indenter–polymer interactions which produce the gouges seen in Figs 5b and 7b. The polymer is torn below the indenter by a rapidly moving crack front (b). With increasing elastic stresses acting on the indenter face, the indenter rides up over the polymer (c).

3.5. Bulk polycarbonate

Figures 7a and 7b show a typical load–depth curve and the corresponding SEM micrograph for scratch tests performed on a bulk polycarbonate sample. As in the polystyrene/polypropylene sample, stick–slip failure seems to be predominant. Similar 'gouges' occur in the polymer, indicating that the schematic diagram shown in Fig. 6 is also applicable to the polycarbonate. The area of each gouge gives the fracture area corresponding to a particular load discontinuity observed in the load–depth curve.

4. DISCUSSION

4.1. Determination of the practical work of adhesion and interfacial fracture toughness

The practical work of adhesion equates the strain energy stored in the film at failure to the energy required to separate the film from the substrate. This energy includes terms involving plastic deformation of the film and substrate. Venkataraman *et al.* [11] showed that the average shear stress ahead of the indenter tip, $\bar{\tau}_{zx}$, at film delamination depends on a number of parameters such as the critical normal load to failure (L_{cr}), the area of film delamination (A_d), the film

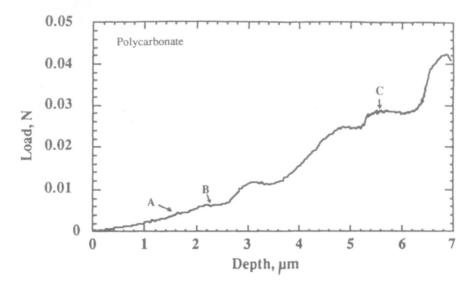

Figure 7. (a) Load–scratched distance plot for scratch tests on bulk polycarbonate showing loading discontinuities. **(b)** Scanning electron micrograph of the scratch test whose scratch loading curve is shown in (a). The chevron-shaped gouges seem to correspond to the loading discontinuities in (a).

thickness (t), the width of the scratch track at failure ($2a$), and the length of the circular or pie-shaped delaminated region from the start to the end of delamination (B) as

$$\bar{\tau}_{zx} = f(L_{cr}, t, a, A_d, B). \tag{1}$$

The strain energy released through the thickness of the film, G_i, can be written in terms of the average stress in the delaminated region as

$$G_i = 2\sum \left(\frac{\bar{\tau}_{ij}^2}{2\mu} t + \frac{\bar{\sigma}_{ij}^2}{2E} t \right), \tag{2}$$

where μ and E are the shear and elastic moduli of the film, respectively. The interfacial fracture toughness, K_i is then given as

$$K_i = \sqrt{\frac{2\mu G_i}{1-\nu}}, \tag{3}$$

where ν is the Poisson ratio of the film material.

This model works well for metal thin films on ceramic substrates, and the strain energy release rate and interfacial fracture toughness for the Cr films on Al_2O_3 substrates are summarized in Table 1. Residual stresses are not considered in this model. However, in similar experiments performed on a Ta/NiP system, residual stresses of ± 1 GPa were found to affect the critical load by only 10–15% [21].

For the DLC films on glass, a section of the film delaminated at the critical load. Using the same model and taking an elastic modulus of 500 GPa for the DLC, the strain energy release rate and the interfacial fracture toughness were calculated as summarized in Table 1. It must be pointed out that the DLC films behaved elastically, as also seen by indentation experiments performed on the same sample. Hence the analysis of Venkataraman *et al.* [11], which is based on an elasticity analysis, is appropriate.

For the DLC/polycarbonate sample, it is only possible to predict qualitatively the adhesion strength in terms of the critical load. The DLC films started cracking but never clearly delaminated from the substrate. This implics that the DLC film adheres well to the polycarbonate. The film periodically cracks as the indenter tip moves into the surface of the sample. Hence the stress built up ahead of the tip is relieved by the cracking of the film and never becomes large enough to delaminate the film.

For the Cu thin films on PET, only a qualitative prediction of the adhesion strength in terms of the critical load was possible, due in part to the deformation behavior of the polymer substrates. In addition, the nature of the intermediate layer in the Cu/PET samples and the lack of a clear delamination region rendered calculation of a strain energy release rate impossible. Earlier results for Cu and Al films on Al_2O_3 had shown that the Cu and Al tended not to delaminate at ambient temperature [13]. This behavior is due to the extremely ductile nature and low yield stress of the Cu and Al, which allow the indenter to plastically deform these metals rather than cause delamination.

Since the fracture behaviors of bulk polycarbonate and of polystyrene film on polypropylene are similar, the two samples were treated identically. Chevron- or heart-shaped fracture areas formed in both cases, and a load corresponding to the fracture area could easily be determined from the scratch loading curves. The above analysis of Venkataraman *et al.* [11] was applied to these samples and the elastic strain energy release rate was calculated at three different fracture areas along the scratch track. For the polycarbonate sample, fracture areas A, B, and C in Fig. 7b were selected. The fracture area, scratch width, and extent of fracture

Table 2.

Calculation of the adhesion strengths for polycarbonate and polystyrene/polypropylene samples. Here L_{cr} is the critical normal load. A_d is the area of delamination, $2a$ is the scratch track width at failure, B is the extent of delamination, G_i is the strain energy release rate, and K_i is the interfacial fracture toughness. A, B, and C are regions of the sample as seen in Figs 5b and 7b

Sample		L_{cr} (mN)	A_d (μm^2)	$2a$ (μm)	B (μm)	G_i (J/m^2)	K_i (MPa$\sqrt{m}$)
Polystyrene/	A	1.5	12	2.8	4	15.5	0.22
polypropylene	B	3.5	45	5.8	9	13.6	0.20
	C	7.0	115	9.5	15.4	10.9	0.18
Polycarbonate	A	5.0	2.2	3.5	2.1	251	0.82
	B	7	22	5.3	7.3	210	0.75
	C	30	101	11.8	12.7	279	0.87

Averages: Polystyrene/polypropylene: $G_i = 13.3$ J/m^2, $K_i = 0.20$ MPa$\sqrt{m}$. Polycarbonate: $G_i = 247$ J/m^2; $K_i = 0.81$ MPa$\sqrt{m}$.

determined using a digital scanner are summarized in Table 2. The critical load corresponding to each of the fracture areas is also presented in Table 2. Since a bulk sample was being tested, the thickness in equations (1), (2), and (3) was equated to the depth of penetration of the indenter tip at that fracture area (depth of fracture). The fracture toughness calculated is then the fracture toughness of the bulk polycarbonate. The strain energy release rate and fracture toughness are also summarized in Table 2.

The calculation of the strain energy release rate and interfacial fracture toughness for the polystyrene/polypropylene sample can be made following a method similar to that used for the polycarbonate. From the shape of the load–displacement curve along with the matched fractography, it was concluded that the film and substrate fractured independently. However, here the elastic strain energy is released through the thickness of the polystyrene film. Three different fracture areas, A, B, and C in Fig. 5b, were selected for the purpose of analysis and the results are summarized in Table 2.

4.2. Significance of the results

The practical work of adhesion is the true work of adhesion plus contributions from plastic deformation of the film and the substrate. No single technique determines a true work of adhesion for metal/ceramic systems, as the energy expended in the deformation of the film and the substrate is also included. The practical work of adhesion for metal/ceramic systems is easily calculated using the continuous microscratch technique. The practical work of adhesion decreases as the thickness of the film decreases, primarily because the energy required for plastic deformation of the film decreases. As the film thickness approaches zero, the practical work of adhesion should approach the true work of adhesion, provided that the amount of inelastic deformation in the substrate is very small— which is the case for sapphire substrates.

For as-evaporated films, the true work of adhesion determined by the continuous microscratch technique is 0.01–0.05 J/m^2 [22]. This range of values is similar to the range calculated for the van der Waals energy of adhesion for metals on ceramics, $0.02-0.08$ J/m^2. The good correlation between these two quantities suggests that the primary bonding for as-evaporated thin metal films on ceramic substrates is due to van der Waals interactions. For as-sputtered samples, however, the adhesion energy depends on the partial pressure of the Ar, increasing as the Ar partial pressure decreases [23]. The thickness of the film has strong effects on the measured practical work of adhesion as the contribution of plastic deformation of the film increases with film thickness. In the Pt/NiO system, for example [22], the practical work of adhesion increased from 0.03 to 4.7 J/m^2 as the film thickness increased from 65 to 1050 nm.

For a comparison, values for the work of adhesion of $0.4-0.7$ J/m^2 for a Cu/Al$_2$O$_3$ system, $0.2-0.8$ J/m^2 for a Ni/Al$_2$O$_3$ system, and 0.5 J/m^2 for a Au/Al$_2$O$_3$ system were determined by Evans and Rühle [24], using an elastic-plastic analysis, the shape and distribution of interfacial flaws, and the flow stress. To obtain values for the work of adhesion for bulk metal/ceramic samples, Agrawal and Raj [25] analyzed the spacing between the cracks formed in the ceramic overlayer when the metal substrate was deformed in tension. This approach yielded values for the work of adhesion of 0.475 J/m^2 for a Cu/Al$_2$O$_3$ system and 0.645 J/m^2 for a Ni/Al$_2$O$_3$ system. Using a blister test, White [26] determined the work of adhesion for metal/polymer systems to be $0.01-0.05$ J/m^2. These independent measurements of the work of adhesion indicate that the model used here for the work of adhesion is not unreasonable for a metal/ceramic system.

5. CONCLUSION

It has been shown that the continuous microscratch test is a simple yet powerful technique for characterizing the adhesion strengths of thin films. Only small volumes of material are required for testing, and testing times are relatively short. The technique is quite quantitative for metal/ceramic systems, where delamination of the film is achieved, and the results obtained correlate well with those obtained by other techniques. For metal/polymer and DLC/polymer interfaces, stresses are relieved by film buckling or cracking so that large-scale delamination of the film is not achieved. In these circumstances, only qualitative prediction of the adhesion strength has been achieved to date. For polymer/polymer interfaces, however, a semi-quantitative prediction of the fracture toughness is possible from the size of the cracked/failed region.

Acknowledgements

We would like to thank Dr. William Dale, Monsanto Company for providing the Cu/PET samples, and Uttandaraman Sundararaj of the Department of Chemical Engineering and Materials Science, University of Minnesota for preparing the polystyrene/polypropylene samples. This research was supported by the Center for Interfacial Engineering at the University of Minnesota under Grant No. NSF/CDR-8721551.

REFERENCES

1. J. W. McPherson and C. F. Dunn, *J. Vac. Sci. Technol.* **B5**, 1321 (1987).
2. B. D. Fabes and W. C. Oliver, *Mater. Res. Soc. Symp. Proc.* **188**, 127 (1990).

3. W. R. LaFontaine, B. Yost and Che-Yu Li, *J. Mater. Res.* **5**, 776 (1990).
4. K. L. Mittal, *Electrocomponent Sci. Technol.* **3**, 21 (1976).
5. D. A. Hardwick, *Thin Solid Films* **154**, 109 (1987).
6. V. Gupta, A. S. Argon, D. M. Parks and J. A. Cornie, *J. Mech. Phys. Solids* **40**, 141 (1992).
7. R. W. Hoffman, *Mater. Res. Soc. Symp. Proc.* **130**, 295 (1989).
8. A. J. Perry, *Thin Solid Films* **107**, 167 (1983).
9. P. A. Steinmann, Y. Tardy and H. E. Hintermann, *Thin Solid Films* **154**, 333 (1987).
10. T. W. Wu, *J. Mater. Res.* **6**, 407 (1991).
11. S. Venkataraman, D. L. Kohlstedt and W. W. Gerberich, *J. Mater. Res.* **7**, 1126 (1992).
12. S. Venkataraman, D. L. Kohlstedt and W. W. Gerberich, *Thin Solid Films* **223**, 269 (1993).
13. S. Venkataraman, M. S. Thesis, University of Minnesota (1992).
14. Y.-C. Lu, S. L. Sass, Q. Bai, D. L. Kohlstedt and W. W. Gerberich, submitted. *Acta Metall.*
15. P. Benjamin and C. Weaver, *Proc. R. Soc. London, Ser. A* **254**, 163 (1960).
16. C. Weaver, *J. Vac. Sci. Technol.* **12**, 18 (1975).
17. M. T. Laugier, *Thin Solid Films* **117**, 243 (1984).
18. S. J. Bull, D. S. Rickerby, A. Matthews, A. Leyland, A. R. Pace and J. Valli, *Surface Coat. Technol.* **36**, 503 (1988).
19. P. J. Burnett and D. S. Rickerby, *Thin Solid Films* **148**, 41 (1987).
20. P. J. Burnett and D. S. Rickerby, *Thin Solid Films* **148**, 51 (1987).
21. R. White, J. C. Nelson and W. W. Gerberich, *Mater. Res. Soc. Symp. Proc.* submitted, Spring 1993.
22. S. Venkataraman, D. L. Kohlstedt and W. W. Gerberich, *Mater. Res. Soc. Symp. Proc.* submitted, Spring 1993.
23. M. Ohring, *The Materials Science of Thin Films.* Academic Press, New York (1992).
24. A. J. Evans and M. Rühle, *MRS Bull.* 46 (Oct. 1990).
25. D. C. Agrawal and R. Raj, *Mater. Sci. Eng.* **A126**, 125 (1990).
26. R. C. White, Paper presented at the International Symposium on Adhesion Measurement of Films and Coatings, Boston (5–7 Dec. 1992).

Adhesion Measurement of Films and Coatings, pp. 175–188
K. L. Mittal (Ed.)
© VSP 1995.

Micro-scratch test for adhesion evaluation of thin films

V. K. SARIN

College of Engineering, Boston University, Boston, MA 02215, USA

Revised version received 21 May 1993

Abstract—In tribological applications, surface properties govern performance, and hence the utilization of coatings to tailor properties has become an essential component of materials technology. The most critical requirement for such coatings is adequate adhesion. Therefore, the need to measure coating/substrate interfacial strength and to characterize the factors influencing it is critical. However, the most common method currently used to characterize adhesion—the scratch test—is inadequate because it does not really measure adhesion. It is indicative of a relative measure of coating durability, which can be useful, at best, for quality control purposes. Consequently, the need for a test that can more accurately determine the adhesion characteristics of thin films is imperative. The development of a micro-scratch test and how it addresses some of the deficiencies of the conventional scratch test are discussed.

Keywords: adhesion; interface; coatings; scratch test; indentation; modelling (scratch test)

1. INTRODUCTION

In recent years, considerable research has been devoted to the development of hard, thin coatings for tribological and structural applications. Such coatings are applied to enhance both the chemical and the physical properties of the substrate material. The underlying objective of this work has been to optimize the coating/ substrate design to achieve desired properties and a reduction of factors inducing failure. Typical of such factors are abrasive wear, chemical wear, thermal shock, and degradation. In order to facilitate optimization, the characteristics of the substrate/coating materials and their interaction must be understood, particularly in the service environment. Many of these properties are either known or may be calculated and empirically verified. However, factors controlling practical adhesion [1], i.e. the actual energy or force required to debond the coating from the substrate, are poorly understood and difficult to quantify.

Adhesion is a macroscopic property that depends on the chemical and mechanical bonding across the interfacial region, the intrinsic stress and stress gradient, and the failure mode. Apart from theoretical evaluation of adhesion from the microscopic viewpoint, and atom-to-atom attractive force, most attempts to characterize thin-film adhesion have been mechanical; that is, physical removal of the coating from the substrate. The particular problem encountered is that of adhesion measurement or, more precisely, its lack of susceptibility to measurement, since all these techniques promote rather complex failure mechanisms, which are influenced by several factors other than just the bond strength between the coating and the substrate.

For decades, researchers have attempted to quantify adhesion. To date, several tests such as the peel test, pull test, scratch test, and many others have been developed. All these tests have been only partly successful at best, and apply to a very specific and limited field of materials. However, none has so far been able to characterize adhesion adequately. The most frequently used scratch test, using the Revetest [Trademarked by LSRH (now CESM)], is strongly influenced by several factors which are not directly related to interfacial strength. Therefore, the critical load (L_c) used as a measure of adhesion in this test yields, at best, an indication of a relative measure of coating durability. Hence, the need to measure the interfacial strength between the coating/substrate interface, and to characterize the factors influencing it, is essential for the systematic design and development of new and/or improved coatings. Furthermore, if coated components are to be used reliably and with confidence, the availability of such information is indispensable.

2. THE SCRATCH TEST

The scratch test in its current form was developed by Benjamin and Weaver [2]. The test is performed by drawing a smoothly rounded indenter point across the coating surface with increasing vertical load until the film is stripped from the substrate. The load on the indenter at which the coating is stripped cleanly from the substrate is defined as the critical load (L_c). The original analysis [2] was based on the proposal that when a rounded indenter is drawn across a coated surface and the adhesion of the indenter to the coating is good, shearing must occur at the weakest point, either at the coating/substrate interface or within the substrate. Assuming that the cohesive forces within the substrate area are greater than the adhesion forces at the interface, shearing should occur at the interface, thus resulting in a measure of the interfacial energy. However, many researchers have investigated the scratch test and found it to be much more complex than the original analysis implied.

Butler *et al.* [3] found that the detached film could be pressed back into the path by the indenter or the film could be thinned to transparency, and thus the critical load could not be determined on the basis of a clear indenter track. They suggested that a much larger number of coating parameters, such as yield stress, surface condition, density, and grain size, were also factors in apparent adhesion differences.

Using scanning electron micrographs of the scratch path that showed hillocks ahead and to the side of the indenter, Weaver [4] concluded that the original theory of detachment by shearing could not be valid. Laeng and Steinmann [5] pointed out the importance of the failure mode and hypothesized that not only does it affect the value of the critical load, but also that some modes do not show a straightforward track clearing and do involve local flaking. Additionally, Laugier [6] found that friction played a critical and complex role in the film-stripping process. Perry *et al.* [7] pointed out that substrate surface roughness may be a factor, further confusing the issue of what the value of L_c really indicated.

There has been some controversy over the effect of film thickness on the critical load (L_c). Originally it was reported that the critical load became constant for films exceeding a certain thickness [2]. The initial sloping part of the load vs. film thickness curve was considered not to be due to a defect in measurement, but rather to be due to the inhomogeneous microstructure of very thin coatings. However, Chopra [8] and Perry and co-workers [9] were among the first of many investigators [10–13] to report that the critical load increased with coating thickness. Additionally, it was reported [9] that the scratch test involved the inducement of a shear stress at the interface as a result of deformation of the system; therefore, a thicker layer might require a greater surface load to attain the same shear stress at the interface. The behavior of thicker films was found to be less reproducible because of the effect of mechanical stress in the films [14].

3. ANALYSES OF THE SCRATCH TEST

In the original analysis, Benjamin and Weaver [2] calculated the shearing force per unit area F, based primarily on the classical equations of Hertz for a flat surface under a hard sphere. Since this assumed an idealized indentation shape and neglected elastic deformation, its accuracy and validity are disputable. Later, Weaver [4] admitted that the scratching process was more complex than originally defined and tried to reconcile observations with the previous successful use of the test. Because vacuum-deposited films are almost always in a state of tensile stress, Weaver postulated that once the perimeter of the contact area was detached, the loosened edge of the film would tend to lift and peel away from the substrate, creating cracks. As the indenter advanced, it would push down the loosened edge, creating a hillock ahead of the indenter.

Optical and electron (SEM) microscopic examination of the topography of the scratch tract [15] usually indicates a groove with a smooth center and raised sides. Often a layer of the coating remains in the center. Gross plastic deformation at the edges of the track, ridge formation due to plastic deformation, and damage to the substrate have all been observed and reported. From these observations, two modes of coating detachments can be considered. The coating can crack and become partially detached after the indenter passes, perhaps because of the failure of the coating and interface to support the stresses resulting from high substrate deformations. Alternatively, the film may become detached in an annulus around the indenter. In some studies, it has been shown that the separation of the film from the substrate usually did not take place directly under the indenter but ahead of it. This observation is in agreement with the theoretical calculations by Hamilton and Goodman [16] for the stress field caused by a loaded spherical tip moving over a flat surface. Laugier [6], using these basic equations [16], suggested a model of the scratch test that included the coefficient of friction as a parameter.

Laeng and Steinmann [5] disproved the original model [2] for hard coatings. To explain their results, they included a dimensionless parameter

$$X = L/(r^2 H),$$

where L is the critical load, r is the tip radius, and H is the substrate hardness. However, since this parameter is itself dependent on the coating thickness and tip radius, it cannot be used to obtain quantitative values of the interfacial strength.

Because interfacial shearing, assumed by many to be responsible for de-adhesion when a critical load is reached, did not always result in film removal, Laugier [17] proposed the use of debonding energy as a criterion. In addition, he suggested that the condition for coating removal ahead of the indenter is that the film be under compression. Hence, for ductile films, material expands and lifts from the surface to relieve the elastic energy stored in the compressively stressed region ahead of the indenter only when this energy is sufficient to overcome the work of adhesion and the work to deform the coating. Laugier hypothesized that this building up and relieving of stress produces folds along the edges of the scratch path. In the case of brittle materials, fragmentation or spalling of the coating occurs ahead of the indenter when the elastic energy is sufficient to result in detachment. He assumed that for both types of films the work of adhesion was much more significant.

When a material undergoes deformation or fracture, some of the elastic energy induced by local stresses can be released in the form of stress waves, which produce small displacements that can be detected on the surface by an appropriate sensor. Change in the amplitude distribution of the acoustic signal has been observed as a precursor to failure. It has been suggested that failure causes the beginning of local cleavage at the crack tip as instability begins [18]. In addition, it has been observed that acoustic emission is dependent on plastic strain [19]. Therefore, a scratch testing system can be equipped with an accelerometer, usually mounted above the scratching indenter that acts as an acoustic detector. Many researchers [5, 9, 10, 12, 20–24] have reported a good correlation between the acoustic emission signal and loss of film adhesion.

Acoustic emission measurements in the 'Revetest' are used to define a critical load (L_c) which measures a complex elastic–plastic deformation response of the coating/substrate combination. At best, the method allows relative comparisons to be made on the same type of coating, provided that such extraneous factors as the indenter conditions and friction between the diamond tip and the coating are factored out.

4. THE MICRO-SCRATCH TEST

The objective was to develop an apparatus and technique to characterize more accurately the adhesion of thick or thin films by overcoming some of the deficiencies of the conventional scratch test. A microhardness tester was modified with the capability of accurately traversing a polished sample in the x–y direction during the loading cycle. A variable speed motor was used to drive the specimen stage at a constant velocity using a precision screw-type mechanism. Stiffness of this mechanism and precision of movement (in micrometer accuracy) were considered to be the two most critical design factors.

A planar surface of the coated article is metallographically polished normal to the interface, exposing a portion of the substrate/coating interface. A static load, via a diamond indenter of predetermined mass, is applied to the substrate or the coating in a direction generally normal to the planar surface. The indenter is moved across the surface from the starting point in a straight line toward and across the substrate/coating interface at an angle relative to the interface between 90° and 20°. The load is incrementally increased until failure is observed (Fig. 1a). The lowest applied load (L_F) at which crack formation is observed at the coating/substrate interface can be used as an empirical value for debonding. However, once the numerical model has been fully developed and experimentally

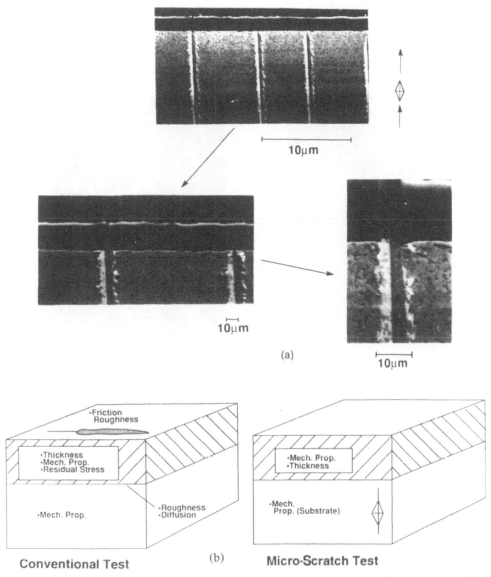

Figure 1. (a) SEM photomicrograph showing the technique of determining L_F (the lowest value of the applied load) at which crack propagation is observed in the micro-scratch test. (b) Schematic diagram comparing the micro-scratch test with the conventional scratch test.

confirmed, such applied load values will be used to evaluate practical adhesion. Figure 1b illustrates schematically the micro-scratch test and lists some properties of the coating/substrate system that influence measurement values as compared with the conventional scratch test. As is evident, several secondary factors that influence L_c do not affect the micro-scratch test, making it a more authentic test of the adhesion characteristics of the coating.

Results on TiC-coated $Si_3N_4 + TiC$ composite ceramic substrates (Fig. 2) were extremely encouraging. It was possible to generate a crack along the coating/ substrate interface and thus actually achieve some measure of the interfacial

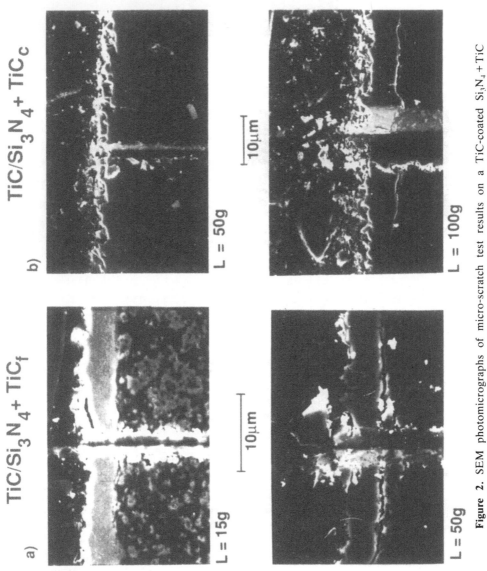

Figure 2. SEM photomicrographs of micro-scratch test results on a TiC-coated Si$_3$N$_4$ + TiC composite substrate containing (a) fine (f) grained and (b) coarse (c) grained TiC particles. L_F values of 50 and 100 g were obtained for the fine and coarse grained materials, respectively.

strength, since an interfacial-type failure is obtained. In fact, the test accurately predicted an increase in practical adhesion due to the increased size of the TiC dispersoid in the composite substrate (Fig. 2b), as has been reported by Sarin and Buljan [25]. However, similar crack propagation at the interface was not obtained in the case of TiC-coated WC–Co cemented carbides (Fig. 3a). Typically, at high loads, cracks were observed to initiate in the substrate and propagate into the coating, resulting in a cohesive failure in the substrate. This has been attributed to the shape and size of the indenter, which was much larger than the thickness of the coating, as shown (Fig. 3b) on scale for a typical scratch track. When the center of the Knoop indenter reaches the coating/substrate interface, shear stress (σ_t) is at a maximum and therefore cracks should propagate along the interface. However, if the length of the indenter exceeds the width of the coating, a bending moment (M), which increases as the indenter traverses through the coating, is obtained (Fig. 3c). Depending on the strength of the interface relative to that of the substrate, crack propagation may occur in the substrate. As can be seen (Fig. 3b), a Vickers indenter under identical loading conditions would be preferred since it would not create such a bending moment (M). However, crack propagation was difficult to obtain using such an indenter. Further confirmation of the effect of the coating thickness on the load to failure (L_F) was observed when a 5 μm sputtered layer of titanium was introduced on two TiC-coated WC–Co substrates. Although cohesive-type failure was obtained. L_F increased in both cases (Table 1). Similarly, in multilayered coatings, higher loads increased as the probability of crack propagation in the WC–Co substrate (Table 2), as opposed to the various interfaces present.

In the conventional scratch test, L_c has been found to increase with coating

Table 1.
Effect of coating thickness and overlayer on the load required for crack propagation (L_F)

Sample/Thickness (μm)	L_F (g)	
	TiC/WC–Co	Ti[a]/TiC/WC–Co
A (TiC = 5 μm)	40	55
B (TiC = 11 μm)	80	90

[a] Thickness of Ti overlayer = 5 μm.

Table 2.
Probability of crack formation in a multilayered coating[a] relative to the applied load

Load (g)	Interface		
	TiN/TiC	TiC/WC–Co	WC–Co (substrate)
25	50%	50%	—
50	—	80%	20%
100	—	15%	85%
150	—	—	100%

[a] The coating consisted of 6 μm TiN/5 μm TiC on a WC–Co substrate.

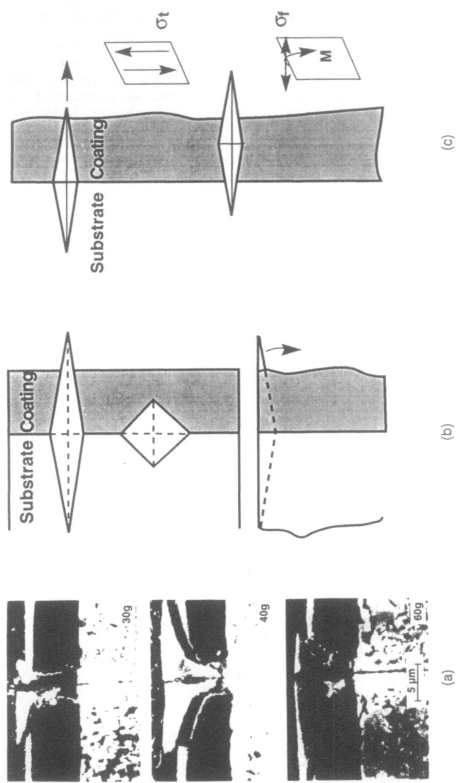

Figure 3. Effect of the indenter size on the crack propagation using the micro-scratch test (a) on a TiC/WC–Co with increasing load. (b) Schematic diagram of a typical size of indentation relative to the coating thickness. (c) Movement created by motion of the indenter beyond the coating/substrate interface.

thickness on cemented carbide and steel substrates [8, 12, 26, 27]. However, our tests show that the L_c decreases with coating thickness on ceramic substrates (Fig. 4a). The reason for this reversal effect of the influence of coating thickness on L_c is hypothesized to be related to the brittle nature of the substrate. Cross-sectional examination of a series of scratch tracks at various loads (Fig. 5) shows that crack propagation in brittle substrates can be obtained even at loads below L_c. Therefore, it is projected that the conventional scratch test will typically result in

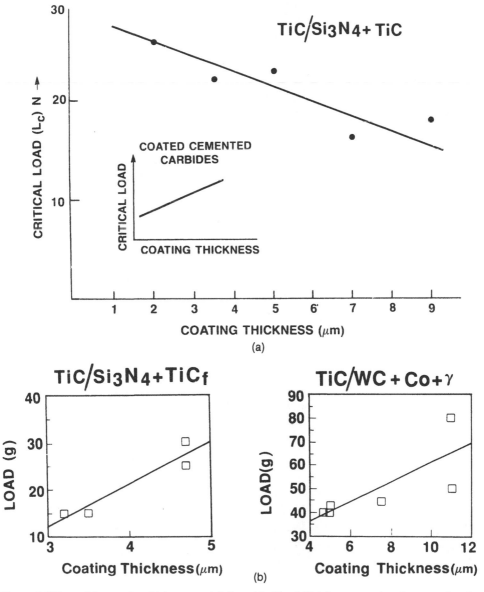

Figure 4. Effect of the coating thickness on (a) the critical load (L_c) for a ceramic substrate using the conventional scratch test. (The insert shows typical results obtained for cemented carbide substrates using the micro-scratch test) and (b) L_F using the micro-scratch test. γ signifies cubic carbide such as TaC, NbC, TiC, etc.

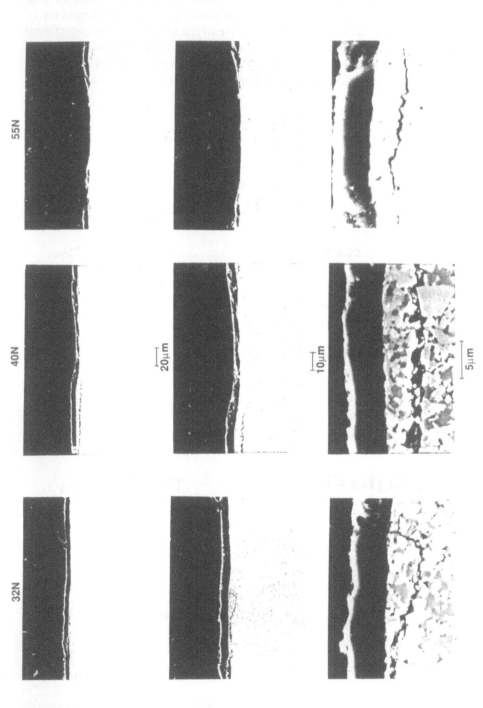

Figure 5. Cross-sections of conventional scratch test tracks taken at several loads before coating failure was observed. Crack formation in the substrate (at 32 N) was observed which could have led to premature failure of the coating and thus a lower critical load L_c (55 N).

failure of the brittle substrate, which will eventually lead to premature failure of the coating.

A linear relationship between increasing coating thickness and L_F, the load required for crack propagation, was observed in all substrate/coating combinations investigated (Fig. 4b). The substrate properties do not seem to influence the micro-scratch test. In addition, it is felt that the observed effect of coating thickness is basically due to the indenter size and geometry. Based on these observations, it is projected that a proper choice of the indenter geometry and size for a specific coating/substrate combination will result in an interfacial failure mode and give a true measure of interfacial adhesion. Indenter shape, size, geometry, and loading conditions are being optimized through a boundary integral method (BIM) numerical simulation model [28].

5. MICRO-SCRATCH TEST MODEL

The primary goal of the model is to quantitatively evaluate the adhesion characteristics between two bonded dissimilar homogeneous elastic/plastic three-dimensional bodies. Debonding is assumed to occur by the propagation of a crack at the flat interface between the two bodies of interest. A new methodology for solution of the related boundary value problem, based on combining the variational and boundary integral approaches, has been developed. It is assumed that a single crack problem for a flaw situated at the interface between two different material combinations, for different indenter geometries, can be solved by evaluating the stress-intensity factors, the energy release rate, and the change in shape of the crack under quasi-static conditions.

The model considers two homogeneous, isotropic, three-dimensional bodies Ω_i and Ω_j bonded along a plane region Γ_{ij} (Fig. 6). A crack is assumed to be situated at an arbitrary position at the interface Γ_{ij} as shown in Fig. 6. The unknown contact zone is assumed to belong to the candidate contact plane region Γ_c.

The boundary value problem governed by the Navier–Cauchy equation of linear elasticity for homogeneous regions Ω_i and Ω_j may be written in the equivalent boundary-integral [29] form as

$$\tfrac{1}{2} u_\alpha^{(\kappa)}(x) + \int_\Gamma T_{\alpha\beta}^{(\kappa)}((x,\xi) u_\beta^{(\kappa)}(\xi) \, \mathrm{d}\Gamma = \int_\Gamma U_{\alpha\beta}^{(\kappa)}(x,\xi) t_\beta^{(\kappa)}(\xi) \, \mathrm{d}\Gamma \qquad (1)$$

where the upper Greek index is used to denote a region, while the lower Greek indices denote spatial directions; in all cases, the summation rule for repeated Greek indices is assumed. In equation (1), u represents a vector for displacements on the boundary; t is the vector of boundary tractions; and $T_{\alpha\beta}(x,\xi)$ and $U_{\alpha\beta}(x,\xi)$ are tensors containing fundamental solutions (kernels) for the tractions and displacements at point ζ due to a unit force at point j [the Kelvin solution for an infinite region may be used in equation (1)].

This integral equation is accompanied by natural boundary conditions for interfacial boundary (continuance of the displacements at the interface), crack region (lack of traction), and unilateral constraints at the candidate contact zone.

It has been shown that the complicated problem above is equivalent to the

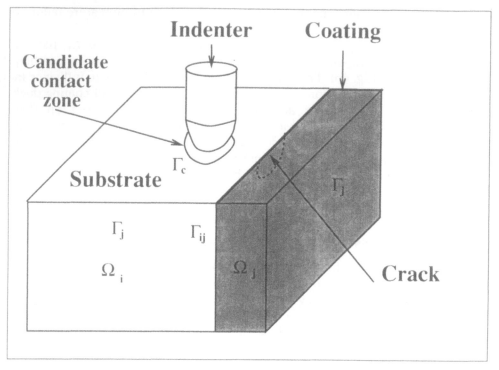

Figure 6. Schematic diagram illustrating the notation for the BIM model under consideration.

minimization of the complementary energy functional:

$$\min_{p \geqslant 0} \Phi(p) = \tfrac{1}{2} \int_{\Gamma_c} p(x) \int_{\Gamma_c} G(x,\eta)p(\eta) \, \mathrm{d}S_\eta \, \mathrm{d}S_x - \int_{\Gamma_c} pg \, \mathrm{d}Sx \, , \qquad (2)$$

where $G(x, \eta)$ is a normal component of a Green's operator related to the finite bodies with given boundary conditions, $p(x)$ is the unknown contact pressure, Γ_c is the candidate contact surface which includes a real unknown contact zone, $p(x) \in H^{(-1/2)}(\Gamma_c)$ is Sobolev's space, and g is a smooth function describing the geometry of the problem: shape of indenter and depth of scratch.

In the case of finite bodies, operator $G(x)$ is unknown and, as mentioned above, the boundary integral technique is being used for numerical evaluation. The generality of the boundary integral method (BIM) allows us to calculate the stress–strain distribution inside the sample and to evaluate different fracture mechanics criteria (energy release rate, stress concentration coefficient) with a reasonable amount of effort. Going into discretization of a volume and making some changes in the governing integral equation, we will be able to implement plasticity as well.

The application of the model as a complementary tool to the experiment permits us to take into account two major fracture mechanisms: steep gradients of the stresses in the proximity of the contact zone resulting in high shear stresses along an interface, and plowing effect of the indenter creating mixed mode I and mode III regimes for interfacial cracks. Since the part of the model included contact problem for laminated elastic bodies is already accomplished, we are able to estimate the effects of coating thickness, indenter geometry, friction, and the rather complex stress–strain distribution due to material inhomogeneity at the interface.

The general scheme for application of the described model to the investigation of the coating adhesion phenomena is as follows. In the framework of a trial-and-error method, we can run a numerical simulation of the micro-scratch test. Based on experimental data from the test (crack geometry), in several iterations we can obtain an approximate estimation for adhesion parameters such as the interfacial energy or energy release rate for interfacial cracks.

6. CONCLUSIONS

If pure interfacial failure can be attained at a coating/substrate interface, the energy or load required to obtain this can be used as a measure of interfacial adhesion. It is felt that the micro-scratch test, with the aid of a numerical model that can optimize the indenter shape and geometry, can attain this objective. On the other hand, our results show that the conventional scratch test will typically result in cohesive failure of the substrate in brittle materials. Once fully developed, the proposed test could prove to be an extremely powerful tool, not only for the development of thin films, but also for the study of interfaces in general.

REFERENCES

1. K. L. Mittal (Ed.), *Adhesion Measurements of Thin Films, Thick Films, and Bulk Coatings*, p. 5. ASTM, Philadelphia (1978).
2. P. Benjamin and C. Weaver, *Proc. R. Soc. London, Ser. A* **254**, 163 (1960).
3. D. W. Butler, C. T. H. Stoddart and P. R. Stuart, *J. Phys. D: Appl. Phys.* **3**, 877 (1970).
4. C. Weaver, *J. Vac. Sci. Technol.* **12**, 18 (1975).
5. P. Laeng and P. A. Steinmann, in: *Proc. 8th Int. Conf. on Chemical Vapor Deposition.* J. M. Blocher, Jr., G. E. Vuillard and G. Wahl (Eds), p. 723. The Electrochemical Society, Pennington, NJ (1981).
6. M. Laugier, *Thin Solid Films* **76**, 289 (1981).
7. A. J. Perry, P. Laeng and H. E. Hintermann, in: ref. 5, p. 475.
8. K. L. Chopra, *Thin Film Phenomena*, p. 313. McGraw-Hill, New York (1969).
9. B. Hammer, A. J. Perry, P. Laeng and P. A. Steinmann, *Thin Solid Films* **96**, 45 (1982).
10. J. H. Je, E. Gyarmati and A. Naoumidis, *Thin Solid Films* **136**, 57 (1986).
11. J. Valli, U. Makela and A. Matthews, *Surface Eng.* **2**, 49 (1986).
12. A. J. Perry, *Thin Solid Films* **107**, 167 (1983).
13. I. S. Goldstein and R. DeLong, *J. Vac. Sci. Technol.* **20**, 327 (1982).
14. M. Laugier, *Thin Solid Films* **79**, 15 (1981).
15. J. Takadoum, J. C. Pivin, J. Chaumont and C. Roque-Carmes, *J. Mater. Sci.* **20**, 1480 (1985).
16. G. M. Hamilton and L. E. Goodman, *J. Appl. Mech.* **88**, 371 (1966).
17. M. T. Laugier, *Thin Solid Films* **117**, 243 (1984).
18. A. A. Pollock, *Non-Destr. Test.* 264 (October 1973).
19. W. Swindlehurst, *Non-Destr. Test.* 152 (June 1973).
20. A. J. Perry, *Thin Solid Films* **78**, 77 (1981).
21. A. J. Perry, in: ref. 5, p. 737.
22. P. K. Mehrotra and D. T. Quinto, *J. Vac. Sci. Technol.* **A3**, 2401 (1985).
23. H. E. Hintermann, *J. Vac. Sci. Technol.* **B2**, 816 (1984).
24. I. Lhermitte-Sebire, R. Colmet, R. Naslain, J. Desmaison and G. Gladel, *Thin Solid Films* **138**, 221 (1986).

25. V. K. Sarin and S. T. Buljan, in: *High Productivity Machining, Materials and Processes,* V. K. Sarin (Ed.), p. 105. ASM (1985). Metals Park, Ohio.
26. T. A. Cruse, *Int. J. Solids Struct.* **5**, 1259–1274 (1969).
27. P. A. Steinmann and H. E. Hintermann, *J. Vac Sci. Technol.* **A3**, 2394 (1985).
28. P. J. Burnett and D. S. Rickerby, *Thin Solid Films* **154**, 415 (1987).
29. V. L. Rabinovich, S. R. Sipcic and V. K. Sarin, *ASME J. Appl. Mech.* Paper No. 93 - WA/APM-20.

Adhesion Measurement of Films and Coatings, pp. 189–215
K. L. Mittal (Ed.)
© VSP 1995.

Mechanics of the indentation test and its use to assess the adhesion of polymeric coatings

R. JAYACHANDRAN, M. C. BOYCE* and A. S. ARGON

Department of Mechanical Engineering, Massachusetts Institute of Technology, Cambridge, MA 02139, USA

Revised version received 1 March 1993

Abstract—The indentation test provides a simple means by which the adhesion of coatings can be qualitatively assessed. One the way to establishing a quantitative measurement of the adhesion strength of coatings and films, it is important that the mechanics of this test are clearly understood. To investigate the influence of factors such as the coating thickness, the indenter radius, and friction during the test, numerical simulations of the indentation of a typical polymeric coating, polymethylmethacrylate (PMMA), bonded to a rigid substrate were conducted by using the finite element method. The stress generated during the indentation test were obtained by employing an accurate constitutive model of the elastic–viscoplastic behaviour of the polymeric coating under consideration. The results of this analysis illustrate the effects of the factors mentioned above on the deformation of the coating during indentation, its confinement, and interfacial shear, and the normal, shear, and hoop stress distributions occurring during indentation. These results provide insight into the possible failure mechanisms operative during the indentation of thin coatings and the important effects of the coating thickness during such tests.

Keywords: Indentation simulation; indentation resistance; adhesion; bond strength; polymeric coating; delamination; debonding; residual stress; interfacial shear strength.

1. INTRODUCTION

Low stiffness polymeric coatings are often used for surface protective applications in the textile, printing, and paper-making industries. They are also used both for functional and for decorative purposes, for example, as paints on automobile bodies and in electronic components such as high density interconnects. In these circumstances, the strong adherence of the coating to the substrate is a basic necessity for successful performance of the coating. In order to design a coating for better adhesion, the measurement of the adhesion strength between the coating and the substrate is of vital importance and the test methods require attention and proper evaluation. The basic

---

*To whom correspondence should be addressed.

requirements of an ideal adhesion test are that the test results should be reproducible and quantitatively reliable.

Mechanical methods such as the scratch test, peel test, lap shear test, and indentation test have been tried by many researchers as a semi-quantitative means for the determination of bond strength (practical adhesion) between the film or coating and the substrate. A very comprehensive bibliography on the adhesion measurement of films and coatings was given by Mittal [1, 2] and a recent review of the methods for the evaluation of coating–substrate adhesion was given by Chalker et al. [3]. Further studies on the current test methods are required because of the lack of a direct correlation of the results with one another.

Indentation testing is becoming increasingly popular, not only because of its simplicity but also due to its ability to induce controlled debonding at the coating/substrate interface. Moreover, it has been shown by Evans and Hutchinson [4] that indentation is unique in that it can measure both the interfacial shear strength and the resistance to interfacial fracture. Ritter et al. [5] and Lin et al. [6] have used the microindentation technique together with Matthewson's analytical expression [7, 8] to measure the interfacial shear strength of thin polymeric coatings on glass substrates. They concluded that the indentation test appears ideal in establishing the true trends in the adherence of a given coating/substrate system. Conway and Thomsin [9] have recently presented an annular plate theory for the determination of debonding strength in conjunction with experiments using a conical indenter. A more precise quantitative determination of the bond strength from the indentation tests requires more accurate analytical models, or highly accurate simulations employing constitutive models that characterize the large strain material behaviour to predict the interfacial strength in shear or tension in combination with the experimental measurements.

We have carried out accurate indentation simulations by the finite element method to obtain the stresses that are generated during the indentation test by employing a large strain constitutive model which models strain softening, strain hardening, strain rate hardening, and the thermal softening response of the glassy polymers. We have taken, as a generic example, a PMMA coating deposited on a rigid substrate for the purpose of this investigation. Moreover, the results from these simulations provide information about the indentation resistance generated by the coating during the test, and particularly the manner in which factors such as the coating thickness, the indenter radius and friction affect the interfacial shear and normal stress distributions.

2. KINEMATICS

Here we give a brief description of the kinematics of large strain elastic–plastic deformation used in this study. More detailed treatments can be found elsewhere [10, 11]. Consider the deformation of an initially isotropic body $\mathcal{B}_0$. The body is loaded to a new state and the current configuration is denoted by $\mathcal{B}_t$. The deformed configuration is determined by the deformation gradient $\mathbf{F}$:

$$\mathbf{F} = \nabla_{\mathbf{X}}\mathbf{x}, \tag{1}$$

where $\mathbf{X}$ and $\mathbf{x}$ represent the initial and the current position of a material point, respectively. The deformation gradient may be multiplicatively decomposed into elastic, $\mathbf{F}^e$, and plastic, $\mathbf{F}^p$, components:

$$\mathbf{F} = \mathbf{F}^e \mathbf{F}^p. \tag{2}$$

Thus, $\mathbf{F}^p$ represents the configuration obtained by complete elastic unloading to a stress-free state. The usual polar decomposition of $\mathbf{F}^p$ gives either the left stretch tensor, $\mathbf{V}^p$, or the right stretch tensor, $\mathbf{U}^p$, and the rotation tensor, $\mathbf{R}^p$, as

$$\mathbf{F}^p = \mathbf{R}^p \mathbf{U}^p = \mathbf{V}^p \mathbf{R}^p. \tag{3}$$

Note that the eigenvalues of the left plastic stretch tensor $\mathbf{V}^p$ are denoted by Λ_i^p, for $i = 1, 2, 3$. The velocity gradient is now obtained by differentiating the product decomposition as

$$\mathbf{L} = \mathbf{F}\mathbf{F}^{-1} = \mathbf{D} + \mathbf{W} = \mathbf{L}^e + \mathbf{F}^e \mathbf{L}^p \mathbf{F}^{e-1}, \tag{4}$$

wherein the elastic and the plastic velocity gradients are introduced. Here $\mathbf{D}$ is the rate of deformation tensor and $\mathbf{W}$ is the spin tensor. Similarly, the velocity gradient of the relaxed configuration is written as

$$\mathbf{L}^p = \mathbf{D}^p + \mathbf{W}^p = \mathbf{F}^p \mathbf{F}^{p-1}. \tag{5}$$

The spin of the relaxed configuration, $\mathbf{W}^p$, is algebraically defined as a result of the symmetry imposed on the elastic deformation gradient. The plastic stretching, $\mathbf{D}^p$, must be constitutively prescribed as detailed in the next section.

3. CONSTITUTIVE MODEL

We use the three-dimensional constitutive model proposed by Boyce *et al.* [10] and later modified by Arruda and Boyce [12, 13] to predict the mechanical behaviour of glassy polymers over a range of temperatures, strain rates, and strain states. This formulation uses a scalar internal variable to model the isotropic plastic resistance and a second-order tensor to model the anisotropic strain hardening behaviour in the form of network resistance against the plastic flow.

Argon [14] developed a kinetic expression for the plastic shear strain rate, $\dot{\gamma}^p$, as a function of the shear stress and temperature:

$$\dot{\gamma}^p = \dot{\gamma}_0 \exp\left[-\frac{As_0}{\theta}\left(1 - \left(\frac{\tau}{s_0}\right)^{5/6}\right)\right], \tag{6}$$

where $\dot{\gamma}_0$ is a pre-exponential factor; As_0 is the zero stress activation free energy divided by Boltzmann's constant; θ is the absolute temperature; s_0 is the athermal shear strength $(= 0.077/(1 - \nu)\mu)$; μ is the elastic shear modulus; ν is Poisson's ratio; and τ is the applied shear stress. This expression has been extended by Boyce *et al.* [10] to include the effects of strain softening, strain hardening, and pressure.

The pressure dependence is modelled by replacing the shear strength s_0 with $s + \alpha p$, where p is the applied pressure and α is the pressure dependence coefficient. Also, the evolution of s is modelled by the following equation:

$$\dot{s} = h\left(1 - \frac{s}{s_{ss}}\right)\dot{\gamma}^{\mathrm{P}}, \tag{7}$$

where h is the softening slope determined based on the macroscopic response of the strain softening of the material. This equation models the decrease in shear resistance with plastic straining until a preferred state represented by s_{ss} is reached.

The anisotropic network resistance which evolves when the molecular chains tend to align themselves along the direction of principal plastic stretch has been modelled by the statistical mechanics approach of rubber elasticity in the form of a back stress tensor **B**. Using the eight-chain network model proposed by Arruda and Boyce, the principal components of **B** are given by

$$\overline{B}_i = C_{\mathrm{R}}\frac{\lambda_{\mathrm{L}}}{3}\mathcal{L}^{-1}\left\{\frac{\Lambda^{\mathrm{P}}_{\mathrm{chain}}}{\lambda_{\mathrm{L}}}\right\}\frac{\Lambda^{\mathrm{p2}}_i - \frac{1}{3}I_1}{\Lambda^{\mathrm{P}}_{\mathrm{chain}}}, \tag{8}$$

where $\overline{B}_i$ are the principal components of the back stress tensor; Λ^{P}_j are the principal network (and plastic) stretch components; $I_1 = (\Lambda^{\mathrm{p2}}_1 + \Lambda^{\mathrm{p2}}_2 + \Lambda^{\mathrm{p2}}_3)$ is the first invariant of stretches; and $\lambda_{\mathrm{L}} = \sqrt{N}$ is the tensile locking network stretch (or natural draw ratio), with N being the statistical parameter related to the limiting value of chain stretch. The term $\Lambda^{\mathrm{P}}_{\mathrm{chain}} = (1/\sqrt{3})I_1^{1/2}$ denotes the plastic stretch on a chain in the network; the hardening modulus $C_{\mathrm{R}} = nk\theta$, where n is the chain density, k is Boltzmann's constant, and θ is absolute temperature. $\mathcal{L}$ is the Langevin function defined by $\mathcal{L}(\beta_i) = \coth\beta_i - 1/\beta_i$, and thus $\beta_i = \mathcal{L}_{-1}(\Lambda^{\mathrm{P}}_{\mathrm{chain}}/\lambda_{\mathrm{L}})$. The model is functionally specified such that $\overline{B}_i$ increases asymptotically as the value of $\Lambda^{\mathrm{P}}_{\mathrm{chain}}/\lambda_{\mathrm{L}}$ approaches unity. The back stress tensor as modelled here is an internal stress state of the material and therefore is non-dissipative.

As discussed in Section 2, $\mathbf{D}^{\mathrm{P}}$ must be prescribed in solving the problem. Therefore, the magnitude of $\mathbf{D}^{\mathrm{P}}$ is prescribed by the plastic shear strain rate, $\dot{\gamma}^{\mathrm{P}}$, given in equation (6), and the tensorial direction of $\mathbf{D}^{\mathrm{P}}$ is specified by the normalized portion of the driving stress tensor as given below:

$$\mathbf{D}^{\mathrm{P}} = \dot{\gamma}^{\mathrm{P}}\mathbf{N}, \tag{9}$$

$$\mathbf{N} = \frac{1}{\sqrt{2\tau}}\mathbf{T}^{*\prime}, \tag{10}$$

where τ is the effective equivalent shear stress given by

$$\tau = \left[\frac{1}{2}\mathbf{T}^{*\prime} \cdot \mathbf{T}^{*\prime}\right]^{1/2}, \tag{11}$$

and $\mathbf{T}^{*\prime}$ is the deviatoric position of $\mathbf{T}^*$ which continues to drive the plastic flow. Moreover,

$$\mathbf{T}^* = \mathbf{T} - \frac{1}{J}\mathbf{F}^e\mathbf{B}\mathbf{F}^e, \tag{12}$$

where $\mathbf{T}^*$ is the driving stress tensor expressed in $\mathcal{B}_t$, $\mathbf{B}$ is the back stress tensor, and J is the volume change given by $\det[\mathbf{F}^e]$. The elastic constitutive relationship gives $\mathbf{T}$, the Cauchy stress tensor [15]:

$$\mathbf{T} = \frac{1}{J}\mathcal{L}^e[\ln \mathbf{V}^e], \tag{13}$$

where $\mathcal{L}^e$ is the isotropic elastic modulus tensor and $\mathbf{V}^e$ is the left elastic stretch tensor. As mentioned earlier, any work associated with $\mathbf{B}$ is non-dissipative.

4. NUMERICAL SIMULATION

4.1. Finite element model

The numerical simulation of the indentation of a thin PMMA coating on a rigid substrate was performed using the finite element method. The physically-based constitutive model discussed in the previous section was used in the simulations. Figure 1 shows the geometry of the indenter, coating, and substrate of the indentation problem. Figure 2 shows the finite element mesh for the coating used in the simulations. The indenter was taken to be a rigid sphere and the problem was modelled using axial symmetry. A finer mesh was chosen near the axis of symmetry where the deformation is expected to be large. The bottommost nodes were constrained in both directions, whereas the nodes along the axis of symmetry were constrained in the radial direction. The sphere was constrained to move only vertically. The centre of the sphere was subjected to a constant displacement, rate, $\dot{\delta}$, in the negative z-direction to simulate the loading condition. In each case, the coating was indented halfway through its

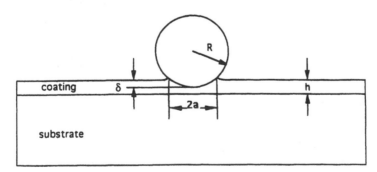

Figure 1. The geometry of the indenter, coating, and substrate.

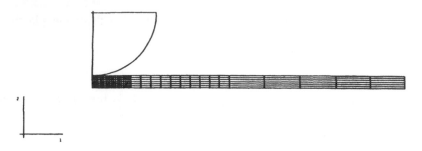

Figure 2. The finite element model used in the simulations.

thickness and then the indenter was withdrawn at the same rate to its initial position for unloading.

The finite element mesh is composed of 8-node, reduced integration, axi-symmetric elements (type CAX8R in ABAQUS [16]). The rigid interface between the indenter and coating was modelled using 3-node interface elements. The interface was taken to be frictionless in most cases. Coulomb's friction law was used whenever friction was considered in the simulations. An optimal mesh was chosen in order to save the computational time without loss of solution accuracy.

4.2. Material properties

The mechanical behaviour of the polymer coating was described using the constitutive law presented in the previous section. The behaviour of PMMA has been shown [13] to depend on the strain rate and temperature. The material constants used in this simulation are taken from Arruda and Boyce [13] as reported based on their experimental compression stress–strain curves. They are given below:

Elastic properties
E 3250 MPa
ν 0.30

Strain rate and temperature-dependent yield
$\dot{\gamma}_0$ $2.8 \times 10^7 \ \text{s}^{-1}$
A 100.6 K/MPa

Strain softening   (14)
h 315 MPa
s_{ss}/s_0 0.82

Pressure-dependent yield
α 0.20

Strain hardening
N (298 K)   2.3
C_R 10.5 MPa

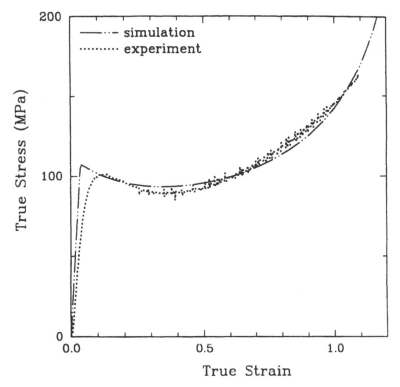

Figure 3. Stress–strain behaviour of PMMA under uniaxial compression conducted at a strain rate of 0.001 s^{-1}.

Figure 3 shows the uniaxial compression stress–strain curve obtained from a simulation using the above material constants for PMMA at a uniaxial compressive strain rate of $\dot{\varepsilon} = 0.001$ s^{-1} along with the experimental data obtained by Arruda and Boyce [13]. As is evident from these curves, the constitutive model accurately models the stress–strain behaviour of PMMA everywhere except at the beginning, where the difference is due to not taking account of the distributed nature of the microstructure. It should be noted that this constitutive model has been shown to capture the strain rate, temperature, and pressure dependence of the yield behaviour. For more details of the experimental verification of this model, refer to Arruda and Boyce [13].

5. RESULTS AND DISCUSSION

The indentation simulations were carried out at a constant displacement rate and in each case, the coating was indented until a depth of 0.5 h was reached, followed by withdrawal of the indenter to its initial position at the same rate as indentation. Most of the simulations were done at a rate of 0.001 h s^{-1}, where h is the coating thickness. A coating thickness of 100 μm and a R/h ratio of 20 were considered for investigation in most cases. The study on the effect of the indenter radius was carried out for $R/h = 5$, 10, and 20. Also, the study on the effect of friction was carried out with a friction coefficient of 0.4 taken at the indenter/coating interface. These results are discussed in the following sections.

5.1. Indentation, loading, and unloading

In order to illustrate material flow under the indenter during the indentation process, we have plotted the history of displaced configurations, the plastic strain rate, shear stress, and normal stress in Figs 4–7, respectively, for the case of isothermal frictionless indentation at a rate of $0.001\,h\,s^{-1}$ and for $R/h = 20$. In these figures, only the polymer coating is included.

Figure 4 illustrates the material flow under the indenter during loading at four different depths of indentation, i.e. $\delta/h = 0.064$, 0.242, 0.422, and 0.5. In the initial stages of indentation, the coating material directly below the indenter is compressed and the material beyond the contact radius shows negligible deformation. As the indentation progresses further, it is seen that the coating material flows laterally towards

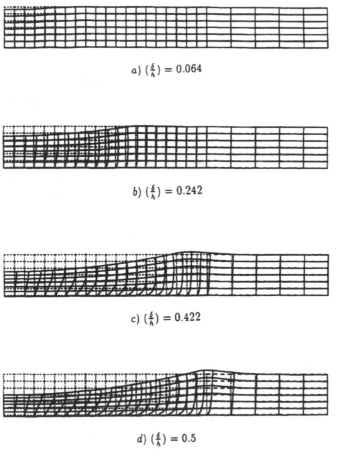

a) $(\frac{\delta}{h}) = 0.064$

b) $(\frac{\delta}{h}) = 0.242$

c) $(\frac{\delta}{h}) = 0.422$

d) $(\frac{\delta}{h}) = 0.5$

Figure 4. A sequence of loaded configurations during indentation. $h = 100\ \mu\text{m}$, $R/h = 20$.

the indenter periphery. This action also induces large shearing of the material at the substrate interface due to the perfect adhesion modelled at the base. Higher depths of indentation necessitate extrusion of the material towards the indenter periphery. We will see later how the interfacial shearing and the material pile-up depend on the rate effects, indenter size, and friction at the indenter interface.

The plastic strain rate contours of Fig. 5 also illustrate the pattern and momentary concentration of material flow during indentation. In the early stages of indentation, the peak plastic strain rate lies directly under the indenter as the material undergoes plastic flow. As the indentation continues, it is noted that the coating material just outside the contact radius is subjected to a considerably high rate of shearing. With the applied displacement rate of $0.001\,h$ s^{-1}, a plastic strain rate of 0.007 s^{-1} is produced at this location. Also note that this region moves outward as indentation progresses, always appearing near the edge of the contact region. This region of localized shearing is a probable site for the initiation of debonding of the coating during the indentation, and a source of possible interface fracture.

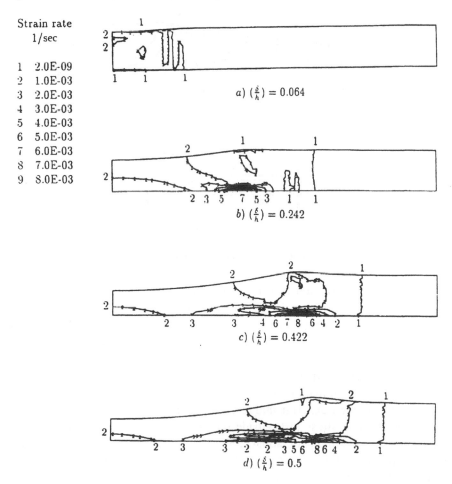

Strain rate
1/sec

1	2.0E-09
2	1.0E-03
3	2.0E-03
4	3.0E-03
5	4.0E-03
6	5.0E-03
7	6.0E-03
8	7.0E-03
9	8.0E-03

a) $(\frac{\delta}{h}) = 0.064$

b) $(\frac{\delta}{h}) = 0.242$

c) $(\frac{\delta}{h}) = 0.422$

d) $(\frac{\delta}{h}) = 0.5$

Figure 5. A sequence of strain rate contour plots during indentation. $h = 100\ \mu$m, $R/h = 20$.

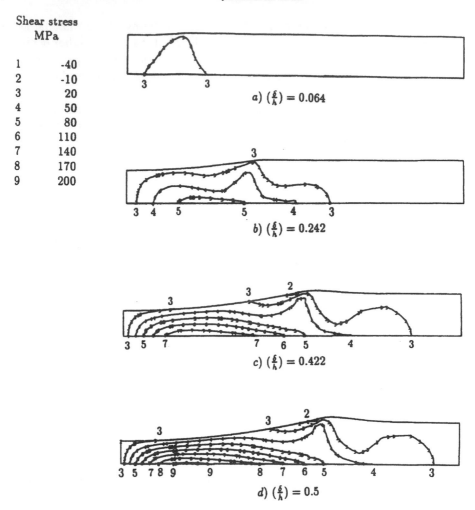

Figure 6. A sequence of shear stress contour plots during indentation. $h = 100\ \mu$m, $R/h = 20$.

Figure 6 shows the evolution of shear stress in the coating during the indentation process. As can be seen from the figure, it is the coating/substrate interface that develops a high shearing stress during loading and this maximum occurs well inside the indenter contact radius. A considerable gradient in shear stress is developed both through the thickness of the coating and in the radial direction. The presence of a high compressive normal stress (Fig. 7) beneath the indenter during loading acts to prevent the coating delamination at this point of maximum shear, as pointed out by Ritter *et al.* [5] from their experimental results. It is important to note that $\delta/h = 0.5$, the magnitude of shear stress just outside the contact radius is only about 80 MPa, as opposed to the maximum of about 200 MPa under the indenter in an annular region close to the centre. Ritter *et al.* [5] observed in their experiments that the delamination initiates at the perimeter of the contact zone and then grows outward on further loading. Our simulations are in agreement with these observations.

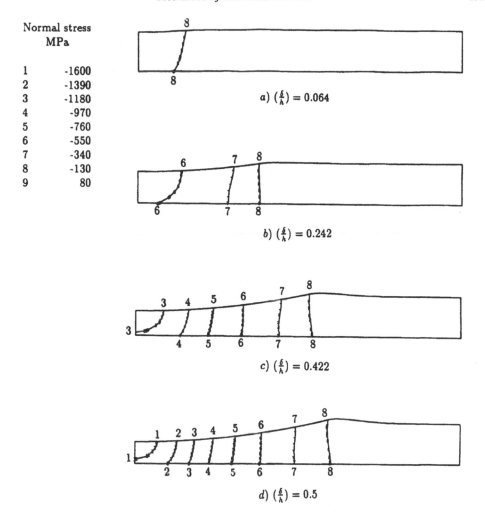

Normal stress MPa
1	-1600
2	-1390
3	-1180
4	-970
5	-760
6	-550
7	-340
8	-130
9	80

a) $\left(\frac{\delta}{h}\right) = 0.064$

b) $\left(\frac{\delta}{h}\right) = 0.242$

c) $\left(\frac{\delta}{h}\right) = 0.422$

d) $\left(\frac{\delta}{h}\right) = 0.5$

Figure 7. A sequence of normal stress contour plots during indentation. $h = 100 \ \mu\text{m}$, $R/h = 20$.

The evolution of normal stress in the coating is depicted in Fig. 7. Under the indenter, the coating is subjected to a high compressive normal stress. The stress is maximum near the axis of symmetry and gradually drops to zero just outside the contact radius where, as pointed out above, shear-induced delamination is often observed. The normal stress distribution is more or less uniform through the thickness of the coating.

As we are more interested in knowing the stress distribution at the coating/substrate interface, we have plotted the evolution of normal and shear stresses in the coating at the integration points close to this interface in Fig. 8. Plotted are the interfacial stress distributions at different relative depths (δ/h) as a function of the radius, where the radius is normalized by the current contact radius a. As the depth of indentation increases, the normal stress increases considerably near the axis of symmetry. How-

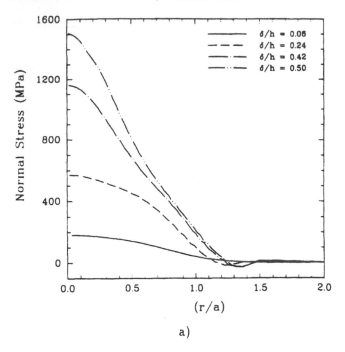

a)

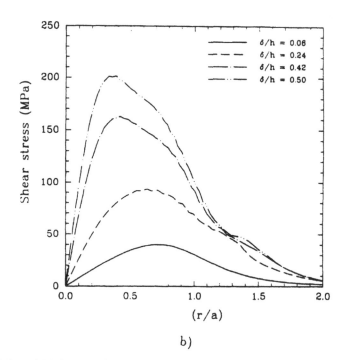

b)

Figure 8. Evolution of (a) the normal stress distribution in the coating near the substrate interface, and (b) the shear stress distribution in the coating near the substrate interface. $h = 100 \ \mu m$, $R/h = 20$.

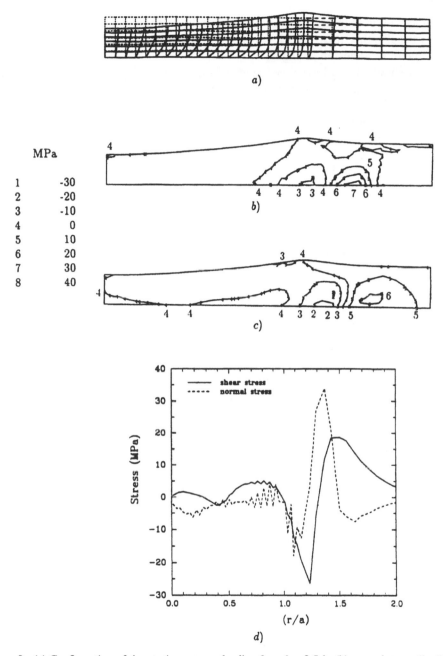

Figure 9. (a) Configuration of the coating upon unloading from $\delta = 0.5\ h$; (b) normal stress distribution in the coating upon unloading; (c) shear stress distribution in the coating upon unloading; (d) normal and shear stress distributions in the coating near the interface upon unloading. $h = 100\ \mu$m, $R/h = 20$.

ever, the rise in the normal stress at the contact radius ($r/a = 1.0$) is diminished with increasing depth of indentation. The shear stress plot also shows that this stress stabilizes to a certain value at the contact radius. Once the magnitude of shear stress at the contact radius is large, it can initiate debonding of the coating from the substrate and it is usually this stress which we seek to find in the indentation test when being used to measure adhesion strength.

Figure 9 shows the unloaded geometry of the coating, the normal and shear stress contours after unloading, and the corresponding stress distribution near the coating/substrate interface. Upon unloading, about 40% of the indentation is recovered. This can be verified from the load vs. displacement plot of the indentation process shown in Fig. 12. Note also that the pile-up appears to increase upon unloading due to elastic recovery of the material beneath it. Negligible normal and shear stresses exist within the indented region of the coating after unloading. However, upon unloading, a rather significant tensile normal stress of about 30 MPa and two peaks in shear stress, −20 and +20 MPa, are observed just outside the final contact radius to which the coating was indented. In cases where delamination occurs during unloading, the mechanism could thus be one of shear-assisted tensile debonding and not simply a shear failure. Clearly, the actual process of debonding and understanding the details of it require additional simulations and experimental observations.

5.2. Effect of the indenter radius

To determine the effect of the indenter radius on the deformation mechanism of the coating, we conducted simulations at $\dot{\delta} = 0.001\,h\,\text{s}^{-1}$ for three different ratios of R/h.

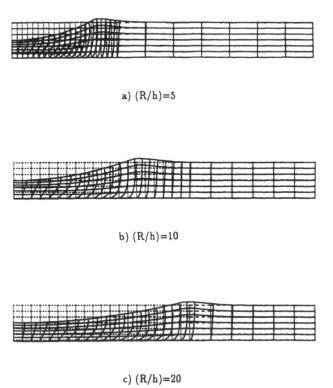

a) (R/h)=5

b) (R/h)=10

c) (R/h)=20

Figure 10. Loaded configurations of the coating for different size indenters. $h = 100\,\mu\text{m}$, $\delta/h = 0.5$.

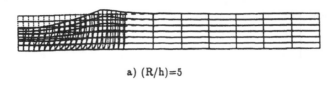

a) (R/h)=5

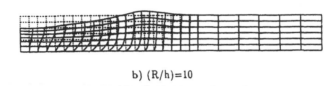

b) (R/h)=10

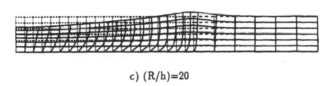

c) (R/h)=20

Figure 11. Configurations of the coating upon unloading from $\delta = 0.5\,h$ for different size indenters. $h = 100\ \mu$m.

We considered a coating thickness of 100 μm and $R/h = 5$, 10, and 20. Pure slip was assumed between the indenter and the coating.

Figure 10 shows the deformed configurations when $\delta/h = 0.5$ during indentation with three different size indenters. For $R/h = 5$, it is seen that the coating material shear at the substrate interface is minimal compared with the case for $R/h = 20$. That is to say, the indentation with a small indenter produces less interfacial shearing of the coating material due to less confinement of the material between the indenter and the substrate. This can be verified by a careful examination of the slopes of the deformed element boundaries shown in Fig. 10.

Figure 11 shows the indentation crater in the coating upon unloading. The permanent plastic indentation left in the coating is greater in the case of $R/h = 5$ than in the case of $R/h = 20$. The elastic recovery is higher when large radius indenters are used because of the hydrostatic nature of deformation. Figure 12 shows the load required by the indenter to carry out the indentation of the coating with three different indenter sizes. Large radius indenters require more load because the indentation is done over a larger area of the coating material and also because the confinement produces a greater pressure in the coating, as also shown in the normal stress contours of Fig. 13. The unloading portion of each of these curves tells us how much permanent indentation is left in the coating.

Figure 13 shows the normal stress distribution in the coating at $\delta/h = 0.5$ and for $R/h = 5$, 10, and 20. A similar pattern of normal stress distribution is seen

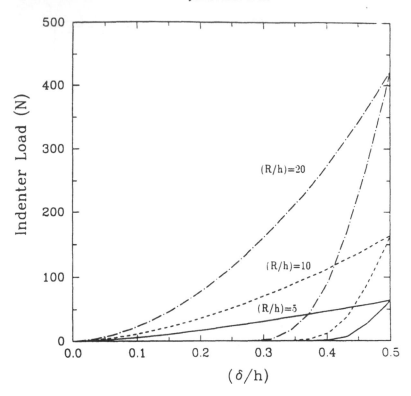

Figure 12. Indenter load vs. displacement curves for different size indenters.

in all cases. As pointed out before, the magnitudes are different because of the different indenter sizes and the corresponding different levels of confinement that each produces. In Fig. 14, we have plotted the normal stress distribution along the radial direction computed at the integration points in the coating material close to the substrate interface. Figure 14a shows the typical variation of normal stress computed at the integration points close to the substrate interface. Replotting the results of Fig. 14a by normalizing the radial distance with respect to the contact radius as in Fig. 14b helps us to visualize the distribution at and beyond the contact radius more easily. Note that the contact radius is different in each of these cases because the indenter sizes are different. In all cases, the normal stress vanishes just beyond the indenter periphery.

Figure 15 shows the shear stress distribution during loading with three different indenters for $\delta/h = 0.5$. Although the indenter size does not affect the pattern of shear stress distribution within the coating, the magnitude of this stress changes depending on the indenter size, and much larger (100% greater) shear stresses are present in the case of $R/h = 20$ as opposed to $R/h = 5$. The variation of shear stress near the substrate interface along the radial direction is depicted in Fig. 16. In Figs 16a and 16b, the shear stress distribution is shown when the radial distance is normalized with respect to the film thickness and with respect to the contact radius, respectively. The shear stress rises from zero at the centre to a peak value in a short distance and

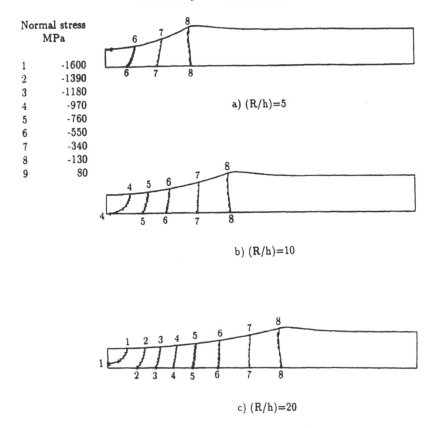

Figure 13. Normal stress distributions in the coating for different size indenters. $h = 100 \, \mu m$, $\delta/h = 0.5$.

then gradually drops to zero beyond the indenter periphery. Although there is a large increase in the peak stress at this interface directly under the indenter with increasing indenter radius, there is an almost negligible increase in the magnitude of the shear stress at the contact radius $(r/a = 1.0)$ as noted in Fig. 16b, where failure would not be prevented by a high compressive normal stress.

Figure 17 shows the pressure distribution in the coating for three different indenter cases under investigation. The peak pressure is about 450 MPa in the case of $R/h = 5$ and about 1200 MPa in the case of $R/h = 20$ when the indentation depth is 0.5 h, illustrating the confining nature of coating indentation which in turn affected the deformation pattern and stress distributions reported above. However, the pattern of pressure distribution remains the same in all three cases.

5.3. Effect of friction

To investigate the effect of friction between the indenter and the coating, we carried out indentation simulations by taking a friction coefficient of 0.4 at the indenter/coating interface to compare with the pure slippage assumed in the earlier cases. For $h = 100 \, \mu m$ and $R/h = 20$, the simulation was conducted at a rate of 0.001 h s^{-1}. Figure 18 shows the deformed configuration when $\delta/h = 0.5$ (a), the unloaded con-

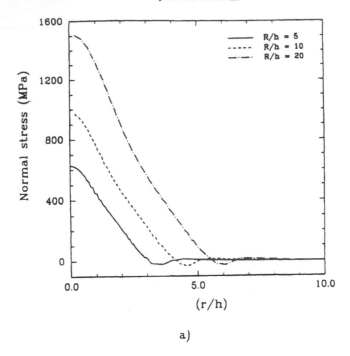

a)

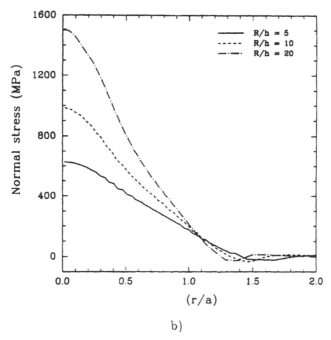

b)

Figure 14. Normal stress distributions in the coating near the substrate interface for different size indenters; (a) when the radial distance is normalized with respect to the film thickness and (b) when the radial
distance is normalized with respect to the contact radius. $h = 100 \ \mu$m, $\delta/h = 0.5$.

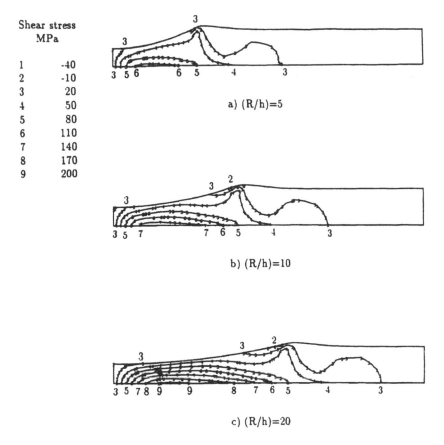

Shear stress
MPa

1	-40
2	-10
3	20
4	50
5	80
6	110
7	140
8	170
9	200

a) (R/h)=5

b) (R/h)=10

c) (R/h)=20

Figure 15. Shear stress distributions in the coating for different size indenters. $h = 100 \ \mu$m, $\delta/h = 0.5$.

figuration of the coating (b), and the load vs. displacement curves of the indenter (c). (Compare these results with the slip results shown in Figs 10c and 11c.) It is clear that the friction at the indenter surface provides an additional constraint to the material flow underneath the indenter. The material pile-up at the indenter periphery is reduced because of this constraint. The material now undergoes an extrusion between the confines of the indenter and the substrate towards the periphery. Figure 18b shows the configuration of the coating upon unloading. The elastic recovery in this case is slightly more than in the corresponding case without friction (see Fig. 11c). Because of the confinement of the coating material underneath the indenter without free flow outwards, more load is required to carry out indentation with friction as noted in Fig. 18c.

The higher load requirement in the case of frictional indentation corresponds to the higher normal stress in the coating (Fig. 19a). But the general pattern of normal stress distribution is affected very little. The strain rate distribution is altered significantly when friction is encountered at the indenter interface. In Fig. 19b, it is seen that a considerable shearing takes place at the indenter interface close to the pile-up region at a rate of 0.004 s^{-1}. However, comparison with the results depicted in Fig. 5d indicates that the maximum value of the strain rate at the substrate interface is reduced

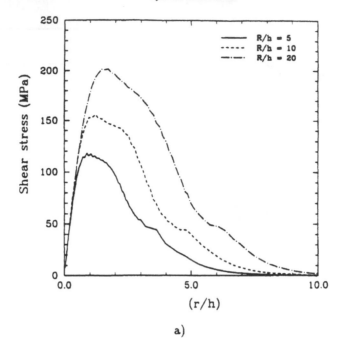

a)

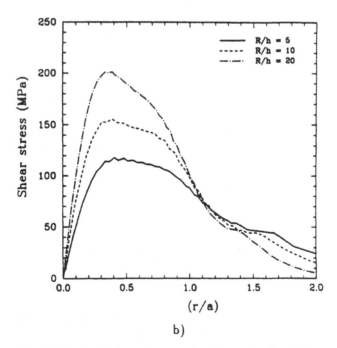

b)

Figure 16. Shear stress distributions in the coating near the substrate interface for different size indenters; (a) when the radial distance is normalized with respect to the film thickness and (b) when the radial distance is normalized with respect to the contact radius. $h = 100$ μm, $\delta/h = 0.5$.

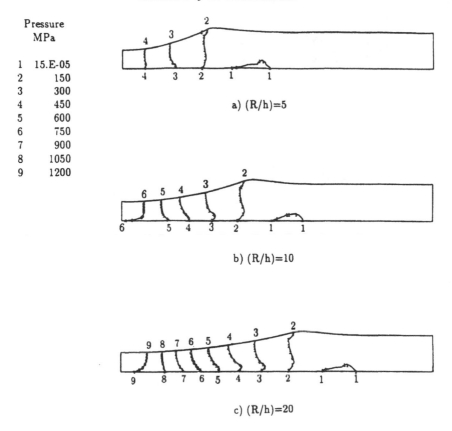

Figure 17. Pressure distributions in the coating for different size indenters. $h = 100$ μm, $\delta/h = 0.5$.

from 0.007 to 0.006 s^{-1}. Figure 19c shows the shear stress distribution in the coating. Both the indenter and the substrate interfaces are subjected to high shearing stresses. Comparison with the result without friction shown in Fig. 6d indicates that the peak shear stress at the interface in the case of friction is more or less the same, approximately, 200 MPa. However, the location of this peak is slightly shifted towards the indenter periphery, as we will see later in Fig. 20.

The hoop stress developed in the coating is depicted in Fig. 19d. It is important to note that because of friction and hence the confinement of the material underneath the indenter, a tensile hoop stress of about 50 MPa develops in the pile-up region. This situation can lead to radial cracking and/or crazing of the material on the surface. PMMA is particularly prone to such a failure because of its tendency to craze in the presence of tensile stresses. The elevated normal stress distribution at the substrate interface due to friction is shown in Fig. 20a. The shear stress distribution at the substrate interface (Fig. 20b) shows that the location of the peak shear stress lifts slightly towards the indenter periphery when friction is considered. A small reduction in the magnitude of shear stress is noticed right at the contact radius ($r/a = 1.0$) because of reduced flow, as noted in the case of the strain rate distribution when friction is included at the indenter interface.

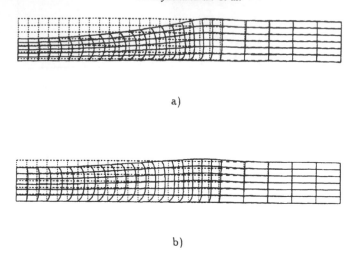

a)

b)

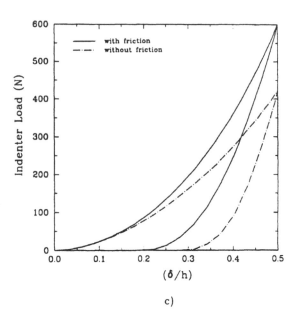

c)

Figure 18. (a) Loaded configuration of the coating at $\delta = 0.5\,h$, with friction at the indenter interface; (b) configuration of the coating upon unloading from $\delta = 0.5\,h$, with friction at the indenter interface; (c) indenter load vs. displacement curves with and without friction at the indenter interface. $h = 100\,\mu$m, $R/h = 20$.

5.4. Comparison with predictions using Matthewson's analysis

For a spherical indenter, a simple elastic–plastic analysis leading to an expression for the interfacial shear stress has been presented by Matthewson [8] to estimate the interfacial shear strength of a coating in terms of the critical indentation load P_c and the corresponding contact radius a_c. In his model, the cylindrical plastic core

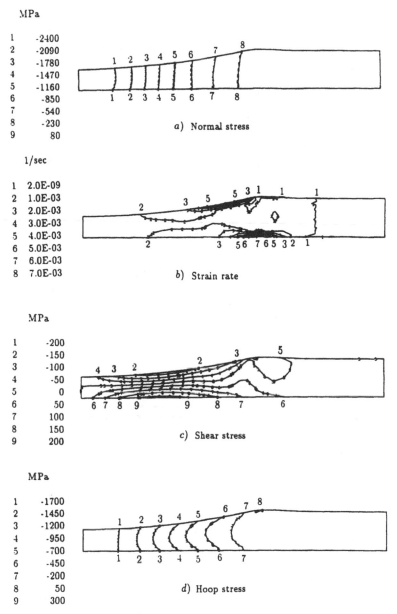

MPa

1	-2400
2	-2090
3	-1780
4	-1470
5	-1160
6	-850
7	-540
8	-230
9	80

a) Normal stress

1/sec

1	2.0E-09
2	1.0E-03
3	2.0E-03
4	3.0E-03
5	4.0E-03
6	5.0E-03
7	6.0E-03
8	7.0E-03

b) Strain rate

MPa

1	-200
2	-150
3	-100
4	-50
5	0
6	50
7	100
8	150
9	200

c) Shear stress

MPa

1	-1700
2	-1450
3	-1200
4	-950
5	-700
6	-450
7	-200
8	50
9	300

d) Hoop stress

Figure 19. Distribution of various stresses and the strain rate in the coating when friction is modelled at the indenter interface. $h = 100 \ \mu m$, $R/h = 20$, $\delta/h = 0.5$.

of radius a is replaced by a pressurized hole with a radial stress to the surrounding elastically deforming coating. Ritter *et al.* [5] used this equation after a slight change in the factor associated with the mean contact pressure in Matthewson's equation based on the assumption that the hardness of the coating is 2.25 times (instead of 3 times) the compressive yield strength of the coating material. Although, from our simulations we cannot predict whether or not the coating material has failed,

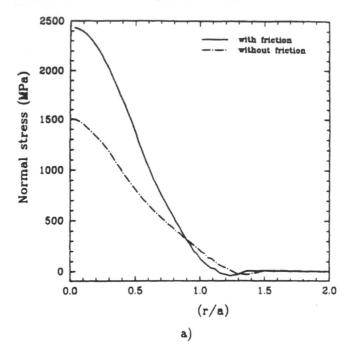

a)

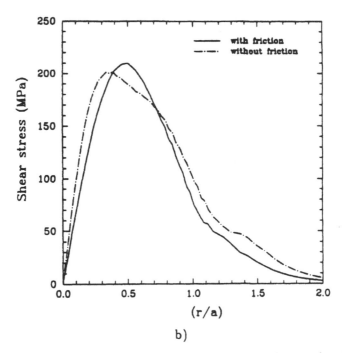

b)

Figure 20. Distribution of (a) normal stress and (b) shear stress in the coating near the substrate interface with and without friction at the indenter interface. $h = 100\ \mu m$, $R/h = 20$, $\delta/h = 0.5$.

we can still pick an indentation load and the corresponding contact radius in order to calculate the interfacial shear stress using Matthewson's analysis. We have used the Matthewson expression with the modification proposed by Ritter *et al.* [5] to calculate the interfacial shear stress at $r = a_c$ as a function of the indentation load and the corresponding contact radius.

Figure 21 shows the comparison of Matthewson's predictions with our results from simulation for various R/h ratios. The interfacial shear stress predicted by Matthewson's analysis is substantially higher than the prediction from the simulation, even for a smaller indentation load or depth of indentation. Whereas in the simulation the shear stress at the contact radius more or less stabilizes to a certain value as

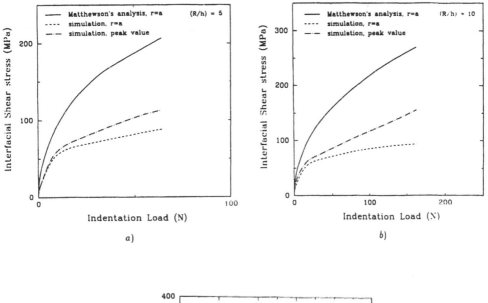

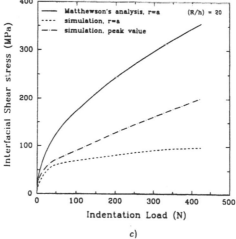

Figure 21. Comparison of interfacial shear stress predictions as a function of the indentation load between Matthewson's analysis and simulation. (a) $R/h = 5$; (b) $R/h = 10$; (c) $R/h = 20$.

the depth of indentation increases, it increases continuously in Matthewson's analysis. Also, the value of peak shear stress that occurs directly under the indenter is well below the value estimated by Matthewson's analysis. The reason for the higher value predicted by the Matthewson model could be the modelling of film surrounding the indenter outside the contact radius as an elastically deforming film where our simulations show the plastic straining which occurs in the critical interface region at and just outside the contact radius. It should also be noted that as indentation progresses, the contact radius keeps changing, and that although the magnitude of the shear stress at $r = a_c$. maintains a constant value, its location keeps moving with the contact radius. Since the failure in the form of debonding is experimentally observed to initiate at $r = a_c$ and keeps moving outwards, the magnitude of the shear stress at $r = a_c$ should adequately represents the adhesion strength of the coating.

6. CONCLUSION

In this study, isothermal indentation simulations were conducted to study the pattern of material flow and the stresses produced under the indenter. A physically based, three-dimensional constitutive law which captures the strain rate, temperature, and the pressure dependence of yield and subsequent orientation-induced hardening of the polymers was used to accurately model the indentation response of polymeric coatings. It was shown that the substrate/coating interface at the contact radius is subjected to an intensive plastic strain rate, indicating the probable site of shear-induced coating delamination during loading. Although the peak shear stress occurs before the contact radius, the compressive normal stress and pressure are large and can act to prevent failure. The shear stress at the contact radius is still large and the normal stress is no longer compressive; therefore failure is likely to occur there. The results also indicate that due to the development of residual tensile normal stress during unloading, there is a possibility of shear-assisted tensile delamination of the coating upon unloading. Due to increasing material confinement with an increase in the ratio of R/h, large pressures, and hence a large normal stress, build up underneath the indenter, thereby increasing the shearing activity at the substrate interface. Also, due to less confinement of the coating material in the case of small indenters or thick coatings, we observe less shearing activity at the substrate interface. The elastic recovery increases with increasing R/h because of the greater confinement and the hydrostatic nature of the deformation. The presence of friction at the indenter interface further confines the material under the indenter, thereby requiring an additional load for indentation, and reduces the material pile-up at the indenter periphery. The tensile hoop stresses generated near the surface due to the indenter friction may cause radial surface cracking/crazing. Thus, the stimulation could be a useful tool for more accurate prediction of the stresses that are generated in the coating and of the interfacial shear strength if the coating material is characterized to include the effects of the factors mentioned above.

Acknowledgements

This work was supported by the Marshall Laboratories of the Du Pont Company. Some computations were performed on the Cray X-MP machine at MIT.

REFERENCES

1. K. L. Mittal, *Electrocomponent Sci. Technol.* **3**, 21–42 (1976).
2. K. L. Mittal, *J. Adhesion Sci. Technol.* **1**, 247–259 (1987).
3. P. R. Chalker, S. J. Bull and D. S. Rickerby, *J. Mater. Sci. Eng.* **A140**, 583–592 (1991).
4. A. G. Evans and J. Hutchinson, *Int. J. Solid Struct.* **20**, 455–466 (1984).
5. J. E. Ritter, T. J. Lardner, L. Rosenfeld and M. R. Lin, *J. Appl. Phys.* **66**, 3626–3634 (1989).
6. M. R. Lin, J. E. Ritter, L. Rosenfeld and T. J. Lardner, *J. Mater. Res.* **5**, 1110–1117 (1990).
7. M. J. Matthewson, *J. Mech. Phys. Solids* **29**, 89–113 (1981).
8. M. J. Matthewson, *Appl. Phys. Lett.* **49**, 1426–1430 (1986).
9. H. D. Conway and J. P. R. Thomsin, *J. Adhesion Sci. Technol.* **2**, 227–236 (1988).
10. M. C. Boyce, D. M. Parks and A. S. Argon, *Mech. Matter.* **7**, 15–33 (1988).
11. M. C. Boyce, D. M. Parks and A. S. Argon, *Int. J. Plasticity* **5**, 593–615 (1989).
12. E. M. Arruda and M. C. Boyce, *J. Mech. Phys. Solids* **41**, 389–412 (1993).
13. E. M. Arruda and M. C. Boyce, *Int. J. Plasticity* (in press).
14. A. S. Argon, *Philos. Mag.* **28**, 839 (1973).
15. L. Anand, *J. Appl. Mech.* **46**, 78 (1979).
16. ABAQUS, HKS, Inc., Providence, RI (1988).

Adhesion Measurement of Films and Coatings, pp. 217–230
K. L. Mittal (Ed.)
© VSP 1995.

Observations and simple fracture mechanics analysis of indentation fracture delamination of TiN films on silicon

E. R. WEPPELMANN,[1,]* X.-Z. HU[2] and M. V. SWAIN[1,3,]†

[1] *CSIRO, Division of Applied Physics, Lindfield, NSW 2070, Australia*
[2] *Department of Mechanical Engineering, University of W.A., Nedlands, WA 6009, Australia*
[3] *Department of Mechanical Engineering, University of Sydney, NSW 2006, Australia*

Revised version received 24 September 1993

Abstract—Ultra micro- or nano-indentation is now widely used to determine the hardness and modulus of thin films on various substrates. In this paper, spherical tipped diamond indenters of small radii are used to investigate the fracture behaviour and delamination of thin TiN films on silicon. The observations clearly establish that the initial behaviour is elastic but at a critical load a sharp discontinuity occurs in the force-displacement curves at the onset of film fracture. Associated with this deformation, the underlying silicon material experiences a pressure-induced phase transformation and film delamination occurs about the contact site. A simple fracture mechanics analysis is presented to estimate the interfacial fracture toughness of the TiN film on the silicon.

Keywords: TiN films; fracture; interfacial toughness; spherical indentation.

1. INTRODUCTION

Thin films are now commonly used for applications as widely disparate as optical sensors to abrasive wear-resistant coatings. Aspects that are particularly important for the long-term reliability of all these films are the presence of residual stresses and the adhesion of such films onto the substrate. As a wear-resistant coating, an additional important parameter is the film's hardness or yield stress.

A number of approaches have been suggested for the measurement of film adhesion, which have been listed in a bibliography by Mittal [1]. However, none of these is entirely satisfactory for routine evaluation and at best they only give a measure of the 'practical' adhesion. Marshall and co-workers [2–4] have shown that indentation with pointed indenters may be used to generate interfacial fracture of films. They indented 2.5–5 μm thick transparent films of ZnO on glass using conventional microhardness instruments. These films enabled the dimensions of the interfacial cracking to be measured easily. Marshall and co-workers [2–4] developed an analysis of such indentation-induced delamination to determine the interfacial toughness and the influence of compressive residual stresses on this cracking. The analysis developed by these authors was based on the elastic–plastic deformation of the ZnO film during indentation

---

*Now at Mechanical Engineering Department, University of Karlsruhe, Karlsruhe, Germany.

† To whom correspondence should be addressed.

that leads to significant compressive radial strains which cause interfacial fracture; it is questionable if this approach is applicable to the current very hard and brittle coatings on a softer substrate.

Some years after the above research, Loubet et al. [5] investigated the cracking behaviour of 8 μm thick TiN films on stainless steel when indented with a Vickers pyramid. Scanning electron micrographs of relatively heavily loaded (10 N) impressions and schematic diagrams presented by these authors suggest that multiple vertical cracking occurred in the films about and beneath the indenter. Loubet et al. [5] claimed that they were able to measure the hardness and stiffness of the TiN films. The authors, although not presenting any experimental evidence for interfacial cracking about the indentation site, proposed an expression for the interfacial fracture toughness very different to the analysis proposed by Marshall and co-workers [2–4].

Since the observations by Marshall and co-workers [2–4], considerable progress in modern indentation instruments has been achieved which now enables the measurement of the hardness and elastic modulus of thin films from very precise force-displacement observations. More recently, Field and Swain [6] proposed the use of small spherical tipped indenters to measure the elastic–plastic/brittle behaviour of materials from which the hardness and modulus with depth, and stress-strain behaviour may be determined. This approach has been successfully applied by Weppelmann et al. [7] to determine the pressures at which phase transformation changes take place during indentation of silicon.

In this paper we explore the use of spherical indentation to investigate the deformation behaviour of TiN films on silicon. The force-displacement data obtained during indentation are initially elastic and show a critical load for film cracking and the influence of film thickness on this load. The extent of interfacial cracking was monitored by SEM observations of cross-sections of the indentation sites and by measuring the size of the uplifted region with a digital interference microscope (WYKO). Apart from interfacial cracking, both Hertzian and radial cracks were observed. A simple fracture mechanics analysis that ignores residual stress is developed to determine the interfacial fracture toughness.

2. THEORETICAL CONSIDERATIONS

2.1. Contact mechanics

The elastic solution for spherical indentation of an isotropic half-space as given by Hertz [8] leads to the following relation for the contact radius, a:

$$a = (4kPR/3E_m)^{1/3}, \tag{1}$$

where P is the applied normal load, R is the indenter radius, and k is a dimensionless constant,

$$k = (9/10)[(1 - \nu_m^2) + (1 - \nu_i^2)(E_m/E_i)], \tag{2}$$

and where E_m, ν_m and E_i, ν_i are the elastic modulus and Poisson's ratio of the half-space and indenter, respectively.

As pointed out by Sneddon [9], the elastic displacements of a plane surface above and below the circle of contact are equal for a completely elastic loading, as shown in

Fig. 1. Therefore, for penetrations which are small compared with the radius of the indenter, the relationship between the depth of penetration, δ, and the contact radius is approximately given by

$$\delta = a^2/R. \tag{3}$$

The penetration of the indenter into the surface of a half-space is given by the combination of equations (1) and (3):

$$\delta = \left(4kP/3E_{\mathrm{m}}\right)^{2/3}(1/R)^{1/3}. \tag{4}$$

The mean indentation pressure, P_{m}, is given by

$$P_{\mathrm{m}} = P/\pi a^2, \tag{5}$$

which when combined with equation (3) leads to

$$P_{\mathrm{m}} = P/\pi\delta R. \tag{6}$$

The influence of a film on the elastic behaviour of an axisymmetric contact has recently been addressed by Gao *et al.* [10]. These authors considered the influence of both single and multi-layered films on the elastic behaviour of an indented half-space. The critical parameter in such tests is the ratio of the contact diameter to the film thickness and the relative moduli of the film(s) and substrate.

Field and Swain [6] have recently extended the spherical elastic contact situation to elastic–plastic contact of materials. This included the provision for strain hardening of the material. Recent papers have shown the applicability of this development along with a novel load-partial unloading indenting procedure to determine the stress-strain behaviour of a range of materials [11].

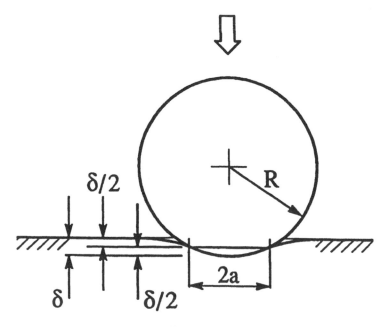

Figure 1. Schematic cross-sectional view of the elastic contact fracture of a sphere and an elastic half-space.

2.2. Indentation-induced delamination

Previous analysis of Vickers pyramid indentation-induced delamination of films by Marshall and Evans [3] was based on the elastic–plastic deformation of the film leading to significant radial strains that caused interfacial fracture. This approach, which assumes that the elastic properties of the film and substrate are identical, leads to the following relationship between the delamination crack size and the interfacial fracture toughness, G_c:

$$G_c = \gamma E P^3 / h H^3 (1 - v^2) c^4, \tag{7}$$

where γ is a constant, h is the film thickness, H is the hardness, and c is the radius of the delamination crack.

An alternative analysis of the interfacial fracture toughness, as mentioned above, has been presented by Loubet et al. [5]. In their model, they suggest that annular cracks develop about the contact area upon the load exceeding a critical value. Subsequent displacement of the material is taken up by the substrate, with debonding occurring when the tensile interfacial stress exceeds the adhesion forces between the film and the substrate. The extent of the delamination is dictated by an energy balance between the strain energy released and the interfacial free energy. A relationship for the interfacial fracture toughness is then proposed, namely,

$$G_c = 3/8 \left[E_c h^3 (d - d_r)^2 \right] / c^4, \tag{8}$$

where $(d - d_r)$ is the displacement of the edge of contact from its original position just prior to cracking and c is the delamination crack radius.

2.2.1. Delamination fracture. Because of the limitations of the above approach, a simple fracture mechanics analysis which is applicable for very hard and brittle films on a softer substrate is now developed. The film is assumed to crack just outside the area of contact which is considered to behave elastically. This assumption is similar to that proposed by Loubet et al. [5] and is consistent with observations to be presented in Section 4. The maximum radial tensile stress which develops in the surface adjacent to the area of contact when loaded with a sphere is given by

$$\sigma_{\text{max-s}} = \frac{1 - 2v}{2} \frac{P_c}{\pi a_c^2}, \tag{9}$$

where the contact diameter a_c, at the critical load P_c, may be calculated from equation (1). An estimate of the maximum radial tensile stress in the film at the load P_c, assuming that no residual stresses exist in the film and assuming the same contact radius, i.e. $a_{cf} = a_c$ (if $h \ll a_c$), may be written as

$$\sigma_{\text{max-f}} = \frac{1 - 2v}{2} \frac{P_c}{\pi a_c^2} f(E_f). \tag{10}$$

However, for the material system in question, $h \leqslant a_c$, and then

$$\sigma_{\text{max-f}} = \frac{1 - 2v}{2} \frac{P_c}{\pi a_c^2} f(h, E_f). \tag{11}$$

An estimate of $f(h, E_f)$ may be obtained by combining equations (9) and (11), i.e.

$$f(h, E_f) = \frac{\sigma_{\text{max-f}}}{\sigma_{\text{max-s}}} = \frac{\sigma_f}{\sigma_s} \tag{12}$$

at or before thin film cracking.

After delamination (Fig. 2a), the change in the strain energy of the thin film is considered to be the energy consumed in delamination (if $h \ll c$), i.e. the interfacial fracture energy, $G_{\text{I-II}}$ (the fracture mode for this situation is likely to be a combination of mode I and mode II loading). This is then given by

$$G_{\text{I-II}}\{\pi(c^2 - a_c^2)\} = \Delta U_f. \tag{13}$$

The strain energy released, ΔU_f, may be calculated from the integration of the stress in the film, i.e.

$$\Delta U_f = \int_0^{2\pi} d\theta \int_{a_c}^c \frac{\sigma_f^2(r)}{2E_f} hr \, dr = \frac{\pi h}{E_f} \int_{a_c}^c \sigma_f^2(r) r \, dr. \tag{14}$$

Therefore, assuming that all the strain energy due to beam deflection of the film is released as surface energy of the delaminated area, the delamination fracture energy may be written as

$$G_{\text{I-II}} = \frac{h}{(c^2 - a_c^2)E_f} \int_{a_c}^c \sigma_f^2(r) r \, dr. \tag{15}$$

As an approximation for the variation of σ_f, we can assume from equation (9) that for $r \geqslant a_c$,

$$\sigma_f(r) = \sigma_{\text{max-f}}\left\{\frac{a_c}{r}\right\}^2. \tag{16}$$

(a)

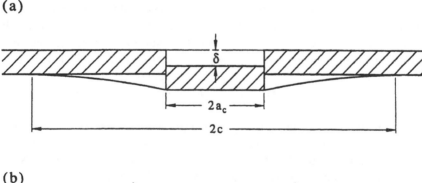

(b)

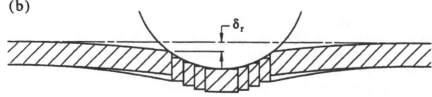

Figure 2. Schematic cross-sections of (a) the elastic contact fracture and delamination of a thin film, and (b) the interference of the spherical indenter in the film deflection after cracking.

Incorporating this expression into equation (15), we have

$$G_{\text{I-II}} = \frac{h\sigma^2_{\text{max-f}}}{12E_f}\left\{\frac{a_c}{c}\right\}^2. \tag{17}$$

The basic assumption with this approach is that the film upon cracking and delamination is free to deflect towards the original surface position, releasing all the indentation-induced strain. However, because of the curvature of the spherical indenter, the deflection is severely restricted. If there exists some residual deflection of the film, i.e. $\delta_r > 0$, the reduction of strain energy in the film is proportional to the square of the residual strain, i.e. ε_r^2, where

$$\varepsilon_r = \varepsilon(1 - \delta_r/\delta). \tag{18}$$

The reduction in the strain energy release rate as shown in Fig. 2b may then be incorporated by multiplying ΔU_f in equation (13) by α. Equation (17) may now be written as

$$G_{\text{I-II}} = \frac{h\sigma^2_{\text{max-f}}}{12E_f}\left\{\frac{a_c}{c}\right\}^2\alpha, \tag{19}$$

with

$$\alpha = \left(\frac{\varepsilon_r}{\varepsilon}\right)^2 = \left(1 - \frac{\delta_r}{\delta}\right)^2,$$

where δ_r is the residual displacement and δ is the actual displacement of the contact circle [i.e. half of δ in equation (3)].

3. EXPERIMENTAL

3.1. Film deposition

The TiN films used in this study were deposited using a physical vapour deposition (PVD) technique. To produce denser films, the PVD system combines an ion bombardment process with a filtered arc device. The system has a small bias voltage of 20–50 V in excess of the cathode potential and the resulting ion beam is steered through a curved magnetic plasma duct to filter out any macro-particles ejected by the arcing process [12]. The combination of the high kinetic energy of the ions on impact with the filtered technique leads to films whose density is close to that of the corresponding bulk material.

The TiN films were deposited on prepolished and slightly oxidized single crystalline silicon discs which were approximately 1.0 mm thick. These discs were such that the surface was aligned parallel to the [100] orientation of the single crystal. During the deposition process, a bias of 50 V was used and the temperature was held at 375°C. These conditions lead to a deposition rate of about 0.35 μm/min. To measure the thicknesses of the TiN films, the samples were broken and the cross-sections examined with an SEM. With this method, the thicknesses were measured and found to be 2.47 and 4.5 μm. Furthermore, X-ray analysis of the lattice parameter of the {222} crystal planes of the TiN revealed that the films were under high residual tensile stresses with values of 0.65 and 0.56 GPa for the thinner and thicker films, respectively.

3.2. Indentation system

All force-displacement measurements were made with a commercially available ultra-micro-indentation system (UMIS-2000). With this system, force-displacement data with an accuracy in the μN and nm range are obtainable for loads up to 800 mN. Details of this instrument have been recently published elsewhere [13]. Indentations were carried out with spherical tipped diamond indenters with radii of 5 and 10 μm. Typically, 60 steps were used during the continuous loading and unloading stages with dwell times of 1 s at each step. The load was varied from 50 to 800 mN and indentations were made in arrays with at least three indentations at each load.

3.3. Examination of the impressions

The impressions made on the TiN films were examined with an SEM and a high precision optical interferometer system (WYKO). With the latter system, very small vertical displacements (a few nm) may be resolved and used to determine the extent of subsurface delamination.

4. RESULTS AND DISCUSSION

Typical force-displacement $(P-\delta)$ observations for indentations made with the 5 and 10 μm radius indenters on the 4.5 μm film are shown in Figs 3 and 4, respectively.

At low loads, for both indenters the force-displacement behaviour is entirely elastic and reversible as shown in Figs 3a and 4a. As the load increases, the film cracks and a break in the force-displacement curve results. With the 10 μm radius indenter, the first crack was observed at loads of $\sim$150 and $\sim$170 mN with the 2.7 and 4.5 μm thick films, respectively, and 130 and 140 mN for the 5 μm radius indenter (see Figs 3b and 4b for the 4.5 μm thick film data). Upon increasing the load, further breaks in the force-displacement curve can be seen which may be associated with further cracking events of the film (Figs 3c and 4c). Using the smaller radius indenter at heavier loads, the extent of plastic deformation was observed to increase (cf. Figs 3c and 4c). For the heaviest load, a change in compliance was observed upon unloading. This behaviour is somewhat similar to the behaviour observed during spherical indentation of the single crystal silicon substrates. The force-displacement curves of the latter materials exhibited a step during unloading if the maximum force had been sufficient to initiate plastic flow. As discussed elsewhere [7], this behaviour was due to a volumetric expanding phase change in silicon that occurs during unloading.

SEM observations of the deformed and cracked region about an indentation carried out with 5 μm and 10 μm radius indenters and at maximum loads of 500 and 767 mN are shown in Fig. 5. The micrograph of the 5 μm radius indentation impression does not show clear evidence of circumferential cracks. However, the scanning electron micrographs of the impressions with the 10 μm radius indenter (Fig. 5b) do exhibit a number of well-defined circumferential cracks. These cracks were present at indentation loads in excess of that necessary to initiate the first step in the load-deflection curve. At heavier loads, the residual impressions were more distinct and radial cracks were also observed. The size of these radial cracks was slightly less than the delamination cracks.

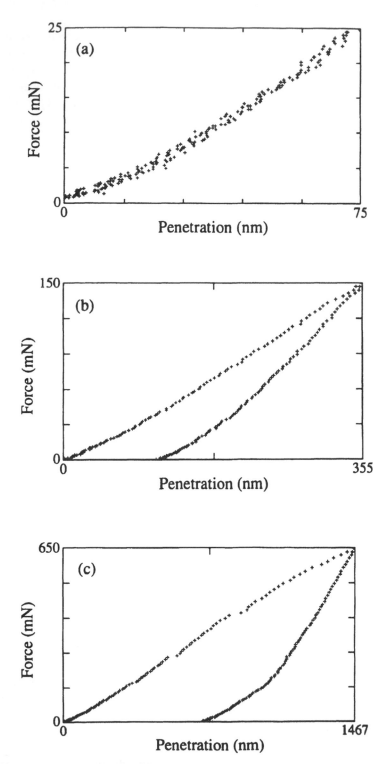

Figure 3. Observations of the force-displacement behaviour with a 5 μm indenter loading a 4.5 μm thick TiN film on silicon. (a) Elastic loading regime; (b) loading just beyond the onset of film cracking; (c) the maximum load applied.

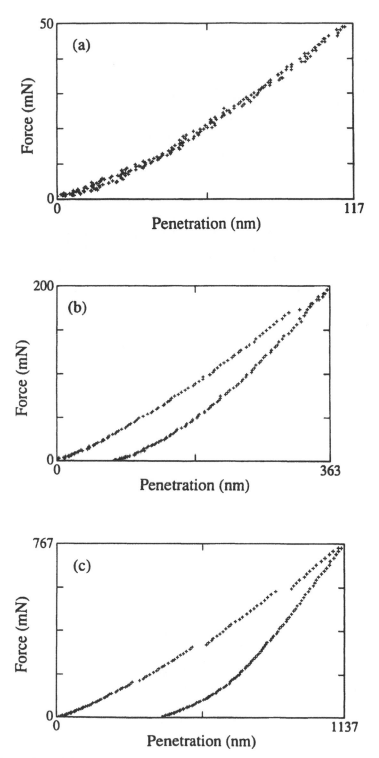

Figure 4. Observations of the force-displacement behaviour with a 10 μm indenter loading a 4.5 μm thick TiN film on silicon. (a) Elastic loading regime; (b) loading just beyond the onset of film cracking; (c) the maximum load applied.

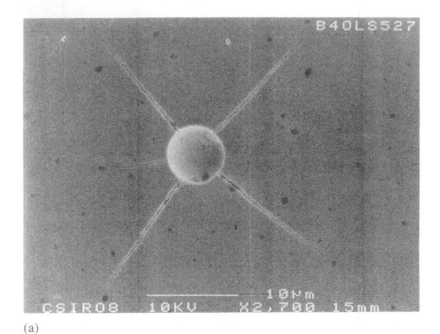

(a)

(b)

Figure 5. Scanning electron micrograph of the residual impression and associated cracking in the 2.7 μm TiN film generated with (a) a 5 μm radius indenter loaded to 500 mN and (b) a 10 μm radius indenter loaded to 767 mN.

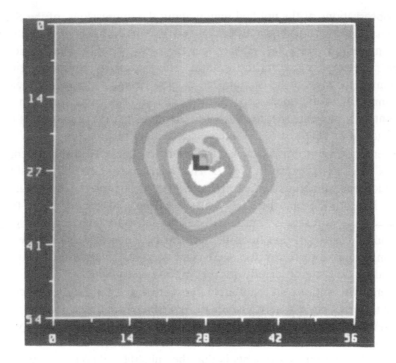

Distance on Surface (μm)

Distance on Surface (μm)

(a)

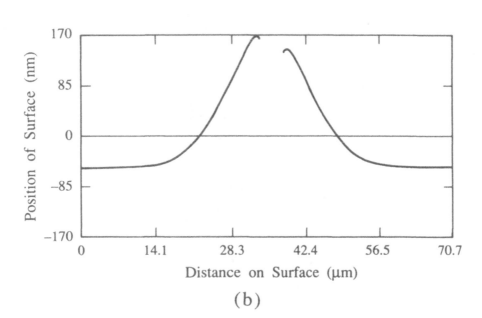

Position of Surface (nm)

Distance on Surface (μm)

(b)

Figure 6. Optical interferometric (WYKO) observations of (a) the contours of uplift about, and (b) cross-section of the impression shown in Fig. 5a.

WYKO observations of the uplifted region about the impressions, of the SEM micrograph shown in Fig. 5a, are shown in Figs 6a and 6b. The contours of the uplifted region about the impressions (Fig. 5a) indicate that the form of the uplifted region partially reflects the radial cracks present in the film. The size of the uplifted region may be used to quantify the size of the delaminated cracks. Recent independent measurements using a scanning acoustic microscope confirm the validity of this approach [14]. Such measurements also indicate that the onset of delamination cracking appears to coincide with film fracture as detected in the force-displacement records.

SEM observation of a cross-section of a 2.74 μm thick TiN film indented with a 5 μm radius indenter is shown in Fig. 7. This micrograph clearly establishes that the film has undergone numerous vertical fractures that are parallel to the columinar grain boundaries. Beneath the film there is a well-established 'plastic' or phase-transformed zone in the silicon substrate. A crack at the film–substrate interface is also observed, which extends some 10–12 μm from the edge of the circle of contact. The size of the delamination cracks observed in the SEM correlates very well with the uplifted zone about the indentation site observed with the WYKO instrument.

The present observations are similar to those reported by Loubet *et al.* [5], who investigated the cracking and delamination of TiN films on stainless steel when indented with a Vickers pyramid. They also observed that the film cracked vertically within the contact area with little or no evidence for plastic deformation of the film. A feature that is different in the present experiments is that the substrate silicon when unloaded after 'plastic' deformation exhibits a 'pop-out' event due to a phase change [7]. This dilation

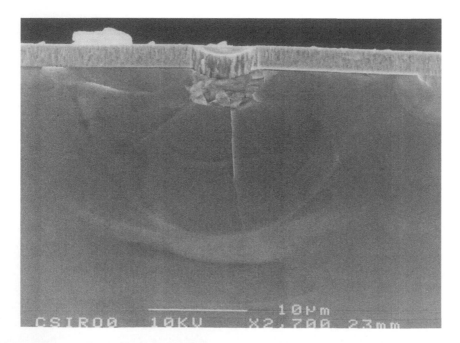

Figure 7. SEM observation of the cross-section of an indented TiN film on a silicon substrate. The vertical faulting of the TiN film, subsurface deformed zone, and associated interfacial cracking are clearly evident. Indenter radius: 5 μm; load: 650 mN; film thickness: 2.7 μm.

of the substrate during unloading may contribute to the further uplift of the film. Such behaviour would be expected to be more substantial at heavier loads when the size of the transformed zone is much greater than at the onset of film cracking.

Estimates of the interfacial fracture toughness can be made with the aid of equations (11), (12), and (17) and with the observations shown in Fig. 7. The relevant dimensions for the calculations are $h = 2.7 \ \mu m$, $E_f = 450$ GPa, $P_c = 650$ mN, $a_c = 3.1 \ \mu m$, and $c = 16 \ \mu m$. Substitution of these values leads to an estimate of G_{I-II} of $\sim 51 \ J\,m^{-2}$, which is greater than the toughness of the substrate silicon and is comparable to that of the TiN material (approximately 6 and 40 $J\,m^{-2}$, respectively). By incorporating the term α, equation (19), to account for the limited recovery, a more reasonable estimate of the interfacial fracture toughness can be obtained. From Fig. 3b the total displacement is $\delta = 1467/2$ nm (half the total displacement at the onset of cracking), the residual displacement $\delta_r = (732 - 120)$ nm, the latter value of 120 nm being typical of the sum of the cracking steps in the loading curve, and using these values to $G_{I-II} = 51(1 - \delta_r/\delta)^2 \simeq 1.2 \ J\,m^{-2}$. These values are likely to be influenced by the presence of substantial residual stresses within the TiN films on silicon.

5. CONCLUSIONS

The present observations have demonstrated that small spherical tipped diamond indenters are suitable for probing the elastic/brittle behaviour of TiN films on silicon and for initiating film delamination. The initial deformation of the system is elastic but upon exceeding a critical load, the film fractures, forming a circular ring crack just outside the area of contact. This brittle behaviour can be observed in the force-displacement record and appears to coincide with the onset of interfacial cracking. With further increasing load, circumferential cracks develop as well as radial cracks and the silicon substrate undergoes a pressure-induced phase transformation. There appears to be minimal evidence for plastic deformation of the film.

A simple fracture mechanics theory to rationalize the extent of interfacial cracking has been proposed. The displacement of the contact area constrained relaxation of the film because indenter interferences are used to estimate the interfacial fracture toughness. SEM observations in support of this approach have been presented and the interfacial fracture toughness between TiN and silicon was estimated to be ~ 1-$2 \ J\,m^{-2}$.

Acknowledgements

We thank P. Martin and A. Bendavid for deposition of the TiN films, T. J. Bell and D. Northcote for their assistance with the UMIS testing, and E. Thwaite and J. Ogilvy for their comments on the manuscript.

REFERENCES

1. K. L. Mittal, *J. Adhesion Sci. Technol.* **1**, 247 (1987).
2. S. S. Chiang, D. B. Marshall and A. G. Evans, in: *Surfaces and Interfaces in Ceramic and Ceramic–Metal Systems*, J. A. Pask and A. G. Evans (Eds), p. 603. Plenum Press, New York (1981).
3. D. B. Marshall and A. G. Evans, *J. Appl. Phys.* **56**, 2632 (1984).
4. C. Rossington, A. G. Evans, D. B. Marshall and B. T. Khuri-Yakub, *J. Appl. Phys.* **56**, 2639 (1984).
5. J. L. Loubet, J. M. Georges and Ph. Kapsa, in: *Mechanics of Coatings*, D. Dowson, C. M. Taylor and M. Godet (Eds), p. 429. Elsevier, Amsterdam (1990).

6. J. S. Field and M. V. Swain, *J. Mater. Res.* **8**, 297–306 (1993).

7. E. R. Weppelmann, J. S. Field and M. V. Swain, *J. Mater. Res.* **8**, 830–841 (1993).

8. H. Hertz, *Hertz's Miscellaneous Papers*, Ch. 5, 6. Macmillan, London (1986).

9. I. N. Sneddon, *Int. J. Eng. Mech. Sci.* **3**, 47 (1965).

10. H. Gao, C.-H. Chiu and J. Lee, *Int. J. Solids Struct.* **29**, 2471–2492 (1992).

11. T. J. Bell, J. S. Field and M. V. Swain, *Mater. Res. Soc.* **239**, 331–336 (1992).

12. P. J. Martin, R. P. Netterfield and T. J. Kinder, *Thin Solid Films* **193/194**, 77–83 (1990).

13. T. J. Bell, A. Bendeli, J. S. Field, M. V. Swain and E. G. Thwaite, *Metrologia* **28**, 463–469 (1991–1992).

14. E. R. Weppelmann, S. Suganomo and M. V. Swain, in preparation.

Adhesion Measurement of Films and Coatings, pp. 231–247
K. L. Mittal (Ed.)
© VSP 1995.

A study of the fracture efficiency parameter of blister tests for films and coatings

YEH-HUNG LAI and DAVID A. DILLARD*

Engineering Science and Mechanics Department, Virginia Polytechnic Institute and State University, Blacksburg, VA 24061, USA

Revised version received 14 December 1993

Abstract—Analytical solutions for circular membranes with different boundary conditions are obtained based on the assumptions of large displacements, small rotations, and small strains. The solutions take into account prestresses of arbitrary magnitude. The maximum stresses and strain energy release rates for the standard and island blister tests are obtained. The fracture efficiency parameter, defined as the ratio between the strain energy release rate and the square of the maximum stress, is then introduced to study the geometric effect on the relationship between the failure along the interface and in the coating. The fracture efficiency parameter is also used to find the optimal specimen dimensions for a given blister test geometry. It is found that the prestresses have a significant effect on the parameter. Although the island blister test is capable of producing very high strain energy release rates at relatively low pressures by decreasing the size of the island, the fracture efficiency parameter does not increase because of the significant increase of stress at the edge of the island. No significant difference in the fracture efficiency parameter is found between the standard and island blister geometries. The result suggests that yielding is likely to take place in all blister test methods at about the same applied strain energy release rate for a given film thickness and properties.

Keywords: Fracture efficiency parameter; test efficiency; blister tests; fracture mechanics; strain energy release rate; films; coatings; membrane solution; stress analysis; standard blister test; island blister test.

1. INTRODUCTION

The blister test was originally proposed by Dannenberg [1] for measuring the adhesion of coatings. To have better control over the debond growth, he confined the blister to form in a narrow groove, thereby resulting in a constant strain energy release rate specimen. Williams [2] utilized a circular debond geometry to measure the fracture energy of elastomeric materials cast to a rigid substrate. Blister tests have been applied to a wide variety of adhering systems, including paints, coatings, elastomers, bonded plates [3], pressure-sensitive adhesive tapes [4], and even adhesion to ice [5]. Blister specimens are quite versatile and have been applied in a number of configurations. The standard blister (Fig. 1) is quite compatible with environmental exposures because the pressurizing medium is contained within the blister region. For circular versions of the blister specimen, the axisymmetric shape minimizes problems associated with edge effects of finite width specimens, and diffusion perpendicular to the debond front eliminates spurious effects on environmental exposure. One of the

---

*To whom correspondence should be addressed.

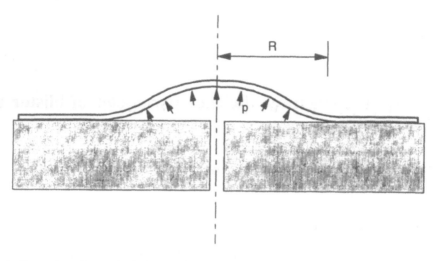

Figure 1. Illustration of a standard blister test.

disadvantages with the standard blister is that the strain energy release rate increases with the debond radius, thus making accurate measurement of the debond radius essential and resulting in an unstable fracture specimen. To minimize this problem, the constrained blister was introduced. The test geometry was proposed by Dillard *et al.* [6, 7] and Napolitano *et al.* [8] and analyzed by Lai and Dillard [9, 10]. While the constrained blister specimen is a constant strain energy release rate specimen only under limiting cases, it does significantly reduce the dependence of the strain energy release rate on the debond radius. To obtain a high strain energy release rate at a relatively low pressure, the island blister (Fig. 2) was proposed by Allen and Senturia [11, 12] for measuring the adhesion of the polyimide films. Allen and Senturia reported that the island blister geometry could be successfully employed to debond prestressed polyimide films, while the standard blister test geometry resulted in film rupture with the same material. Although the island blister test is capable of producing a high strain energy release rate, the debonding process is unstable. To have a stable debonding process while retaining the advantage of the high strain energy release rate, the peninsula blister was introduced by Dillard and Bao [13]. Strain energy release rates for both membrane and plate solutions were derived, indicating that this test geometry is a truly constant strain energy release rate specimen over a relatively large test section. Both the peninsula and the island blister specimens derive high strain energy release rates from the fact that the debond front is reduced to a very small length. A moderate increase in compliance is produced by a relatively small increase in debond area, thereby causing large strain energy release rates. Although these two test geometries are capable of producing high strain energy release rates at relatively low applied pressures, there are questions about whether high strain energy release rates could be produced at lower coating stresses than other test geometries [14].

In all fracture tests for adhesion strength, the most important property is the ability to debond the specimen without causing rupture or gross yielding. Since most films and coatings have relatively strong adhesion and a low load-bearing capacity due to the small thickness, measurement of the adhesion strength of films and coatings may require

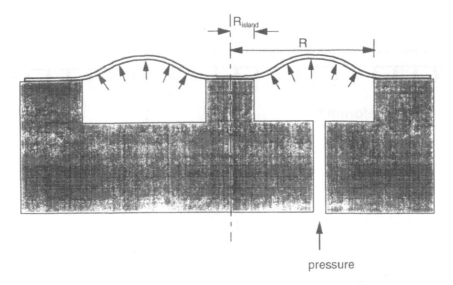

Figure 2. Illustration of an island blister test.

tests that are capable of producing high strain energy release rates at relatively low stresses. The applied force, test geometry, material properties, prestress, and specimen dimensions are the factors which determine the magnitude of applied strain energy release rates and stresses. Before performing blister tests to determine the adhesion strength of a specific material system, one needs to consider all the above factors and possibly to optimize the tests to successfully debond the film without rupturing it. In this study, a new parameter called the 'fracture efficiency parameter' will be defined and used to optimize these factors. Although there are a variety of blister tests, we will discuss two circular versions, the standard blister test and the island blister test. These will be sufficient to demonstrate the salient features while avoiding the complications of the three-dimensional peninsula blister specimen.

2. MATHEMATICAL FORMULATIONS FOR THE STANDARD BLISTER TEST AND THE ISLAND BLISTER TEST

2.1. Circular membrane solutions

For a circular membrane of which the cross-section is shown in Fig. 3, the constitutive equations at the middle surface of the membrane are [17]

$$N_r = \frac{Eh}{(1 - \nu^2)} \left(\bar{\varepsilon}_r + \nu \bar{\varepsilon}_\theta \right) + \bar{\sigma}_0 h, \tag{1}$$

$$N_\theta = \frac{Eh}{(1 - \nu^2)} \left(\bar{\varepsilon}_\theta + \nu \bar{\varepsilon}_r \right) + \bar{\sigma}_0 h, \tag{2}$$

where $\bar{\sigma}_0$ is the prestress in the membrane, N_r is the meridian stress resultant, N_θ is the circumferential stress resultant, E is Young's modulus, ν is Poisson's ratio, h is the thickness of the membrane, $\bar{\varepsilon}_r$ is the meridian strain, and $\bar{\varepsilon}_\theta$ is the circumferential strain.

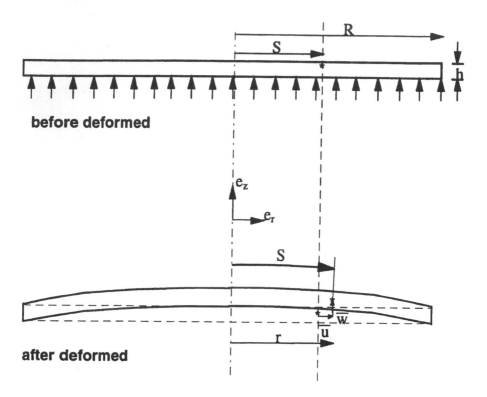

Figure 3. Cross-section of a pressurized membrane.

It should be noted that N_r and $\bar{\varepsilon}_r$ are not the radial stress resultant and radial strain, but rather are allowed to rotate with the middle surface. The prestress may arise during the curing process due to mismatch of thermal expansion coefficients, or may be due to moisture- or solvent-induced swelling of the film, etc.

The meridian strain and circumferencial strain are given by [17]

$$\bar{\varepsilon}_r = \frac{d\bar{u}}{dr} + \frac{1}{2}\left(\frac{d\bar{w}}{dr}\right)^2, \tag{3.1}$$

$$\bar{\varepsilon}_\theta = \frac{\bar{u}+r}{r} - 1 = \frac{\bar{u}}{r}, \tag{3.2}$$

where $\bar{u}$ and $\bar{w}$ are the displacements of a point at the middle surface of the membrane in the radial and normal to radial directions, respectively.

Equilibrium equations in the meridian direction and transverse direction are

$$(rN_r)_{,r} = N_\theta \tag{4.1}$$

and

$$\frac{1}{r}(rN_r\bar{w}_{,r}) + p = 0, \tag{4.2}$$

where the comma denotes derivatives with respect to the indices which follow.

Substituting the equations for strain/displacement relations into the constitutive equations and then into the equilibrium equations, we can obtain the governing equations for a circular membrane as

$$
3\overline{w}_{,rr}\overline{w}_{,r}^2 + \frac{\overline{w}_{,r}^3}{r} + 2\overline{w}_{,rr}\left(\overline{u}_{,r} + \nu\frac{\overline{u}}{r} + \frac{1-\nu^2}{E}\overline{\sigma}_0\right)
$$
$$
+ 2\overline{w}_{,r}\left(\overline{u}_{,rr} + \frac{\overline{u}_{,r}}{r}(1+\nu) + \frac{1-\nu^2}{E}\frac{\overline{\sigma}_0}{r}\right) + 2\frac{1-\nu^2}{Eh}p = 0,
\tag{5.1}
$$

$$
\overline{u}_{,rr} + \frac{\overline{u}_{,r}}{r} - \frac{\overline{u}}{r^2} + \overline{w}_{,r}\overline{w}_{,rr} + \frac{(1-\nu)\overline{w}_{,r}^2}{2r} = 0.
\tag{5.2}
$$

By introducing the nondimensional notations x, w, u, σ_0, σ_r, and σ_θ, which are expressed as follows:

$$
x = \frac{r}{R},
\tag{6.1}
$$

$$
w = \frac{\overline{w}}{R}\left(\frac{pR}{Eh}\right)^{-1/3},
\tag{6.2}
$$

$$
u = \frac{\overline{u}}{R}\left(\frac{pR}{Eh}\right)^{-2/3},
\tag{6.3}
$$

$$
\sigma_0 = \frac{\overline{\sigma}_0}{E}\left(\frac{pR}{Eh}\right)^{-2/3},
\tag{6.4}
$$

$$
\sigma_r = \frac{\overline{\sigma}_r}{E}\left(\frac{pR}{Eh}\right)^{-2/3},
\tag{6.5}
$$

and

$$
\sigma_\theta = \frac{\overline{\sigma}_\theta}{E}\left(\frac{pR}{Eh}\right)^{-2/3},
\tag{6.6}
$$

equations (5.1) and (5.2) can be rewritten as

$$
3w_{,xx}w_{,x}^2 + \frac{w_{,x}^3}{x} + 2w_{,xx}\left(u_{,x} + \nu\frac{u}{x} + (1-\nu^2)\sigma_0\right)
$$
$$
+ 2w_{,x}\left(u_{,xx} + \frac{u_{,x}}{x}(1+\nu) + (1-\nu^2)\frac{\sigma_0}{x}\right) + 2(1-\nu^2) = 0
\tag{7.1}
$$

and

$$
u_{,xx} + \frac{u_{,x}}{x} - \frac{u}{x^2} + w_{,x}w_{,xx} + \frac{1-\nu}{2}\frac{w_{,x}^2}{x} = 0.
\tag{7.2}
$$

Nondimensional stresses are expressed as

$$
\sigma_r = \frac{1}{(1-\nu^2)}\left(u_{,x} + \frac{1}{2}w_{,x}^2 + \nu\frac{u}{x}\right) + \sigma_0
\tag{8.1}
$$

and

$$\sigma_\theta = \frac{1}{(1-\nu^2)} \left(\frac{u}{x} + \nu u,_x + \frac{\nu}{2} w,_x^2 \right) + \sigma_0. \tag{8.2}$$

From equations (7.1) and (7.2), it is noted that one set of solutions, $\{w, u, \sigma_r, \sigma_\theta, \bar{\varepsilon}_r, \bar{\varepsilon}_\theta\}$ exists for each set of $\{\nu, \sigma_0,$ boundary conditions$\}$. It should also be noted that solving equations (7.1) and (7.2) requires neither knowledge of the pressure, Young's modulus, membrane thickness, nor the size of the outermost radius. The only geometric parameters involved are the ones in the boundary conditions. For the island blister test, it is the nondimensional island radius, a. For the constrained blister test, it is the nondimensional constraint height. For the standard blister test, there is no geometric parameter involved when solving the nondimensional governing equations, equations (7.1) and (7.2).

The volume under the deformed membrane can be presented in two different forms, the true volume, $\bar{V}$, and the nondimensional volume, B, which are expressed as follows:

$$\bar{V} = 2\pi \int \bar{w} r \, dr \tag{9.1}$$

and

$$B = 2\pi \int w x \, dx. \tag{9.2}$$

Since equations (7.1) and (7.2) are two simultaneous nonlinear partial differential equations, closed form solutions are not available. The non-prestressed version of equations (7) has been solved by several researchers using different numerical techniques [15–17]. In this study, a nonlinear relaxation technique by Kao and Perrone [16, 17], which is an iterative approach used in conjunction with finite difference approximations, will be used to solve the above nonlinear equations. Although the closed form solution is not available, only a minimal numerical effort is needed to determine the effects of each parameter on the nondimensional form of the governing equations, equations (7.1) and (7.2).

The boundary conditions for the standard blister test are

$$w,_x = u = 0, \qquad \text{at} \quad x = 0, \tag{10.1}$$

$$w = u = 0, \qquad \text{at} \quad x = 1. \tag{10.2}$$

The solutions $w(x)$ and $u(x)$ depend on Poisson's ratio, ν, and the nondimensional prestress, σ_0. There is no geometric parameter involved in the nondimensional solution. In the case without prestress, i.e. $\sigma_0 = 0$, the solutions $w(x)$ and $u(x)$, and therefore σ_r and σ_θ, depend on Poisson's ratio only.

The boundary conditions for the island blister test are

$$w = u = 0, \qquad \text{at} \quad x = a, \tag{11.1}$$

$$w = u = 0, \qquad \text{at} \quad x = 1, \tag{11.2}$$

where a is the nondimensional island radius R_{island}/R.

The solutions $w(x)$ and $u(x)$ depend on Poisson's ratio, ν; the nondimensional island radius, a; and the nondimensional prestress, σ_0. For the case without prestress, i.e. $\sigma_0 = 0$, the solution $w(x)$ and $u(x)$, and therefore σ_r and σ_θ, depend only on Poisson's ratio and the nondimensional island radius.

2.2. Formulation of the strain energy release rate

It should be noted that the strain energy release rates for the standard and island blister tests found in the literature are either for the case without prestress [4] or for the case with dominating prestress [11, 12]. The strain energy release rates derived in this section for both test geometries will consider an arbitrary magnitude of prestress.

By using the energy balance equation

$$\mathcal{G}\delta A = p\delta\overline{V} - \delta\mathcal{U}, \tag{12}$$

where $\mathcal{G}$ is the strain energy release rate, δA is the change in debond area, $\delta\overline{V}$ is the change in volume of the blister in the debonding process, and $\delta\mathcal{U}$ is the change in strain energy of the blister in the debonding process, and performing some algebra, the strain energy release rate for the standard blister test with an arbitrary magnitude of prestress can be obtained as

$$
\mathcal{G} = \frac{(pR)^{4/3}}{3(Eh)^{1/3}} \left\{ \frac{1}{\pi} \left[5B - \sigma_0 \frac{\partial B}{\partial \sigma_0} \right] \right.
$$
$$
- \left(5 - \sigma_0 \frac{\partial}{\partial \sigma_0} \right) \int_0^1 \left[\left((\sigma_r - \sigma_0)^2 - 2\nu(\sigma_r - \sigma_0)(\sigma_\theta - \sigma_0) + (\sigma_\theta - \sigma_0)^2 \right) \right. \tag{13}
$$
$$
\left. \left. + 2(1 - \nu)\sigma_0 (\sigma_r + \sigma_\theta - 2\sigma_0) \right] x \, dx \right\}.
$$

It should be noted that the energy dissipation due to plasticity and viscoelasticity is assumed to be negligible.

In the case without prestress, one can simply substitute zero for σ_0 and obtain the strain energy release rate without prestress. Alternatively, one can obtain $\mathcal{G}$ for the case without prestress using a generalized compliance approach proposed by Lai and Dillard [9]:

$$\mathcal{G} = \frac{5B}{4\pi} \frac{(pR)^{4/3}}{(Eh)^{1/3}}. \tag{14}$$

The strain energy release rate for the island blister test with an arbitrary magnitude of prestress can be obtained as

$$
\mathcal{G} = \frac{(pR)^{4/3}}{2a(Eh)^{1/3}} \left\{ \frac{1}{\pi} \frac{\partial B}{\partial a} \right.
$$
$$
- \frac{\partial}{\partial a} \int_a^1 \left[\left((\sigma_r - \sigma_0)^2 - 2\nu(\sigma_r - \sigma_0)(\sigma_\theta - \sigma_0) + (\sigma_\theta - \sigma_0)^2 \right) \right. \tag{15}
$$
$$
\left. \left. + 2(1 - \nu)\sigma_0 (\sigma_r + \sigma_\theta - 2\sigma_0) \right] x \, dx \right\},
$$

where

$$a = \frac{R_{\text{island}}}{R}, \tag{16}$$

and $\mathcal{G}$ in the case without prestress using the generalized compliance approach is

$$\mathcal{G} = \frac{(pR)^{4/3}}{(Eh)^{1/3}} \frac{3}{8\pi} \frac{\partial B}{a \partial a}. \tag{17}$$

2.3. Maximum tensile stress

With the knowledge of membrane stresses and the strain energy release rate for a crack at an interface, the bending moment at the crack front takes the form [18]

$$M = \left[\frac{\mathcal{G}Eh^3}{6(1 - \nu^2)} - h^2 N^2/12 \right]^{1/2}, \tag{18}$$

where N is the difference of the stress resultants at the two sides of the crack tip. It should be noted that the above equation can be applied to the blister test on the condition that the debonding radius of the standard blister and the island radius of the island blister are much larger than the thickness of the coating [19]. By introducing a nondimensional form as in equations (6), the bending stress of the film at the debond front can be obtained as

$$\sigma_b = (\sigma_{\text{mt}} - \sigma_0) \left[\frac{6g}{1 - \nu^2} - 3 \right]^{1/3}, \tag{19}$$

where σ_{mt} is the membrane stress at the crack front, and

$$g = \left\{ \frac{1}{\pi} \left[5B - \sigma_0 \frac{\partial B}{\partial \sigma_0} \right] - \left(5 - \sigma_0 \frac{\partial}{\partial \sigma_0} \right) \int_0^1 \left[\left((\sigma_r - \sigma_0)^2 - 2\nu(\sigma_r - \sigma_0)(\sigma_\theta - \sigma_0) \right. \right. \right.$$

$$\left. \left. \left. + (\sigma_\theta - \sigma_0)^2 \right) + 2(1 - \nu)\sigma_0(\sigma_r + \sigma_\theta - 2\sigma_0) \right] x \, dx \right\} \left[\sigma_{\text{mt}}(\nu, \sigma_0) - \sigma_0 \right]^{-2}$$

for the standard blister test and

$$g = \frac{1}{2a} \left\{ \frac{1}{\pi} \frac{\partial B}{\partial a} - \frac{\partial}{\partial a} \int_a^1 \left[\left((\sigma_r - \sigma_0)^2 - 2\nu(\sigma_r - \sigma_0)(\sigma_\theta - \sigma_0) + (\sigma_\theta - \sigma_0)^2 \right) \right. \right.$$

$$\left. \left. + 2(1 - \nu)\sigma_0(\sigma_r + \sigma_\theta - 2\sigma_0) \right] x \, dx \right\} \left[\sigma_{\text{mt}}(\nu, \sigma_0) - \sigma_0 \right]^{-2}$$

for the island blister test.

The nondimensional maximum tensile stress can then be found at the crack tip as

$$\sigma_{\text{max}} = \sigma_{\text{mt}} + \sigma_b. \tag{20}$$

2.4. Fracture efficiency parameter

To answer questions such as 'Under what conditions will the specimen debond without yielding?' and 'Why does the same material rupture in one test geometry while it does not rupture in another one?', we need to study the relation between the adhesion strength and the cohesive strength of the films and coatings. One of the most used failure theories for fracture is the 'total strain energy release rate theory', which states that debonding of a bonded system will begin when the total strain energy release rate reaches a limiting value, namely, the interfacial fracture energy. Three frequently used failure theories for material strength on a continuum level are the maximum principal stress criterion, von Mises criterion, and the maximum shear stress theory, or Tresca criterion. The most appropriate criterion may be selected based on brittleness of the material.

It will be shown in the Section 3 that the maximum stress occurs at the boundary, $x = 1$ for the standard blister and $x = a$ for island blister, where the crack tip stress singularity is located. If the singular region is very small or the yielding is confined to a small distance from the crack tip, typically, a distance less than a layer thickness from the boundary, the maximum stress such as the one obtained in equation (20) could be used as the parameter to indicate whether gross yielding or rupturing will occur in the coating. Since the debonding is determined by the magnitude of the strain energy release rate (which is related to the singular stress field) and the gross yielding of the coating is determined by the magnitude of the maximum effective stress (which is related to the stress field other than the singular stress field), whether the specimen will debond or yield depends on the magnitudes of two competing quantities — the strain energy release rate and the maximum effective stress.

In the original Griffith criterion [20], the strain energy release rate in a cracked infinite body is found to be a function of both the material properties and the crack length and is proportional to the square of the stress at infinity. Likewise, the strain energy release rate in an adhesion test can be expressed as the product of the square of maximum stress, which can be the maximum tensile stress, maximum shear stress, or maximum effective stress, and a function of the material properties, test geometry, and loading condition. The function which is the ratio between the strain energy release rate and the square of maximum stress can then be used to evaluate the 'efficiency', or a test's ability to produce a maximum strain energy release rate at a certain maximum stress level. This function is defined in the present study as the 'fracture efficiency parameter', T_e, and is expressed as

$$T_e = \frac{\mathcal{G}}{\bar{\sigma}^2_{max}}, \tag{21}$$

where $\bar{\sigma}_{max}$ can be the maximum tensile stress, maximum shear stress, or maximum effective stress in von Mises criterion.

For simplicity, the maximum tensile stress is used in this study, which suggests that the results are more appropriate for brittle materials. It should be noted that the fracture efficiency parameter can be applied to more general materials by choosing an appropriate effective maximum stress.

In the case without prestress, the fracture efficiency parameters for the standard and island blister tests are as follows:

$$\text{Standard blister test: } T_e = \frac{5}{4\pi}\frac{h}{E}\frac{B}{\sigma_{max}^2}, \tag{22.1}$$

$$\text{Island blister test: } T_e = \frac{3}{8\pi}\frac{h}{E}\frac{1}{\sigma_{max}^2}\frac{1}{a}\frac{\partial B}{\partial a}. \tag{22.2}$$

For the case with prestress, the fracture efficiency parameters are

Standard blister test:

$$
\begin{aligned}
T_e = \frac{h}{E}\Bigg\{ &\frac{1}{\pi}\left[5B - \sigma_0\frac{\partial B}{\partial\sigma_0}\right] \\
&- \left(5 - \sigma_0\frac{\partial}{\partial\sigma_0}\right)\int_0^1 \left[\left((\sigma_r - \sigma_0)^2 - 2\nu(\sigma_r - \sigma_0)(\sigma_\theta - \sigma_0)\right.\right. \\
&\left.\left. + (\sigma_\theta - \sigma_0)^2\right) + 2(1 - \nu)\sigma_0(\sigma_r + \sigma_\theta - 2\sigma_0)\right]x\,dx \Bigg\}\sigma_{max}^{-2},
\end{aligned}
\tag{23}
$$

Island blister test:

$$
\begin{aligned}
T_e = \frac{h}{2Ea}\Bigg\{ &\frac{1}{\pi}\frac{\partial B}{\partial a} \\
&- \frac{\partial}{\partial a}\int_a^1 \left[\left((\sigma_r - \sigma_0)^2 - 2\nu(\sigma_r - \sigma_0)(\sigma_\theta - \sigma_0) + (\sigma_\theta - \sigma_0)^2\right)\right. \\
&\left. + 2(1 - \nu)\sigma_0(\sigma_r + \sigma_\theta - 2\sigma_0)\right]x\,dx \Bigg\}\sigma_{max}^{-2}.
\end{aligned}
\tag{24}
$$

Several comments about the fracture efficiency parameter can be made here:

(a) The fracture efficiency parameter defined in equation (21) is a function of the material properties, test geometry, specimen dimensions, and prestress. In linear systems and some nonlinear systems such as the membrane blister without prestress, it is independent of the applied load. In cases such as the membrane with prestress, the fracture efficiency parameter is weakly dependent on the pressure, as seen in the nondimensional form of the prestress in equation (6.4).

(b) For the standard blister specimen without prestress, T_e depends on three parameters, E, ν, and h. It is interesting to note that the fracture efficiency parameter is independent of the radius of the blister; i.e. if the blister yields before debonding at a specific radius, it will yield for any blister size despite the fact that increasing radius can significantly increase the strain energy release rate for a given pressure. The only way to avoid yielding for a specific material is to increase the thickness of the coating, since the fracture efficiency parameter is linearly proportional to the thickness.

(c) For the blister specimens with prestress, T_e depends on the nondimensional prestress, σ_0. It should be noted that

$$\sigma_0 = \frac{\bar{\sigma}_0}{E}\left(\frac{pR}{Eh}\right)^{-2/3},$$

so changing the prestress, pressure, and outer blister radius can change the fracture efficiency parameter. Increasing pressure or blister radius has the same effect as decreasing prestress; therefore, the fracture efficiency parameter is no longer linearly proportional to the thickness.

(d) It is possible to define a critical fracture efficiency parameter as the radio between the interface toughness and the square of the coating strength, so a fracture efficiency parameter smaller than this value indicates debonding of the coating without coating failure, and vice versa.

3. RESULTS AND DISCUSSION

Figure 4 illustrates the nondimensional membrane stress distributions in the standard blister test without prestress along the meridian direction for different Poisson's ratios. It was found that for both cases with and without prestress, the maximum membrane stress was located at the center of the blister for Poisson's ratios ranging from 0 to 0.5. However, if the bending stress is taken into consideration, the maximum tensile stress is located at the edge of the debond front for both cases with and without prestress. The yielding will most likely occur at the edge of the debond front. It should be noted that the bending moment obtained from equation (19) has only a 0.5% difference compared with the result obtained from a boundary layer solution by Jensen [19]. It should also be noted here that the maximum stress in the standard blister test is dependent on only two nondimensional parameters, ν and σ_0. Since the strain energy release rate is proportional to the nondimensional blister volume in the case without prestress as seen in equation (14), it is possible to formulate an equation for the strain energy release rate in the standard blister specimen by curve-fitting the data of nondimensional blister volume as a function of Poisson's ratio. By choosing a second-order polynomial, the strain energy release rate can be written as

$$\mathcal{G} = \frac{5}{4\pi}(1.207 - 0.336\nu - 0.253\nu^2)\frac{(pR)^{4/3}}{(Eh)^{1/3}}. \tag{25}$$

Equation (25) offers a short cut to obtain the strain energy release rate as long as the material properties, blister radius, blister thickness, and critical pressure are known. That is, if one knows the material properties and thickness of the blister, the only measurements needed are the critical pressure and debonding radius. The simplicity eliminates laborious numerical or experimental effort. In order to check the accuracy of the fitted curve, Gent and Lewandowski's data [4] were used. The difference between the fitted

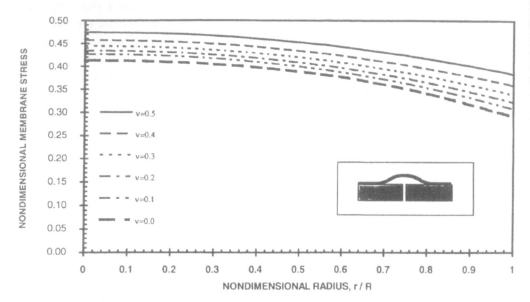

Figure 4. Nondimensional membrane stress distribution in the standard blister test without prestress for different Poisson's ratios.

value and Gent's data is within 0.3% for a Poisson's ratio of 0.5. In the Poisson's ratio range between 0.25 and 0.5, which includes the values of most polymers, the percent difference in strain energy release rate is less than 12%. The relatively small variation in strain energy release rate across a wide range of Poisson's ratios suggests that one may choose an approximate value for Poisson's ratio, perform a simple tensile test to obtain Young's modulus of the film, and a standard blister test to obtain the critical pressure, and then use equation (14) to obtain the critical strain energy release rate with a reasonable accuracy.

A typical nondimensional membrane stress distribution in the island blister test is illustrated in Fig. 5 for the case with a nondimensional prestress of 0.64 (as was found in the polyimide film studied by Allen and Senturia [11, 12]), a nondimensional island size of 1/150, and a Poisson's ratio of 0.34. To check the convergence of the nonlinear relaxation iteration technique used in this study, several finite difference meshes with 21, 41, and 81 nodes were used. All meshes showed satisfactory results except for the stresses at the edge of the island. An axisymmetrical nonlinear finite element (FE) model of 150 elements was also used for comparison with the results from analytical solutions. The FE mesh consisted of three layers of elements through the thickness. Since only membrane stress was to be compared with the analytical solution, the mesh was not refined enough to obtain the singular behavior of the stress field very near the crack tip. The finite element program used was ABAQUS [21] and the element type used was an eight-node quadrilateral element. Very good agreement was seen between the finite element and analytical results for different finite difference mesh sizes except in the region near the island. A high stress concentration was seen near the edge of the island. To find the maximum membrane stress at the edge of the island, the finite difference mesh was refined near the edge of the island until a convergent value was found. In the region away from the island, the membrane stress was slightly higher than the prestress, which suggests that the prestress was high enough to dominate

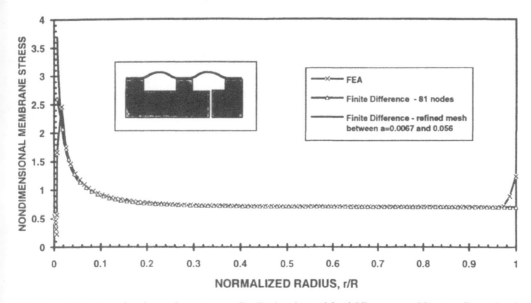

Figure 5. Nondimensional membrane stress distribution in an island blister test with a nondimensional prestress of 0.64 and a Poisson's ratio of 0.34. The nondimensional island radius is 1/150.

the stress field in the region far away from the island. From all the island blister analyses, the maximum membrane stress was found to be located at the edge of the island.

Figure 6 illustrates the nondimensional strain energy release rate and nondimensional maximum membrane stress versus the nondimensional island radius for an island blister test without prestress. It is seen that both the strain energy release rate and the maxi-

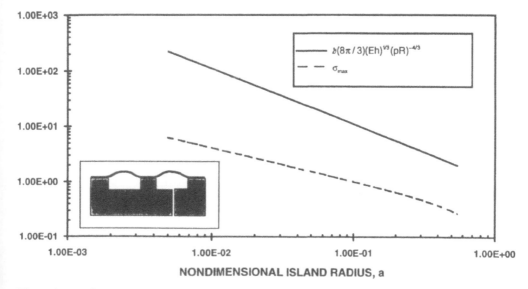

Figure 6. Nondimensional strain energy release rate and nondimensional maximum membrane stress versus the nondimensional island radius for an island test without prestress. The Poisson's ratio is 0.3.

mum membrane stress increase significantly as the island size decreases. Whether gross yielding will occur during the debonding process depends on whether the maximum stress reaches the coating strength. The results also suggest that the island blister test is capable of producing a very high strain energy release rate under a relatively low pressure.

Figures 7 an 8 illustrate the nondimensional fracture efficiency parameter versus the prestress and nondimensional island radius for the standard and island blister tests with

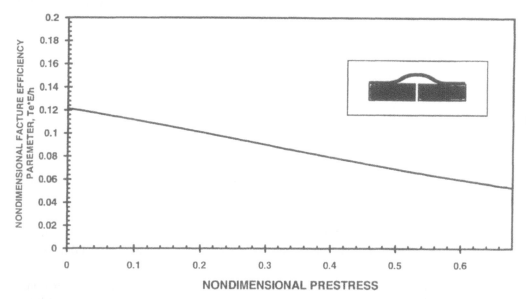

Figure 7. Nondimensional fracture efficiency parameter versus the nondimensional prestress for a standard blister test with a Poisson's ratio of 0.3.

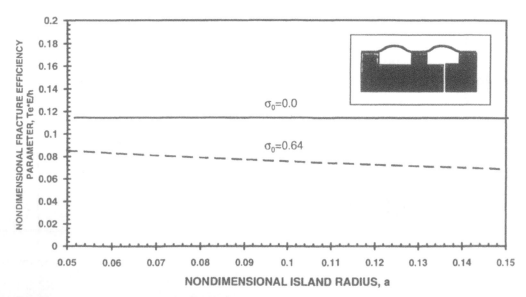

Figure 8. Nondimensional fracture energy parameter versus the nondimensional island radius for an island blister test with two typical nondimensional prestresses. The Poisson's ratio is 0.3.

nondimensional prestresses of zero and 0.64. The Poisson's ratio is 0.3. The fracture efficiency parameter of the standard blister test decreases as the nondimensional prestress increases. Since decreasing nondimensional prestress is equivalent to the increase of the blister radius, the results in Fig. 7 suggests that once the debonding of the blisters can be initiated without rupturing or yielding, it will continue if the pressure can be reduced accordingly. In Fig. 8, it is seen that although the strain energy release rate increases significantly as the island radius decreases under constant pressure, the fracture efficiency parameter increases only slightly. The analytical results of the island blister test suggest that if gross yielding occurs, the island radius needs to be decreased to have possible debonding initiated without yielding of the film. However, the variation is so small that it is unlikely that changing the size of the island will prevent yielding in the experiments. The nondimensional fracture efficiency parameter for a standard blister test without prestress was found to be 0.118 as shown in Fig. 7, which is very close to that found in the island blister test with a different island radius shown in Fig. 8. The result of the case without prestress suggests that if films yield without debonding in the standard blister test, it will also be likey to do so in the island blister test even though the island blister test needs a much lower pressure. It should be noted that the maximum stress used in the fracture efficiency parameter is not the maximum membrane stress, but the maximum value of the sum of the bending stress and membrane stress. It is interesting to see in Fig. 8 that increasing pressure and blister radius can increase the fracture efficiency parameter of the standard blister test with prestress, while the fracture efficiency parameter is independent of the pressure and blister radius for a standard blister test without prestress. It should be noted that as the bending moment is determined at the debonding front by using equation (18), the phase angle can be easily determined [18]. It was found that the phase angles were about $-50°$ and that they differed by less than 4% for the standard and island blister tests for the case with the same material properties for the coating and bottom substrate. Thus, it is appropriate to directly compare the fracture efficiency parameter between the two tests without considering the effect of mode mixity. In order to compare the fracture efficiency parameter between the standard and island blister tests, the same real prestress, rather than the nondimensional prestress, should be chosen for both tests. In the case of an island blister test with a nondimensional prestress of 0.64 and a nondimensional island size of 0.1, the equivalent nondimensional prestress for the standard blister test is 0.38. It can be seen from Figs 7 and 8 that the standard blister test has a nondimensional fracture efficiency parameter of 0.081, while the island blister test has one of 0.075. The results suggest that it is unlikely to induce film yielding in one test without inducing film yielding in another, which is indeed what was observed in Allen and Senturia's experiment. Allen and Senturia also found that their specimens ruptured in the standard blister tests but not in the island blister tests after gross yielding of films in both tests [22]. In order to predict rupture after yielding, elastic–plastic analysis is needed and awaits further study. The comparison of the fracture efficiency parameters is not capable of predicting rupture after large-scale yielding. The analytical solutions obtained in this study may be used to explain the experimental observation by Allen and Senturia. Figures 4 and 5 indicate high stresses throughout the standard blister specimen and low stress in the island blister specimen, except for the high stress concentration at the edge of the island. Statistically, it may be more likely to rupture the standard blister specimen than the island blister specimen due to the high stress level all over the standard blister specimen.

4. CONCLUSIONS

In this study, large deformation, analytical solutions for the standard and island blister specimens were obtained. The solutions took into account prestresses of arbitrary magnitude. Maximum stresses and strain energy release rates for the standard blister test and island blister test were obtained. The fracture efficiency parameter, defined as the ratio between the strain energy release rate and the square of the maximum stress, was then used to determine whether a particular coating would yield before debonding in one test geometry and yet debond without yielding in another. A preliminary study of the relationship between yielding and debonding was done for both the island and the standard blister tests with and without prestress. It was shown that test parameters such as the radius, thickness, and pressure can be optimized for a given blister test geometry. It was found that the prestress has a strong effect on the fracture efficiency parameter. No significant difference in the fracture efficiency parameter was found between the standard and the island blister geometries. This result suggests that yielding is likely to take place in all blister test methods at about the same applied strain energy release rate for a given film thickness and properties, which supports the observations made by Allen and Senturia.

Acknowledgements

We would like to acknowledge the Center for Adhesive and Sealant Science at Virginia Tech and the National Science Foundation's 'High Performance Polymeric Adhesives and Composites' Science and Technology Center (Contract DMR 9120004) for their support of this work.

REFERENCES

1. H. Dannenberg, J. Appl. Polym. Sci. **5**, 125–134 (1961).
2. M. L. Williams, J. Appl. Polym. Sci. **13**, 29–40 (1969).
3. M. Fernando and A. J. Kinloch, Int. J. Adhesion Adhesives **10**, 69–76 (1990).
4. A. N. Gent and L. H. Lewandowski, J. Appl. Polym. Sci. **33**, 1567–1577 (1987).
5. K. R. Jiang and L. S. Penn, J. Adhesion **32**, 203–226 (1990).
6. D. A. Dillard and Y. S. Chang, CASS/ESM-87-8, Report, Virginia Polytechnic Institute, Blacksburg, VA (1987).
7. Y. S. Chang, Y. H. Lai and D. A. Dillard, J. Adhesion **27**, 197–211 (1989).
8. J. Napolitano, A. Chudnovsky and A. Moet, J. Adhesion Sci. Technol. **2**, 311–323 (1988).
9. Y. H. Lai and D. A. Dillard, J. Adhesion **31**, 177–189 (1990).
10. Y. H. Lai and D. A. Dillard, J. Adhesion **33**, 63–74 (1990).
11. M. G. Allen and S. D. Senturia, J. Adhesion **25**, 303–315 (1988).
12. M. G. Allen and S. D. Senturia, J. Adhesion **29**, 219–231 (1989).
13. D. A. Dillard and Y. Bao, J. Adhesion **63**, 253–271 (1991).
14. Y. Bao, Master's Thesis, Department of Engineering Science and Mechanics, Virginia Polytechnic Institute and State University, Blacksburg, VA (1992).
15. H. Hencky, Z. Math. Phys. **63**, 311–317 (1915).
16. N. Perrone and R. Kao, J. Appl. Mech. **38**, 172–178 (1971).
17. R. Kao and N. Perrone, Int. J. Solids Struct. **7**, 1601–1612 (1971).

18. J. W. Hutchinson and Z. Suo, *Adv. Appl. Mech.* **29**, 63–191 (1992).
19. H. M. Jensen, *Eng. Fract. Mech.* **40**, 475–486 (1991).
20. A. A. Griffith, *Philos. Trans. R. Soc. London, Ser. A* **A221**, 163–197 (1921).
21. Hibbit, Carlsson and Sorenson, Inc., ABAQUS User Manual, Version 4-7-1, Providence, RI (1992).
22. M. G. Allen, PhD Dissertation, Massachusetts Institute of Technology (1989).

Adhesion Measurement of Films and Coatings, pp. 249–264
K. L. Mittal (Ed.)
© VSP 1995.

Adhesion (fracture energy) of electropolymerized poly(*n*-octyl maleimide-co-styrene) coatings on copper substrates using a constrained blister test

JENG-LI LIANG,[1] JAMES P. BELL[1,*] and ASHIT MEHTA[2]

[1] *Polymer Science Program, Institute of Materials Science, University of Connecticut, Storrs,
CT 06269, USA*
[2] *IBM Corp., Endicott, NY 13760, USA*

Revised version received 29 March 1993

Abstract—The fracture energy required to separate electropolymerized *n*-octyl maleimide/styrene polymer film from various copper substrates was measured using a constrained blister test. A G value (strain energy release rate) of 75 J/m^2 at 0.5 mm/min debonding rate was found for separation of the *n*-octyl maleimide/styrene polymer film from a smooth cooper surface. This value is slightly higher than the G value (60 J/m^2) of pressure-sensitive adhesive tape on copper, and much higher than the estimated 4.4 J/m^2 of Kapton® polyimide on a similar copper surface. It was found that rough surfaces having a regular pattern substantially increased the adhesion strength of the polymer coatings. The polymer films adhered well to the rough surfaces; attempts to separate the polymer film from the copper surface resulted in cohesive failure.

Keywords: Electropolymerized coating; copper foil electrode; poly(*n*-octyl maleimide-co-styrene); adhesion; constrained blister test; fracture energy.

1. INTRODUCTION

Several poly(*N*-substituted maleimide-co-styrene) films were electropolymerized onto copper substrates and reported previously [1] as possible candidates for electronic applications. Iroh *et al.* [2] have electropolymerized poly(3-carboxyphenyl maleimide-co-styrene) onto AS-4 graphite fibers from aqueous monomer–dilute sulfuric acid solutions. The preparation of polymer-coated metal samples in the present study is based on an electropolymerization technique similar to that developed for graphite fiber composites. Graphite fiber electrodes are replaced by metal sheets in this study. We believe that the starting monomer solutions are of very low viscosity and can easily wet the metal substrates, and that polymerization occurs in the proximity of the metal surface. The resulting adhesion to the metal may therefore be improved.

---

*To whom correspondence should be addressed.

Teng and Mahalingam [3] and Garg *et al.* [4] have investigated the use of electropolymerization to coat metal electrode surfaces with acrylate polymers. These polymers have rather low glass transition temperatures and therefore appear to be unsuitable as candidates for electronic applications where high temperature resistance is required.

Recently, a number of blister techniques have been reported for flexible and thin adhesive/adherend systems, including the standard blister test applied to a polyurethane elastomeric adhesive by Briscoe and Panesar [5] and to two incompatible polymers by Chu and Durning [6], and the island blister test produced by Allen and Senturia [7]. To minimize the variation in the strain energy release rate using this standard blister technique, Napolitano *et al.* [8, 9] and Chang *et al.* [10] introduced the constrained blister test to evaluate the adhesion of thin pressure-sensitive adhesive tape on metal surfaces, but we found no reports using this test for non-pressure-sensitive polymer films on metal substrates. Stain energy release rate measurements are reported in this paper.

The 4-carboxylphenylmaleimide/styrene (4-CMI/ST) copolymer films investigated to date have fractured before debonding from the copper substrates. At the time of fracture, a strain energy release rate larger 60 J/m^2 was found, using a roughened copper surface. To study electropolymerized coatings of the maleimide/styrene type without fracture, a more ductile *n*-octyl maleimide/styrene copolymer with a lower glass transition temperature (115°C) and a flexible side-chain was prepared. Work is continuing on the strain energy release rate measurements for the high T_g 4-CMI/ST coatings.

2. EXPERIMENTAL

2.1. Materials and electropolymerization

n-Octyl maleimide (OTMI) monomer was prepared according to the method of Sauers [11] using reagent-grade ingredients as described previously [1]. Monomer purity was confirmed by IR spectroscopy and differential scanning calorimetry (DSC). The IR absorbance for NH bending (amide II band) at 1550 cm^{-1} was absent after the recrystallization of the *N*-substituted maleimides studied. The DSC thermogram of these monomers showed only one sharp melting peak. Styrene (99.9%, Aldrich Chemical Co.) was extracted twice with 10% sodium hydroxide solution and washed several times with distilled water to remove inhibitors.

The electropolymerization procedure was performed as described previously [1]. In this process, monomers are electropolymerized onto copper sheets which function as the cathode in the reaction cell. The monomer–electrolyte solution was prepared by dissolving the desired amounts of *N*-substituted maleimide and styrene in dimethylacetamide (DMAc). The resulting solution and aqueous sulfuric acid (0.025 M) were then mixed in a 50:50 volume ratio. The mixture was introduced into the central chamber of a three-compartment electrochemical cell [1b]. A copper sheet (5 cm × 15 cm) of 25 μm thickness was degreased by soaking in acetone for 24 h. The sheet was then dried at room temperature and centrally positioned in the central chamber. Aqueous sulfuric acid solution (0.0125 M) was placed in the two outer counter-electrode (anode) chambers. The anodes were tantalum substrates coated with magnesium oxide. Additional information about the cell has been given previously [1a]. Electropolymerization was done at a current density of 0.6 mA/cm^2 sheet area. After a given time, the coated copper sheet was withdrawn from the cell, rinsed thoroughly with a DMAc/water mixture to remove the monomer residue, and dried at room temperature overnight. The partially dried copper sheet was vacuum-dried at 250°C to remove any residual DMAc.

2.2. Characterization

A number of copper substrates provided by IBM Corp. were characterized in terms of surface morphology and composition. Surface morphology was examined by scanning electron microscopy (SEM). Elemental analysis of the surface region was done simultaneously with SEM coupled to energy dispersive X-ray analysis (EDXA). Information concerning surface roughness was provided from Talysurf-S-120 profilometer measurements.

The thickness of the dried polymer film was determined by dividing the weight gain by the polymer density. The glass transition temperatures of the polymer were obtained using a DuPont 9900 DSC, under a nitrogen environment at a heating rate of 10°C/min.

X-Ray photoelectron spectroscopy (XPS) data were collected for 9 min using a Perkin-Elmer 5000 XPS system with an Al K_α X-ray source. A sampling angle of 62° was used.

A constrained blister test apparatus (Fig. 1a), as developed by Napolitano *et al.* [8, 9], and Chang *et al.* [10] and Lai *et al.* [12], was connected to an image analyzer (Contextvision microGOP system, Struers Vision Co., Sweden) and gas pressure measurement device. This image analyzer can take and store images automatically, and perform very sophisticated enhancing and sharpening functions to improve the blister images using a SUN® computer system. A schematic diagram of the set-up is shown in Fig. 1b. The debonded area $A(t)$, debonded radius a, and the distance d between contact points of the film with the substrate and the constraint, which we will refer to as the 'horizontal suspended distance', were measured by the image analyzer as a function of time at a given constant gas pressure. It should be noted that term 'horizontal suspended distance' is specialized, and is not the same as the debonded distance. Substituting into the following Chang, Lai, and Dillard equation [10], we can obtain the strain energy release rate G:

$$G = phq$$
$$q = \left[1 - (d/2a)\right] + \left[(d/3a) - (1/2)\right](\partial d/\partial a),$$

where p and h are the applied pressure and the constraint height, respectively. $\partial d/\partial a$ is the partial derivative of the horizontal suspended distance vs. debond radius relation. The constrained blister test has the advantages of (1) a relatively constant fracture

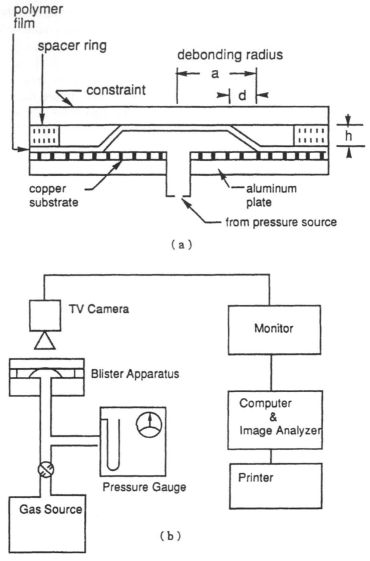

Figure 1. (a) Geometry of the constrained blister specimen. (b) Schematic diagram of the blister test set-up.

energy G as the crack grows, (2) less deformation of the film than a free blister test, and (3) avoidance of the fracture and tearing that are common for such thin films (~ 25 μm) when used with a peel test or other tests with larger deformation [13].

A circular area of 6 mm diameter was etched out the copper side, using 30 wt% ferric chloride solution, from a 3 cm $\times$ 3 cm polymer-coated copper substrate. This copper specimen was adhered to 3 mm thick aluminum plate using epoxy adhesive (Devcon 5-Minute® Epoxy, Devcon Co., USA). The bonded specimen was then baked in an oven at 60°C for 12 h and then slowly cooled to room temperature before testing.

3. RESULTS AND DISCUSSION

3.1. Copper morphology

Scanning electron micrographs of three different substrates described previously [1b] are shown in Fig. 2. It was found that copper D is rather smooth and is purely copper (Fig. 2, D). On the contrary, copper E with a pink color is rough and has a rather uniform peak to valley ratio of 5 μm (Fig. 2, E). Copper F, with a brown color, also has a rough surface and a higher peak to valley ratio of 10 μm (Fig. 2, F). However, dendritic structures were found only on copper F at the peaks of its surface profile. The dendrites were examined by EDXA; the spectra showed a strong signal associated with magnesium.

n-Octyl maleimide (OTMI) was copolymerized with styrene onto copper surfaces D, E, and F. The thicknesses were 16, 17, and 11 μm on copper surfaces D, E, and F, respectively.

The glass transition temperature for OTMI/styrene was found to be 115°C by DSC at a heating rate of 10°C/min.

3.2. Polymer–substrate interactions

The X-ray photoelectron spectra of the uncoated and OTMI/styrene-coated copper surface D are shown in Fig. 3. On the uncoated surface not only were the expected Cu $3s$ and Cu $3p$ found, but also C $1s$ was found due to contamination (Fig. 3a). However, there was very little contamination (C $1s$) left after sputter-etching in the XPS vacuum chamber for 30 s (Fig. 3b). Figure 3c presents the XP spectrum of polymer-coated copper after tetrahydrofuran (THF) solvent extraction for 24 h. N $1s$ corresponds to imide nitrogen. These data indicate that the electropolymerized polymer film was not completely removed from the copper surface by solvent extraction. The polymer signal remained even after sputter-etching for 30 s (Fig. 3d). Although it is not clear whether or not chemical bonding exists, the XPS data demonstrate that the thin film on the copper substrate is rather well attached.

A scanning electron micrograph of an electropolymerized polymer film surface after dissolving copper E by means of ferric chloride solution is presented in Fig. 4b. The surface morphologies match the surface contour of copper E (Fig. 4a) exactly. This result demonstrates excellent mechanical interlocking due to both wetting by the starting monomer solution and the fact that electropolymerization occurs in the proximity of the metal surface, which should lead to improved adhesion.

3.3. Fracture energy measurement

A typical image of a polymer film inflated by pressure in the constrained blister test is shown in Fig. 5. Two annuli are clearly observed. The small annulus was reflected under light as a result of the inflated polymer film compressed by a polymethylmethacrylate constraint. The large annulus was reflected under light in the same way when the inflated polymer film peeled from the copper substrate. Therefore, the small

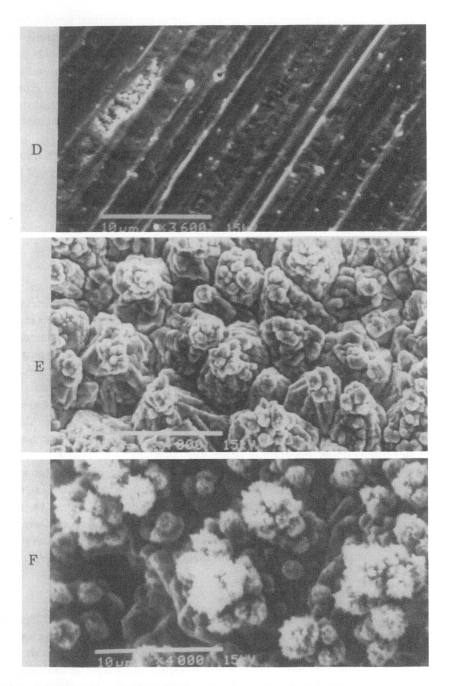

Figure 2. Scanning electron micrographs [1a] of copper surfaces D, E, and F.

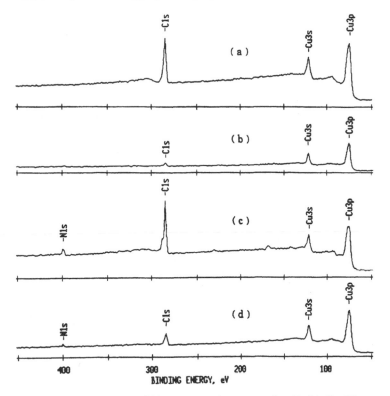

Figure 3. X-Ray photoelectron spectra of (a) an uncoated copper surface D, (b) after 30 s sputter-etching, (c) polymer-coated copper surface after 24 h THF extraction, and (d) after 30 s sputter-etching.

annulus and large annulus correspond to the suspended zone and the debond zone, respectively. The areas of the two zones were measured by the image analyzer to determine the debond radius a and the horizontal suspended distance d. Using the model of Chang *et al.* [10], the strain energy release rate G is calculated by substituting the values of a and d into Chang, Lai, and Dillard's equation mentioned in Section 2.

Since the horizontal suspended distance d changes only slightly as the debond grows, the term $\partial d/\partial a$ appears to be negligible for the cases of the octyl maleimide/styrene copolymer (Fig. 6). Table 1 shows the average q value and debonding rate for copper

Table 1.

Average q value and debonding rate for copper D, E, and F at three different gas pressures

Gas pressure (kPa)	Average q value for copper			da/dt (mm/min)		
	D	E	F	D	E	F
57	0.78	0.71	0.71	0.26	0.21	0.19
68	0.85	0.82	0.81	0.35	0.22	0.21
82.5	0.90	0.86	0.85	0.51	0.26	0.24

Constraint height, h, is 1 mm.

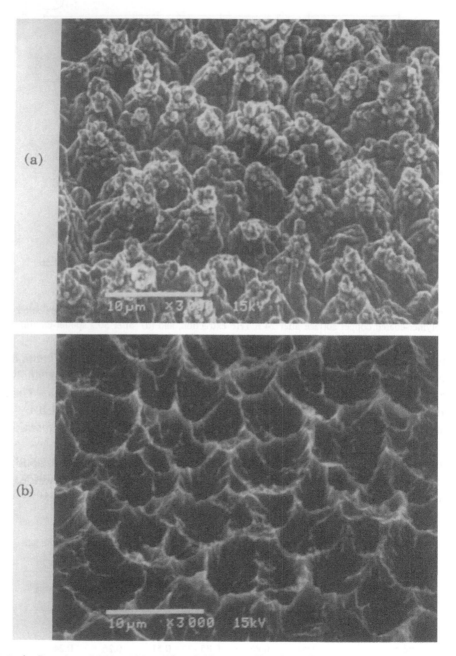

Figure 4. Scanning electron micrographs of (a) copper E and (b) the exposed polymer surface after removing the copper substrate.

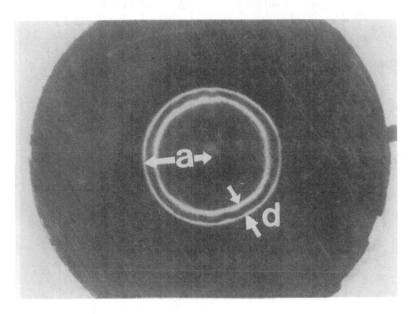

Figure 5. The blister image of an inflated polymer film.

D, E, and F at three different constant gas pressures. In the case of copper surface E, a and d were measured at 57 kPa gas pressure; the correction factor q can be calculated and varies from about 0.7 at the beginning of the test to 0.73 at the end. The average q value of 0.71 was used for calculating the strain energy release rate G at a given pressure. The debond rate of n-octyl maleimide/styrene polymer coatings as a function of the strain energy release rate is plotted for copper surfaces D, E, and F in Fig. 7; the strain energy release rate was independently varied by changing the gas pressure. The error bars in Fig. 7 represent the range of two measurements; the symbols are at the mean. The expected phenomenon is observed, i.e. an increasing debond rate occurs as the strain energy release rate is increased. The larger the slope, the greater the tendency for cracking between the polymer films and the copper substrates. It was found that copper surface D has the highest debond rate, and copper surfaces E and F have similar debond rates at the same desired strain energy release rate. The order of the results is not surprising, since copper D has the smoothest surface, giving little contribution from mechanical interlocking to retard crack growth. The rougher copper E and F would be expected to have higher bond strenghts due to the mechanical interlocking (Fig. 4b) associated with the surface roughness; this is best discussed in terms of the strain energy release rate G required to propagate a given crack.

Table 2 shows a comparison of the G values at 0.5 mm/min debonding rate for different material/copper systems. A G value of 60 J/m^2 at a debonding rate of

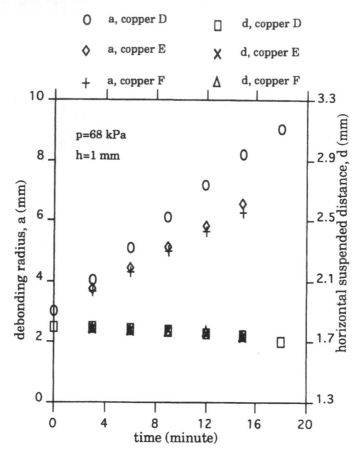

Figure 6. The debonding radius and the horizontal suspended distance vs. time for different polymer-coated copper substrates.

Table 2.
Comparison of the G values (in J/m^2) for different polymer/copper systems at 0.5 mm/min debonding rate

Pressure-sensitive tape on copper [a]	OTMI/ST on copper surface D	4-CMI/ST on copper surface E [b]	Kapton polyimide on copper [c]	Electrolessly plated copper on Kapton polyimide
60	75	> 60	4.4	< 4.0 [d] 70 [e]

[a] Data from Napolitano and Moet [14].

[b] The failure pressure was used to estimate a 'low limit' G value.

[c] Calculated from the fracture mechanics analysis [16] of Chang's data [15], and then using a linear extrapolation as was done by Chang et al. [10].

[d] Calculated from the fracture mechanics analysis [16] of Baum et al.'s data [17], and then using a linear extrapolation as was done by Chang et al. [10].

[e] After treatment of the Kapton® with aqueous NaOH and then with an aqueous solution containing $K_3Fe(C_2O_4)_3$ and $Pd(NH_3)_4Cl_2$ [17].

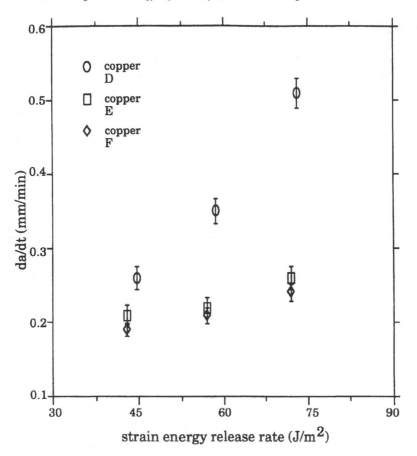

Figure 7. The debonding rate vs. applied strain energy release rate for different polymer-coated copper substrates.

0.5 mm/min for a pressure-sensitive adhesion tape on a copper substrate was reported by Napolitano and Moet [14]. The high G value was attributed to energy dissipative processes (cavitation, flow) during interfacial failure, however. For a similar debonding rate, the G value of OTMI/styrene polymer film on copper D was 75 J/m^2.

An empirical relationship reported by Chang *et al.* [10] shows that the G values from the free membrane blister test, the constrained blister test, and several standard peel tests with different peel angles are all in good agreement with each other. The debonding rates show a linear relationship over a large range of G values. With that result, the peel energy of a DuPont Kapton® polyimide film (prepared by imidization of a polyamic acid solution on the copper substrate [15]) was calculated for comparison with the result of the electropolymerized OTMI/styrene system. A fracture energy (180°-peel test) of 87.5 J/m^2 (calculated from the fracture mechanics analysis of the peel test [16]) at a debonding rate of 10 mm/min (derived from the reported crosshead speed) was chosen. Based on a linear extrapolation, a G value of 4.4 J/m^2 at 0.5 mm/min for the Kapton® polyimide-untreated copper system was obtained.

Following the same procedure, a fracture energy value of less than 4.0 J/m^2 was obtained for the electrolessly plated copper film/Kapton® polyimide (without base hydrolysis of the polyimide) using the data of Baum *et al.* [17]. The G values from these Kapton® polyimide-untreated copper systems are in good agreement with each other, although admittedly the linear extrapolation is not rigorous. In contrast, a G value of 70 J/m^2 was obtained in the same study for a sample of the polyimide that had been treated with aqueous NaOH and then with an aqueous solution containing $K_3Fe(C_2O_4)_3$ and $Pd(NH_3)_4Cl_2$. A G value greater than 60 J/m^2 for the 4-CMI/styrene polymer film on the roughened copper surface E [1a] is also listed in the same table for comparison. The 4-CMI/styrene polymer film adhered well to the copper surface. However, an undesired fracture of the polymer film resulted before debonding from the copper surface began. Because of this fracture, the debonding rate could not be measured.

Representative fracture surfaces of debonded areas from the adherend surfaces are shown in Fig. 8. Fracture surfaces on copper surface D show both exposed smooth copper surfaces and fractured polymer pieces stuck to the copper surface (Fig. 8a). The fractured pieces align parallel to each other but are perdendicular to the debonding direction (left to right). The existence of exposed copper surfaces indicates poor bonding between the copper surface and polymer film. The relatively low G value for copper surface D at a given da/dt is probably due to this fracture pattern. On the other hand, the fracture surfaces of both copper E and copper F demonstrate a different morphology (Figs 8b and 8c). It is difficult to find exposed copper on surfaces E and F. The aligned polymer pieces remaining on the surfaces are perpendicular to the debonding direction (left to right). Both copper E and F surfaces appear to have higher G values due to failure occurring within the polymer matrix.

EDXA elemental analysis of the debonded surface provides additional evidence for the above hypothesis. The debonded surfaces of both the n-octyl maleimide/styrene polymer film side and the metal side of the copper surface D specimens were scanned. The EDXA spectra of the metal side showed that almost half of the area is copper; this was indicated by the relative copper peak height to that of the bare copper surface (Figs 9a and 9b). The gold (Au) element is present from the gold which was deposited to avoid charging of the samples. The elemental analysis of the polymer side showed substantially less copper than on the metal side, although a notable amount of copper was still detected (Fig. 9c). The EDXA scans and scanning electron micrographs strongly suggested that failure was occurring close to the metal region. Similar EDXA spectra were obtained on the debonded surfaces of both the metal and the polymer sides of the copper surface E sample (Fig. 10) and F sample (not shown). In this case, however, the metal side showed a greatly decreased amount of copper (Fig. 10a) and the polymer side showed no sign of copper (Fig. 10b). The small copper signal on the copper side may come from underneath the surface. Since the sampling depth of EDXA is more than 5 μm, the copper electrons can still scatter through a thin top polymer layer. These data tend to support the hypothesis that failure was occurring within polymer matrix.

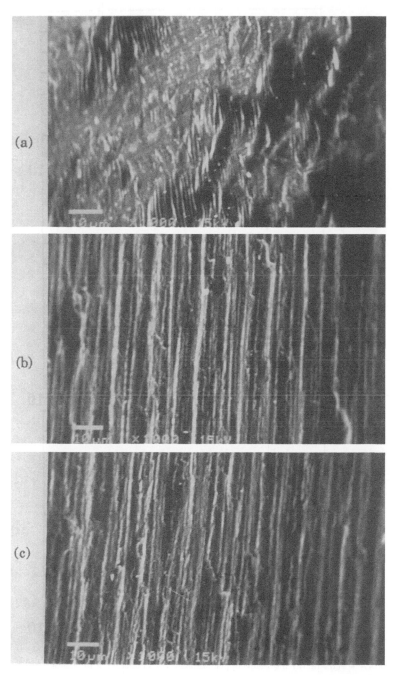

Figure 8. Scanning electron micrographs of the debonded (a) copper D surface, (b) copper E surface, and (c) copper F surface.

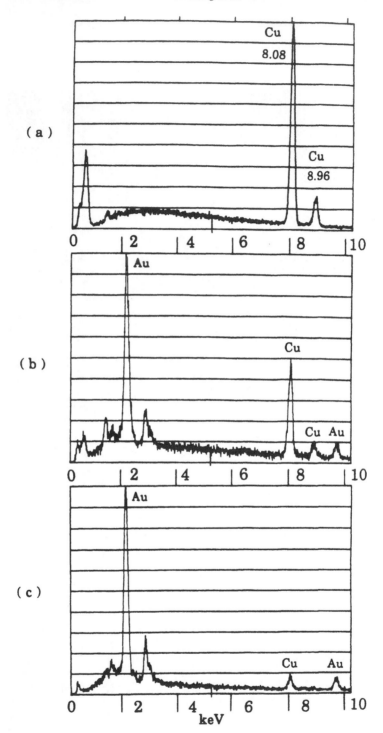

Figure 9. EDXA spectra of (a) the bare copper surface, (b) debonded copper side, and (c) debonded polymer side of copper D.

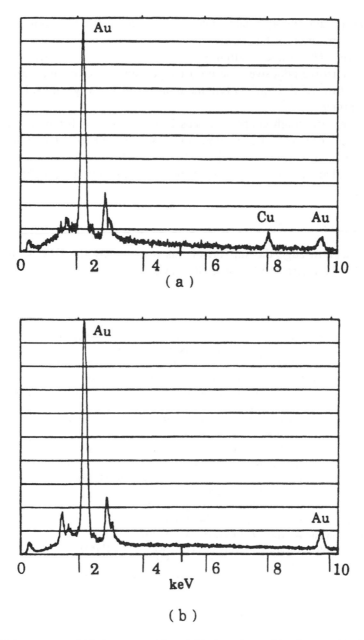

Figure 10. EDXA spectra of (a) the copper side and (b) the polymer side of copper E, after debonding.

4. CONCLUSIONS

A constrained blister test used with an image analyzer was found to be suitable for the characterization of the adhesion (fracture energy) of thin electropolymerized films to supporting materials. The fracture energy in terms of the strain energy release rate was measured by this technique for octyl maleimide/styrene copolymer films.

A G value (strain energy release rate) of 75 J/m² at 0.5 mm/min debonding rate
was found to separation of the n-octyl maleimide/styrene polymer film from a smooth
copper surface. This value is slightly higher than the G value (60 J/m²) reported
for pressure-sensitive adhesive tape on a copper surface, and much higher than the
estimated 4.4 J/m² of Kapton® polyimide imidized from polyamic acid solution onto
the copper surface.

The G values for rough copper surfaces E and F are much higher than the value
for smooth copper D, and EDXA measurements indicated cohesive failure within the
polymer coating on surfaces E and F. The reason for the better adhesion is likely
due to the mechanical interlocking contributed by the surface roughness of copper
surfaces E and F.

Acknowledgements

We wish to acknowledge the support of this research by the IBM Corp., Endicott,
and especially the assistance by Dr C. R. Davis.

REFERENCES

1. (a) J.-L. Liang, PhD Thesis, University of Connecticut, Storrs (1991); (b) J.-L. Liang, J. P. Bell,
   J. O. Iroh and A. Mehta, *J. Appl. Polym. Sci.* **48**, 465–476 (1993).
2. J. O. Iroh, J. P. Bell and D. A. Scola, *J. Appl. Polym. Sci.* **41**, 735–749 (1990).
3. F. S. Teng and R. Mahalingam, *J. Appl. Polym. Sci.* **23**, 101–113 (1979).
4. B. K. Garg, R. A. Raff and R. V. Subramanian, *J. Appl. Polym. Sci.* **22**, 65–87 (1978).
5. B. J. Briscoe and S. S. Panesar, *Proc. R. Soc. London, Ser. A* **433**, 23–43 (1991).
6. Y. Z. Chu and C. J. Durning, *J. Appl. Polym. Sci.* **45**, 1151–1164 (1992).
7. M. G. Allen and S. D. Senturia, *J. Adhesion* **29**, 219–231 (1989).
8. M. J. Napolitano, A. Chudnovsky and A. Moet, *Polym. Mater. Sci. Eng.* **57**, 755 (1987).
9. M. J. Napolitano, A. Chudnovsky and A. Moet, *J. Adhesion Sci. Technol.* **2**, 311 (1988).
10. Y. S. Chang, Y. H. Lai and D. A. Dillard, *J. Adhesion* **27**, 197 (1989).
11. C. K. Sauers, *J. Org. Chem.* **34**, 2275 (1969).
12. Y. H. Lai and D. A. Dillard, *J. Adhesion* **33**, 63–74 (1990).
13. K. L. Mittal, *J. Adhesion Sci. Technol.* **1**, 247 (1987).
14. M. J. Napolitano and A. Moet, *J. Adhesion* **33**, 149–167 (1991).
15. K. P. Chang, Masters Thesis, University of Connecticut, Storrs (1985).
16. A. N. Gent and A. J. Kinloch, *J. Polym. Sci., A-2* **9**, 659 (1971).
17. T. H. Baum, D. C. Miller and T. R. O'Toole, *Chem. Mater.* **3**, 716 (1991).

Adhesion Measurement of Films and Coatings, pp. 265–281
K. L. Mittal (Ed.)
© VSP 1995.

An experimental partitioning of the mechanical energy expended during peel testing

RICHARD J. FARRIS[1, *] and JAY L. GOLDFARB[2]

[1]*Polymer Science and Engineering Department, University of Massachusetts, Amherst, MA 01003, USA*
[2]*Raychem Corporation, 300 Constitution Drive, Menlo Park, CA 94025-1164, USA*

Revised version received 25 March 1993

Abstract—Peeling of polyimide coatings bonded to aluminum substrates was analyzed from a thermodynamic perspective with the intent of determining how the energy expended in separating the bonded materials is consumed. The mechanical work expended and the heat dissipated during peeling were simultaneously measured using deformation calorimetry. The surfaces exposed by peeling were analyzed by electron microscopy and electron spectroscopy. The thermodynamics of tensile drawing for polyimide were studied using deformation calorimetry and thermomechanical analysis. When polyimide coatings were peeled from aluminum substrates at a peel angle of 180°, almost all of the mechanical energy was consumed in propagating the bend through the coating being peeled. The fraction of peel energy dissipated as heat was 48 ± 1.3% and nearly all of the remainder was stored as latent internal energy in the peeled polyimide. When the bend is propagated through aluminum, which has a limited capacity to store latent internal energy, 97–100% of the mechanical energy is dissipated as heat.

Keywords: Adhesion; peel energy; calorimetric measurements; thermal dissipation; inelastic deformation; energy balance.

1. INTRODUCTION

In this work, adhesion is analyzed in terms of a global thermodynamic energy balance. From a global thermodynamic perspective, a complete detailed analysis of the micro-mechanisms and fracture mechanics of separating bonded materials is unnecessary because the intent is purely to determine how the mechanical energy expended in separating the bonded materials, the adhesion energy, is consumed. Globally partitioning the adhesion energy shows which aspects of bonded structures are most important in determining the total mechanical energy required to separate the structures.

The bonded systems were separated by peeling, which is used extensively to characterize the adhesion strength of flexible tapes and coatings. In a peel test, the force

---

*To whom correspondence should be addressed.

required to pull a flexible film away from a substrate, to which it is bonded, is measured. In the absence of energy dissipation due to plasticity or viscoelasticity, the energy required to peel the film from its substrate is a direct measure of the interfacial adhesion. However, for polyimide coatings, almost all of the work expended in peeling is consumed by dissipative processes accompanying deformation of the test sample. The peel test is extremely sensitive to energy dissipative mechanisms. In peel tests, energy dissipation can occur throughout the bulk of the test sample, and bond strengths can be measured which are outrageously high when compared with values obtained by other test methods.

The experimental approach to decomposing the adhesion energy was to simultaneously measure the mechanical work expended and heat dissipated during peeling of bonded materials. This accomplished partitioning of the mechanical work into heat and an internal energy change which is the difference between the internal energy of the peeled and bonded materials. If mechanisms of energy dissipation other than heat flow are negligible, for example, acoustic and light emission, conservation of energy requires that the mechanical work done on the sample during peeling, which does not appear as heat, raises the internal energy of the test specimen. While much of the mechanical energy expended during peeling is dissipated as heat, not all of the mechanical work which is consumed by dissipative mechanisms during peeling appears as heat. Most of the mechanical work of peeling, which does not appear as heat, is also consumed by dissipative processes but is stored as latent internal energy in the peeled materials rather than dissipated as heat.

The energy of peeled materials may be raised by a substantial amount when glassy polymeric films are deformed during peeling. Peeling causes molecular rearrangements in regions of the film exposed to bendings strains similar to that which occurs when the film is subjected to homogeneous tensile or compressive deformation. These deformations are frozen into the deformed film, leaving it in a high-energy thermodynamic state.

Energy is consumed in creating surfaces exposed by peeling, raising the internal energy of the test specimen. This is the energy required to reversibly separate the bonded materials, commonly referred to as the thermodynamic work of adhesion. In cases where the bond fails at the interface, it is logical to attempt to relate the mechanical peel energy to the thermodynamic work of adhesion, which reflects the energetic contribution of the intrinsic adhesion forces acting across the interface. Relating the thermodynamic work of adhesion to bond strengths measured by mechanical tests is complicated because these measurements contain indeterminate contributions from dissipative energy losses in the coating and substrate. Although the intrinsic adhesion forces acting across the coating–substrate interface may affect the bond strength, they are usually completely obscured by other contributions.

Other researchers have applied energy balance analysis to peel test data. The primary objective of developing an energy balance for peeling has been to calculate the energy consumed by plastic and viscous dissipation and thereby determine the intrinsic adhesion from peel data by subtracting the dissipative contribution. The dissipation comes from two sources. One is the near tip stress analyzed by Kaelble [1] and Crocombe and Adams [2, 3], who formulated a numerical finite element solution of the elastoplastic peel problem, calculating the stress distribution ahead of the interfacial crack. The other is the plastic deformation caused by bending strains imposed on the film during peeling. For peeling thin, stiff films strongly adhered to rigid substrates, plastic deformation induced by bending usually predominates. Therefore, energy balance approaches to peeling have concentrated on estimating the energy consumed by plastic and viscous processes due to propagating the bend in a film. Gent and Hamed [4, 5] calculated the energy required to propagate a bend in an ideal elastic–plastic strip using elementary beam bending theory. Kim *et al.* [6–9] have developed a more detailed analysis of bending dissipation during peeling and have applied it to ideal elastic–plastic, strain-hardening, and linear viscoelastic materials, all on elastic substrates. Their analysis also includes the effects of reverse plastic bending required to straighten the peeled film.

Complete stress analysis is almost impossibly difficult for peeling of strain-hardening or viscoelastic materials. A great deal of information can be gained from these studies; for example, they have provided explanations for the effects of test conditions such as the thickness of the coating layer and the peel angle. Direct measurements of the energy dissipated during peeling are presented in this work, eliminating the need for such complex analysis. Regardless of the method of obtaining the dissipation, measurements, or calculations, peel energy data may never be useful for determining intrinsic adhesion. The accuracy of the intrinsic adhesion determined from the peel energy is only as good as the accuracy to which the dissipation is known. In most cases of practical interest, the peel energy is at least 100 times larger than the intrinsic adhesion, thus completely obscuring it. Alternatively, the intrinsic adhesion energy may be estimated from peeling by extrapolation of the peel energy to experimental conditions where dissipation is negligible, for example, extremely low peel rates. Accurate extrapolation of the small energies associated with intrinsic adhesion requires data at very low peel forces. The large magnitude of the peel energy at easily accessible experimental conditions and the inaccuracies associated with measuring small peel forces make this a difficult proposition.

2. PEELING OF POLYIMIDE-COATED ALUMINUM

The polyimide used in this study, PMDA-ODA, is not soluble and is thus applied as a poly(amic acid), DuPont Pyralin 2540, polymerized from pyromellitic dianhydride and oxydianiline, which is soluble in N-methyl-2-pyrrolidone (NMP). The peel test samples were prepared by spin-coating a solution of polyamic acid onto 100 μm thick aluminum substrates using a Headway Research EC101 photoresist spinner. The aluminum substrates were etched in chromic acid, bonded to glass plates with a high temperature adhesive to prevent curling, and solvent-wiped with acetone prior to coating. The spun-on films were then thermally imidized by baking in several steps with a final temperature of 360 °C, totally converting the polyamic acid to polyimide.

To produce thick coatings, additional layers were applied on top of partially imidized polyamic acid coatings which had been cured to 150 °C. For coatings spun at 1000 rpm, the thickness was approximately 15–20 μm per layer. Coatings were prepared having thicknesses ranging from 28 to 95 μm. Internal tensile stresses in the films limited the coating thickness to less than 120 μm, a thickness at which stress-driven spontaneous delamination occurred during post-curing cooldown. After the final curing step, the aluminum was easily separated from the glass plates and peel test samples were prepared by cutting strips from the polymer/metal sheets. When removed from the glass, the sample would curl into the polyimide side, indicative of tensile stress in the polyimide layer.

Two types of calorimeter peel test sample were fabricated from the polyimide/aluminum sheets. A photograph of both is displayed in Fig. 1. A rigid steel wire was bonded to the back of the aluminum strip in the top sample, which prohibited the aluminum from bending while a 180° bend was propagated through the polyimide film to peel it from the aluminum. A rigid steel wire was bonded to the back of the polyimide film in the bottom sample, prohibiting it from bending while the aluminum was peeled from the polyimide.

The technique of deformation calorimetry was adapted to measure the mechanical work and heat of peeling. Deformation calorimetry is a technique which can measure the heat and work associated with deforming a solid material. If processes for transferring energy out of a sealed system other than heat flow, for example, acoustic and light emission, are negligible, the internal energy change associated with the deformation is also measured. The instrument used in this work was designed and built by Lyon and Farris [10]. The instrument operates by measuring pressure changes in a gas surrounding the sample, which is contained in a sealed chamber, relative to a sealed reference chamber. It is designed such that any change in differential gas pressure is only due to the emission or absorption of heat by the sample. The area under the time-dependent differential pressure between the calorimeter cells is found to be proportional to the total heat absorbed or emitted during a process. A mechanical tester

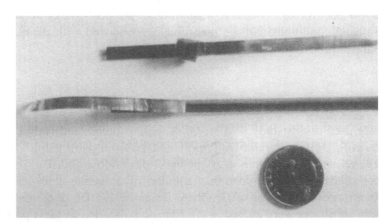

Figure 1. Calorimeter peel test samples, fabricated from the polyimide/aluminum sheets, are shown next to a quarter to indicate relative size. A rigid steel wire was bonded to the back of the aluminum strip on the top sample, which prohibited the aluminum from bending while a 180° bend was propagated through the polyimide film to peel it from the aluminum. A rigid steel wire was bonded to the back of the polyimide film in the bottom sample, prohibiting it from bending while the aluminum was peeled from the polyimide.

equipped with a load cell and a variable displacement transducer is connected to the sample in the sample cell by a tungsten wire passing through a gas-tight mercury seal in the top of the sample chamber. The mechanical work is calculated from the force and displacement data. Details of the operation and construction of the calorimeter may be found elsewhere [10, 11]. A minimum heat flow of 84 μW is required to produce a pressure deflection equal to twice the signal-to-noise ratio and the minimum detectable heat is about 0.4 mJ [10].

Peeling experiments on polyimide/aluminum in the deformation calorimeter have been reported previously [12]. During peeling in the calorimeter at a steady peel rate, heat is evolved and the pressure in the sample chamber rapidly increases. The peeling proceeds in a steady continuous manner with relatively little fluctuation in the peel force. After peeling, the peeled film is unloaded, negating the effect of thermal expansion and releasing the nominal elastic energy in the peeled film. The rate of thermal dissipation and work expenditure appear to be constant during peeling, increasing proportionally with the peeled area. Table 1 gives mean values of the work and heat, the number of measurements, and the standard deviations for peeling 63 μm polyimide films from rigid aluminum substrates in the deformation calorimeter.

The standard deviations in the mean heat values range from 3 to 17% of the mean. While large, the source of these deviations is not necessarily the method of measurement, but variability in the peeling experiments as reflected by the standard deviations of the mean work values which are almost equal to those of the heat. The systematic errors of the work measurements are known to be smaller than those of the heat. The combined effect of all systematic errors inherent in the heat measuring technique is best evaluated from calibration experiments done with wire heating elements, and the systematic errors in the work measuring technique are best evaluated from calibration experiments done by moving a weight through a known distance. The systematic errors in the heat and work measured with the deformation calorimeter, where the magnitudes of the work and heat are similar to those in the peeling experiments, are estimated to be less than 5% and 1%, respectively. The large standard deviations in the experimental populations are indicative of random errors due to the poor repeatability of peel experiments.

The heat and work of peeling aluminum from rigid polyimide, as shown in the bottom sample in Fig. 1, were measured using the deformation calorimeter at several peel rates. The thermodynamic data are presented in Table 2.

Table 1.

Measurements of the work, W, and heat, Q, of peeling 63 μm polyimide films from rigid aluminum substrates in the deformation calorimeter

Peel rate (cm/min)	No. of peels	W (J/m^2)	Q (J/m^2)
0.68	4	640 ± 50	−290 ± 50
2.71	5	660 ± 30	−330 ± 40
6.67	6	640 ± 40	−330 ± 50
13.4	3	620 ± 10	−300 ± 10

Table 2.

Measurements of the work, W, and heat, Q, of peeling 100 μm aluminum films from rigid polyimide substrates in the deformation calorimeter

Peel rate (cm/min)	No. of peels	W (J/m^2)	Q (J/m^2)
0.68	3	1050 ± 100	-1000 ± 90
2.71	5	800 ± 70	-740 ± 80
6.67	4	803 ± 50	-830 ± 90
13.4	4	900 ± 90	-870 ± 90

Table 3.

The work W, heat Q, internal energy change ΔU, and ratio of heat to work Q/W of peeling determined from combining measurements at all peel rates

Peeled layer	No. of measurements	W (J/m^2)	Q (J/m^2)	ΔU (J/m^2)	Q/W (%)
Polyimide	18	625 ± 9	-302 ± 9	323 ± 13	48 ± 1.3
Aluminum	16	847 ± 35	-852 ± 42	-5 ± 55	100 ± 2.7

The thermodynamic data for peeling polyimide or aluminum does not show. any significant rate dependence. The ratio of the highest to lowest peel rates is only 20. This small change in deformation rate would not significantly affect the viscoelastic response of a glassy, elastic polymer like polyimide unless the combination of test rate and temperature caused the polymer to behave as if it were in the vicinity of its glass transition. It is justified to combine the thermodynamic measurements, done at different peel rates, to obtain a larger experimental population and, therefore, more precise mean values for the thermodynamic quantities. The mean of the combined population was obtained by weighting the individual mean values from each peel rate by the reciprocal of the standard deviations squared, and the standard deviation of the combined population was computed from the sum of the reciprocals of the individual standard deviations squared. The resulting mean work and heat values are presented in Table 3. The internal energy change and ratio of work to heat were computed from the mean values for work and heat. The standard deviations of these quantities were computed from the variances using the propagation of errors formula.

The peeled polymer dissipates approximately half of the work of peeling as heat. In contrast, the peeled metal dissipates all of the work of peeling as heat. These samples were cut from the same plate and their interface chemistry is identical. If the locus of separation and the surfaces exposed by peeling are the same for both types of peel experiment, the true adhesion strength of the interface should also be equal.

X-Ray photoelectron spectroscopy (XPS) of the peeled side of the film and substrate surfaces exposed in the polyimide/aluminum peel tests was used to provide information on the elemental compositions at the surfaces exposed by peeling. From this information, the molecular composition of the failure surfaces and locus of failure were determined. A Perkin Elmer PHI 5100 ESCA instrument equipped with a

Mg K_α source was used. XPS has been widely applied to this type of study and has been used previously by other investigators [13–15] to determine the elemental composition and interface chemistry of polyimide coatings peeled from metal and semiconductor substrates.

The locus of failure and surface compositions are nearly identical for the metal and polymer exposed when the bend is propagated through the aluminum or the polyimide during peeling. The XPS data show that separation occurs in the polyimide film close to the interface such that the organic layer retained on the metal is approximated, from the electron sampling depths, to be of the order of 10 nm [11]. This thin layer is suggestive of a boundary layer in the film, near the interface, with inferior strength compared to the bulk film. The same conclusion was reached by Anderson *et al.* [13] in their XPS study of PMDA-ODA peel adhesion. While the surfaces exposed by peeling may appear to be smooth, they are actually microscopically rough such that the retention of a uniform layer of polyimide on the aluminum is not realistic. The surfaces exposed by peeling polyimide-coated aluminum were examined using electron microscopy. A JEOL scanning electron microscope was used to examine the surfaces which were coated with gold. Electron micrographs showed that the surface of the peeled aluminum consists of a carpet of broken polyimide fibrils attached to the aluminum surface [11]. On the exposed surface of the aluminum side of the peeled samples, XPS detected that the surface contained more polyimide than aluminum. As the sampling depth increased, the ratio of aluminum to polyimide increased. This may reflect the distribution of the distances that the polyimide fibrils extend normal to the aluminum surface. The XPS data indicates that the projection of the polyimide fibrils, normal to the aluminum surface, is of the order of 10 nm. The principal reason for examining the surfaces exposed by peeling with microscopy and XPS was to determine whether there were any major differences between the surfaces exposed by propagating the bend through the polyimide or aluminum during peeling. No obvious differences in the peeled surfaces were detected.

It has now been shown that the peeled polymer dissipates approximately half of the work of peeling as heat and, in contrast, the peeled metal dissipates nearly all of the work of peeling as heat. The interface chemistry of these samples has been proven to be identical, as is the locus of failure and the appearance of the surfaces exposed by peeling. If the interface and exposed surfaces are identical the true adhesion strength of the interface must also be equal. The true adhesion strength should be considered to be the reversible work of creating the surfaces exposed by peeling, the thermodynamic work of adhesion and the irreversible work expenditure confined to the vicinity where the actual separation of the layers occurs during peeling. Irreversible work expended in deforming the test sample, outside of the vicinity of separation, is not part of the true adhesion. In 180° peeling, irreversible work is expended in propagating the bend in the peeling film. For elastic bending, the bending energy is released when the film is unloaded. When the force required to separate the bonded layers is large, the peeled film may exceed its elastic limit curvature during peeling and mechanical energy will be irreversibly consumed by plastic and viscous processes associated with propagating the bend in the peeling film. As a consequence of inelastic bending, the peeled polymer is tightly curled as seen in the top sample shown in Fig. 1. The peeled metal also shows visible evidence of plastic deformation, but it has a higher modulus and lower yield stress, resulting in little residual curvature. The reinforced substrate does not inelastically deform. The only difference between peeling polymer from metal and metal from polymer is that inelastic deformation occurs in the bulk of the polymer film in the former and in the bulk of the metal in the latter. In both cases, separation occurs cohesively in the polyimide, so dissipation and irreversible work expenditure will occur in a thin layer in the polyimide, compared to the film thickness, near the surfaces exposed by peeling.

During 180° peeling with the bend propagated through polyimide, the actual observed bending radius of the polyimide approaches the film, greatly exceeding its elastic limit curvature. In this case, the maximum tensile strain in the bent film is one half the thickness of the film divided by the radius of the film or 0.5 at the outside edge, and the maximum compressive strain in the bent film is −0.5 at the inside edge. The yield strain of polyimide films in tension is approximately 0.01. Thus, inelastic deformation due to bending occurs in approximately 98% of the film volume during peeling. The surfaces exposed by peeling are nearly smooth, so that the irreversible work associated with the advance of the crack probably occurs in a layer of the order of the height of the retained polyimide fibrils on the aluminum side which is 10 nm. In contrast, irreversible work associated with propagating the bend in the peeling film, when the bending radius approaches the film thickness, will occur throughout the volume of the film. This is approximately 10 000 times greater than the volume of material in which irreversible work associated with the advance of the peel crack occurs.

When polyimide film was peeled from rigid aluminum, 46.7–49.3% of the mechanical energy consumed by peeling was dissipated as heat. In contrast, when aluminum was peeled from rigid polyimide, 97.3–100% of the mechanical energy was dissipated as heat. Separation occurs cohesively in the polyimide. When polyimide is peeled from aluminum, the aluminum does not inelastically deform and most of the thermal dissipation is caused by inelastic deformation associated with propagating the bend in the polyimide film or with advancing the crack through the polyimide layer. When the aluminum is peeled from polyimide, energy is dissipated due to inelastic deformation caused by propagating the bend in the aluminum and by advancing the crack through the polyimide. The volume of polyimide being inelastically deformed due to advancing the crack when aluminum is peeled is estimated to be 10 000 times smaller than the volume of aluminum film inelastically deformed due to bending. Therefore, the majority of energy dissipation, when aluminum is peeled, occurs in the aluminum. The mechanism of energy dissipation in aluminum is plasticity, which is a flow process and does not change the internal energy. When aluminum is peeled from polyimide, the internal energy of the peeled and bonded samples is identical within experimental error.

When polyimide is peeled from aluminum, 52% of the mechanical energy expended in peeling is consumed by raising the internal energy of the peeled sample. Two of the possible mechanisms for creating the internal energy of the peeled samples appear to be energetically insignificant. These are the creation of the surfaces exposed by peeling and the release of elastic energy from the stressed polyimide film.

The mechanical energy expended in creating the surfaces exposed by peeling would raise the internal energy of the peeled samples. The locus of failure and the appearance of the fracture surfaces were shown to be identical for both types of peel experiments, so that the work of creating the surfaces exposed by peeling should be the same. The fracture surfaces were found to be smooth and consisted of polyimide. Therefore, the change in internal energy of the peel samples due to surface creation can be estimated to be approximately twice the surface free energy of PMDA-ODA polyimide. The surface free energy of Kapton PMDA-ODA has been determined from liquid contact angle measurements to be 0.05 J/m^2 [16]. This resulting change in internal energy, 0.1 J/m^2, is nearly four orders of magnitude smaller than the mechanical energy of peeling and cannot be resolved within the precision of the peeling experiments. The bonded polyimide contains residual tensile stresses which disappear when it is peeled from the aluminum. The stresses become apparent when the samples are removed from the glass plates and they curl into the polyimide side. The elastic energy associated with these stresses is released upon peeling, reducing the internal energy of the sample, not increasing it. Therefore, elastic energy associated with the residual stresses cannot contribute to the observed internal energy increase.

The key to resolving the cause of the large internal energy change in peeling polyimide is the difference in the thermodynamics of peeling polymer from metal and metal from polymer. Although the mean values for the work and heat of peeling are equal when the bend is propagated through the aluminum, it is statistically possible and logically probable that the internal energy of the sample increases with peeling. The internal energy could be stored as latent energy in the deformed aluminum, in the deformed polymer in the region where cohesive failure occurred, and as the free energy of the surfaces exposed by peeling. The capacity of the aluminum to store deformation energy as latent internal energy is minimal when compared with the polymer. Regardless of the mechanism, the value of the internal energy change for peeling aluminum from polyimide is likely to be less than the standard deviation of the measurement, or 55 J/m^2. When the polyimide is peeled, the internal energy change is 323 ± 13 J/m^2. Since the chemistry of the interface, location of separation, and the appearance and composition of the fracture surfaces are identical for both types of sample, the reversible work consumed in breaking the bonds must be the same. Therefore, no more than 55 J/m^2 can be consumed by thermodynamically reversible processes, such as creating surfaces, when the polyimide is peeled. Thus, the energy consumed by irreversible processes when the polymer is peeled is nearly double the energy actually dissipated as heat. The remainder must be stored in the peeled polymer as latent internal energy.

3. THERMODYNAMICS OF POLYIMIDE DEFORMATION

Evidence of the ability of polyimide to convert mechanical deformation energy into latent internal energy is provided by the behavior of polyimide in tensile deformation. Most of the mechanical work expended in peeling the polyimide-coated aluminum is consumed by bending. The inelastic deformation due to bending during peeling is analogous to homogeneous tensile or compressive deformation.

For interpreting thermodynamic data on polymer deformation, it is convenient to classify the molecular events which take place in polymers in response to an applied stress as energy elastic, entropic elastic or pure flow processes. Energy elastic processes result in latent internal energy storage equal to the work which they consume. Isothermal entropy elastic processes result in heat flows equal to the work consumed, resulting in no internal energy storage, but the entropy of the system changes. Pure flow processes irreversibly dissipate all of the energy consumed as heat and do not change the internal energy of the system.

Polymer molecules may respond to stress by distortion of covalent bond angles or by changes in covalent bond lengths. These are energetically elastic processes. Rotation about covalent bonds may also occur, changing the conformation of the molecules in the chain. Through conformational changes, the end-to-end dimensions of the molecules may change in response to an applied stress. The extended molecules may also orient in the direction of the applied stress. Conformational changes are entropically elastic processes and do not directly contribute to the latent internal energy change, but the molecular motions may cause changes in morphology or crystallinity which will cause changes in the intermolecular and intramolecular interactions which are energetically elastic. Inelastic volume dilatation during deformation is indicative of structural reorganization, which results in changes in the intramolecular and intermolecular interactions within the solid.

The primary forces which act on the macromolecules to keep them frozen are the intermolecular interactions. These include hydrogen bonding, van der Waals forces, and dipole–dipole interactions. These forces create secondary bonds which are weak relative to primary covalent bonds. These secondary bonds can also deform in response to an applied stress without breaking, storing latent internal energy. Physical anchorage points between the chains due to entanglements, crosslinks, or crystal domains may also restrict motion. When sufficient stress is applied to cause yielding, the elastic energy in some of the chain segments, between anchorage points, is sufficient to cause breakage of secondary bonds, facilitating motion. These bonds reform when the stress is removed, leaving the polymer frozen in the deformed state. The process of breaking and reforming secondary bonds conserves energy because the broken and reformed bonds have identical energies, but the elastic energy in deformed chain segments is dissipated as heat by the thermodynamic cycle of stretching and retracting of chain segments to statistically more favorable dimensions. Upon recoil, the elastic energy in a stretched chain would be released. Some of this energy should be irreversibly dissipated as heat by frictional forces as the broken chains and those entangled with them slip by other chains into lower stress positions. The frictional dissipation would not change the internal energy of the material. If there are no frictional forces restricting the recoil of the entropically stretched chain segments after the bonds break, no work is done by the chains during recoil. In this case, the extension, scission, and retraction of single chains during fracture could be considered as adiabatic stretching followed by free adiabatic recoil. An amount of work, ΔW, is done on the sample during stretching. If the process is adiabatic, the heat flow,

ΔQ, is zero and the internal energy of the system increases by ΔW, resulting in a temperature rise. During adiabatic free recoil, no work is done by the system and there is no heat flow. The temperature and internal energy of the system would not change during recoil. Successive adiabatic stretch, break, and retract events would effectively act as an internal energy pump, progressively increasing the temperature and internal energy of the system. While the molecular events in a polymeric material undergoing fracture can be considered as adiabatic on a time scale commensurate with the deformation and fracture of single chains, the system is not truly adiabatic and will eventually attain thermal equilibrium with its surroundings dissipating the internal energy. Therefore, polymeric materials can dissipate mechanical energy in the absence of frictional forces.

The energy elastic mechanisms can account for large latent internal energy changes when polymers are deformed. The distortion of primary covalent and secondary bonds is primarily responsible for the latent internal energy changes. Viscous flow is not required to produce thermal dissipation during the post-yielding deformation of glassy polymers. Thermal dissipation could be caused by decreasing the entropy of the solid polymer. If viscous flow were minimal, it would suit the exceptional ability of these materials to recover their original shape when heated. Dimensional recovery would be facilitated by the application of thermal energy to the system by increasing the entropically elastic restoring force of stretched molecular chain segments between anchorage points while providing sufficient activation energy to break some of the secondary bonds which are preventing the molecules from returning to their undeformed, thermodynamically more stable dimensions.

The molecular mechanisms of yielding in PMDA-ODA are a combination of entropy and energy elastic processes which are consistent with thermal dissipation, latent internal energy storage, and dimensional recovery upon heating. Figure 2 shows the work, heat, and internal energy change measured with the deformation calorimeter

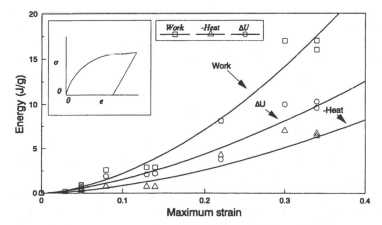

Figure 2. The work, heat, and internal energy change ΔU were measured during uniaxial drawing of spun-cast polyimide film in the calorimeter at a strain rate of 0.011 s^{-1}. The films were drawn to the maximum strain and then the force on the samples was immediately retracted. The loading–unloading path is illustrated in the inset.

during uniaxial drawing of spun-cast polyimide films from which the aluminum sub-strate was peeled. The samples were drawn in the deformation calorimeter at $25\,^\circ C$ and a constant strain rate of $0.011\ s^{-1}$ to a maximum strain and then retracted im-mediately until the force on the samples was removed. The loading–unloading cycle is illustrated by the approximate stress–strain curve in the inset of the figure. Each set of work, heat, and internal energy change points on the graph corresponds to an individual experiment done with a different sample. At extensions beyond the yield strain, more than half of the energy under the stress–strain curve is stored in the deformed material. At a strain of 0.34, the average ratio of heat to work is 0.4.

Drawn polyimide exhibits remarkable dimensional recovery when heated. Polyimide drawn at room temperature begins to shrink at temperatures $300\,^\circ C$ below the glass temperature and almost recovers to its original dimensions when held at $300\,^\circ C$ for 30 min [11]. Analogously, the tightly curled peeled film recovers to a nearly flat state when heated to $300\,^\circ C$ for 30 min. Therefore, it is likely that the heat production measured during tensile drawing of polyimide is caused by reductions in entropy, not flow processes.

When the deformation of a polymeric material results in fracture, as in peeling, the possibility exists that changes in the internal energy of the system can result from the breakage and reformation of chemical bonds. When a polymeric sample undergoes fracture, a new surface is created by a combination of chain scission, requiring the severance of primary (covalent) bonds, and chain pullout, requiring the severance of secondary (van der Waals or hydrogen) bonds. Breakage of secondary bonds will be followed by reformation of similar bonds in the fractured sample, producing no change in internal energy. In contrast, broken covalent bonds will produce reactive radicals which could recombine with many different species, producing internal energy changes. The extent to which each type of bond is broken depends on the polymer being tested and the test conditions. Highly crosslinked polymers cannot be fractured without the breakage of primary bonds. In the fracture of thermoplastics, it may be possible for molecules to slide past one another by breaking secondary bonds only. Chain entanglements and crystalline regions in semi-crystalline polymers function as physical crosslinks in high molecular weight thermoplastics, providing sufficient anchoring of the chains to require some breakage of primary covalent bonds during fracture. However, the density of chemical crosslinks in a thermoset polymer is much higher than that of the physical crosslinks in a thermoplastic. Therefore, primary bond breakage is far more important in thermoset fracture.

Primary covalent bonds will be in a highly stressed state before breaking. Chain breakage should facilitate molecular reorganization within the polymer, producing latent internal energy changes. It would be impossible to separate internal energy changes due to covalent bond breakage, measured in the deformation calorimeter, into chemical bond energy and molecular rearrangement components. However, the local stress at the molecular level which would cause secondary bond breakage is probably small compared with that required for breaking primary covalent bonds. Therefore, the molecular rearrangement resulting from the breakage of a single primary bond would certainly be more extensive than that resulting from the breakage of a secondary bond. Latent internal energy changes in the range of $2–10\ J/g$ are measured for drawing, not fracture, of thermoplastics, where bond breakage is confined to secondary bonds. For comparison, the chemical energy needed to break primary bonds during the fracture of highly oriented Nylon 6 fibers has been estimated from ESR data to be 0.1 J/g [17]. For the scission of a covelant bond, it would seem likely that the latent internal energy change associated with reorganization of the material following breakage would obscure the change in chemical bond energy.

Plastic deformation in polyimides, being commercially important polymers, has been extensively studied [18, 19]. PMDA-ODA polyimide is composed of rigid monomer units 18 Å long interconnected by a flexible diphenyl ether linkage [20]. The rigid monomer units have been found to rotate freely around the ether linkage. Bond angle distortions about the imide nitrogens are also possible. However, the long rigid monomer and the limited mobility of the polymer create an unusual structure for the solid polymer. Films of the material are not crystalline but are highly ordered. The structure of PMDA-ODA in the bulk can best be described as smectic ordering, where there is lateral alignment of the chain segments with the positions of the phenyl ether linkages coordinated [19]. For thin polyimide films, prepared on substrates, the polymer chain axes are preferentially aligned parallel to the plane of the substrate. Polyimide is tough and can be elongated 50–70% before break at room temperature. Ressell and Brown [19] have studied PMDA-ODA deformation with X-rays. The X-ray studies have shown that PMDA-ODA exhibits a conformational change, becoming extended in the direction of stretching and contracting parallel to it. Bundles of chains ordered in a smectic manner orient as a unit maintaining the lateral alignment of the chain segments. If the stretching direction is parallel to the chain axis, then the projection of the monomer unit onto the chain axis can increase by bond angle distortions at the imide nitrogens and at the diphenyl linkages. Before substantial bond angle distortion can occur, PMDA-ODA chains must orient in the stretching direction. X-Ray studies provide evidence of such orientation. The observed change in d-spacing, parallel to the stretching direction, for elongations of 70% is 1.1 Å. From theoretical calculations, the energy required to distort the bond angles sufficiently to cause this change in length is 3–3.5 kcal/mol or 31.5–36.8 J/g [19]. The total area under the stress–strain curve of PMDA-ODA films at an elongation of 70% is approximately 70 J/g. Upon release of the applied stress, only a small amount of dimensional recovery is observed, showing that the internal deformations are frozen into the material. The bond distortional energy is approximately half of the total mechanical work of deformation. Bond angle distortion is an energetically eleastic process, so that the work consumed by it would remain in the material as stored latent internal energy. The mechanical work expended in changing the conformations of the polymer molecules would produce most of the heat when the material is deformed. Deformation calorimetric measurements show that approximately one half of the mechanical energy input is dissipated as heat and the other half is stored as latent internal energy in the deformed material. The recovery behavior of the stretched PMDA-ODA fits the deformation mechanism well. Substantial length recovery is observed starting 300°C below T_g. Elastic energy in covalent bond angle and length distortion would be fully recovered when the thermal energy in the system was sufficient to break the secondary bonds, freezing the system in the deformed state. The entropically stretched polymer chains would also retract to statistically more favorable dimensions. The mechanisms of deformation proposed by Russell and Brown [19] are qualitatively and quantitatively consistent with the deformation calorimetric and thermomechanical data for PMDA-ODA films.

The experimental and theoretical evidence presented in this section leaves little doubt that almost all of the mechanical energy expended in separating polyimide-coated aluminum by peeling, including that which is manifested as an internal energy change, is consumed by inelastic deformation caused by propagating the bend in the peeling film. Furthermore, almost all of the mechanical peel energy which is not dissipated as heat when the polymer is peeled is stored as latent internal energy in the peeled polymer. An indistinguishable amount of the mechanical energy expended in peeling is consumed in creating the surfaces exposed in peeling and it is therefore impossible to accurately determine the intrinsic adhesion from peel test data for this system.

4. SEPARATION BY STRESS-DRIVEN DELAMINATION

If most of the mechanical energy required to separate polyimide-coated aluminum by peeling is consumed by inelastic bending in the bulk of the sample, there must be a way to separate the layers with less energy. The bonded polyimide contains residual tensile stresses which develop upon cooling from elevated cure temperatures due to the thermal expansion mismatch between the polyimide coating and aluminum/glass laminate substrate. A film or coating bonded to a substrate and under a state of residual biaxial tensile stress will delaminate if the elastic energy from the residual stress is high compared with the adhesion. Delamination is likely to initiate from an edge or a crack in the film because shear stresses are concentrated near the intersection of the coating–substrate interface and the edge or crack. The estimated elastic energy per unit area, U, in a film with an equal biaxial stress state is

$$U = \frac{t\sigma^2}{E}(1 - \nu), \tag{1}$$

where t is the film thickness, σ is the stress, E is Young's modulus, and ν is the Poisson ratio. For a given film stress, the elastic energy will increase with the film thickness. Polyimide coatings thicker than 120 μm spontaneously delaminated from the aluminum, which was still bonded to the glass plate, during post-cure cooldown. The maximum strain energy in the 120 μm film is estimated to be 23 J/m^2 [11].

Figure 3 shows the peel energy for polyimide films of varying thickness peeled from rigid aluminum substrates at 180°. The films were prepared as described in Section 2. While the stored elastic energy is small compared with the peel energy, it is sufficient to cause stress-driven delamination. Extrapolating the peel energy to the thickness at which spontaneous delamination occurs in Fig. 3 predicts a peel energy of approximately 500 J/m^2. Residual stress-driven delamination requires less energy than peeling because there is much less plastic deformation than with peeling.

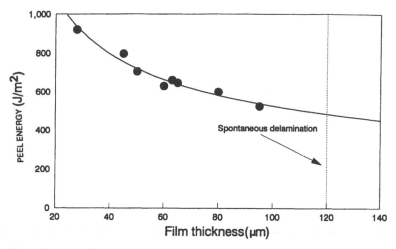

Figure 3. The peel energy of polyimide films peeled from rigid aluminum substrates at 180° is plotted vs. the film thickness. Polyimide films exceeding 120 μm in thickness spontaneously delaminate from aluminum substrates during post-curing cooldown.

The resistance of the interface to residual stress-induced delamination is far less than that predicted by peel testing. Spontaneous delamination, driven by residual stress, has been used to measure adhesion [11, 21]. Since spontaneous delamination occurs with little energy dissipation, tests based on this phenomenon should give values of adhesion closer to the intrinsic energy required to reversibly separate the bonded materials.

5. CONCLUSIONS

During 180° peeling, the peeled material must be bent through 180° to bring stress to bear on the interface. When polyimide coatings were peeled from aluminum substrates, almost all of the mechanical energy was consumed by propagating the bend in the peeling coating. The fraction of the peel energy dissipated as heat was 46.7–49.3% and nearly all of the remainder was stored as latent internal energy in the peeled polyimide. When the bend was propagated through aluminum, which has a limited capacity to store latent internal energy, 97.3–100% of the mechanical energy was dissipated as heat. Therefore, it is unlikely that more than 2.7% of the peel energy could have been consumed by other mechanisms of energy consumption, such as surface formation, which could not be detected within experimental precision.

When a stiff coating which strongly adheres to a rigid substrate is peeled, most of the mechanical energy is consumed by deforming the bulk of the peeling film. For a proper characterization of the adhesion energy, it is necessary that inelastic deformation and dissipation be caused directly by the crack tip stresses associated with the propagating peel crack, and not with bending stress in the film. If separation occurs cohesively within one or both of the bonded layers, it is impossible to separate the bonded materials without crack tip specific deformation and dissipation. When the majority of the mechanical work is consumed by bending, the peel energy outrageously overestimates the adhesion energy, which is therefore totally useless for predicting failure of bonds under loading conditions other than peeling. The fact that most of the peel energy of stiff coatings is consumed by bending creates a serious deficiency of the test for use on coatings containing tensile stresses. When a crack is introduced in the coating, a shear stress, which acts to delaminate the coating, is concentrated near the intersection of the crack and the coating–substrate interface. This process occurs with little bending deformation as compared with peeling and requires considerably less energy. As a result, spontaneous delamination was observed for coatings exceeding a critical thickness, while slightly thinner coatings had high peel strengths. The spontaneous delamination energy of a 120 μm thick polyimide coating from aluminum is estimated to be less than 23 J/m^2, while the extrapolated value of the peel energy at that film thickness is 500 J/m^2. The peel test is not capable of detecting the detrimental effect of tensile stresses on the adhesion of coatings. From an energetics point of view, the energy consumed to create the surfaces exposed by peeling is insignificant since it accounts for an indeterminately small fraction of the total peel energy. However, the peel energy must depend on the chemistry and physics of the interface because the bonded materials can only be subjected to stresses during peeling, which results in dissipation, if the interface is strong.

Acknowledgements

We gratefully acknowledge the support of the Materials Research Laboratory at the University of Massachusetts.

REFERENCES

1. D. H. Kaelble, *Trans. Soc. Rheol.* **9**, 135–163 (1965).
2. A. D. Crocombe and R. D. Adams, *J. Adhesion* **12**, 127–138 (1981).
3. A. D. Crocombe and R. D. Adams, *J. Adhesion* **13**, 241–248 (1982).
4. A. N. Gent and G. R. Hamed, *J. Appl. Polym. Sci.* **21**, 2817–2831 (1977).
5. A. N. Gent and G. R. Hamed, *Polym. Eng. Sci.* **17**, 462–466 (1977).
6. J. Kim, K. S. Kim and Y. H. Kim, *J. Adhesion Sci. Technol.* **3**, 175–187 (1989).
7. K. S. Kim, *J. Eng. Mater. Technol.* **110**, 266–273 (1988).
8. N. Aravas, K. S. Kim and M. J. Loukis, *Mater. Sci. Eng.* **A107**, 159–168 (1989).
9. K. S. Kim, *Mater. Res. Soc. Symp. Proc.* **119**, 31–41 (1988).
10. R. E. Lyon and R. J. Farris, *Rev. Sci. Instrum.* **57**, 1640–1646 (1986).
11. J. L. Goldfarb, PhD Thesis, University of Massachusetts, Amherst, MA (1992).
12. J. L. Goldfarb and R. J. Farris, *J. Adhesion* **35**, 233–244 (1991).
13. H. R. Anderson, Jr, M. M. Khojasteh, T. P. McAndrew and K. G. Sachdev, *IEE Trans. CHMT-9* No. 4, 364–369 (1986).
14. H. J. Leary and D. S. Campbell, *Surface Interface Anal.* **1**, 75–82 (1979).
15. L. P. Buchwalter and A. I. Baise, in: *Polyimides: Synthesis, Characterization and Applications*, K. L. Mittal (Ed.), Vol. 1, pp. 537–548. Plenum Press, New York (1984).
16. J. A. Kreuz, *Dimensional Stability and Adhesion Studies of Kapton Polyimide Film for Flexible Circuitry*, Internal Publication. E. I. duPont de Nemours & Co.
17. H. H. Kausch, *Polymer Fracture*. Springer-Verlag, New York (1978).
18. A. S. Argon and M. I. Bessonov, *Philos. Mag.* **35**, 917–933 (1977).

19. T. P. Russell and H. R. Brown, *J. Polym. Sci. B* **25**, 1129–1148 (1987).
20. T. P. Russell, *J. Polym. Sci. B* **22**, 1105–1117 (1984).
21. R. J. Farris and C. L. Bauer, *J. Adhesion* **26**, 293–300 (1988).

Adhesion Measurement of Films and Coatings, pp. 283–298
K. L. Mittal (Ed.)
© VSP 1995.

Comparison of finite element stress analysis results with peel strength at the copper–polyimide interface

R. H. LACOMBE,[1] L. P. BUCHWALTER[2,][*] and K. HOLLOWAY[3]

[1] *IBM Technology Products, Route 52, Hopewell Jct., NY 12533, USA*
[2] *IBM Research, T. J. Watson Research Center, Yorktown Heights, NY 10598, USA*
[3] *IBM Technology Products, 1000 River Road, Essex Jct., VT 05452, USA*

Revised version received 21 June 1993

Abstract—An exploratory stress analysis has been performed on dense wiring structures, such as those used for multichip modules, with the objective of anticipating possible delamination failure mechanisms. The structure consists of an array of parallel copper lines imbedded in a polyimide insulator, all of which is supported by an underlying silicon wafer. Different versions include thin layers of silicon nitrides on top of and at the base of the copper lines. Using the finite element technique, several models were constructed to explore where delamination could occur in this structure by calculating the driving force (strain energy release rate) which could act on any preexisting microcrack or delamination. It was found that delaminations propagating from the base of the copper line up the sidewall were a strong possibility. The predictions of the stress analysis were substantiated by transmission electron microscopy photographs of real structures which showed the same type of delamination as that anticipated by stress analysis. For these parts, the adhesion between the copper line and the polyimide insulator was mediated by a liner of tantalum, which gives weak adhesion to copper under certain circumstances. X-ray photoelectron spectroscopy and peel test analyses were used to verify the suspected failure modes.

Keywords: Adhesion; copper; polyimide; stress.

1. INTRODUCTION

With increasing circuit densities and correspondingly finer feature sizes, the design and fabrication of the next generation of computer devices and their packages are becoming more complex and difficult. Problems of increased sensitivity to minute amounts of contamination and subtle phenomena such as electromigration are well known and require the utmost attention of the device or package designer. However, a more treacherous problem lies in assessing the thermal-mechanical stability of any particular wiring design and a choice of materials. The dual specters of delamination or cracking are ever present and often obscured by complexities of the process flow. In

---

*To whom correspondence should be addressed.

this paper, the focus is on estimating the potential for delamination in a densely-packed microcircuit wiring structure, basing the calculation solely on the design geometry and estimated material properties. A detailed stress analysis is performed using the finite element method. The driving force for suspected delamination modes is then calculated using fracture mechanics techniques.

The purely computational aspects of this problem are well established and known to be reliable. The least secure part of this type of calculation comes from estimating the material properties of the metals and insulators which comprise the wiring structure. These properties, e.g. elastic constants, expansion coefficients, etc., determine the basic constitutive relations which underlie any stress calculation.

Under ideal circumstances, we would have direct measurements of all the elastic properties of each material used in our structure. Furthermore, every material sample would have been fabricated under conditions closely corresponding to those of the actual manufacturing environment. Needless to say, this ideal state of affairs is almost never realized in practice. The process of carefully measuring the elastic properties of thin film samples is altogether too complex and time-consuming to be successfully carried out. In practice, only a few of the elastic properties are known on the materials from direct measurements on physically relevant samples. For the remaining properties, one needs to rely on the available literature and on 'engineering judgment'.

The imperfect knowledge of the basic material properties imposes a number of constraints on the type of stress calculations which can be carried out. Almost invariably it will be limited to linear elastic models which ignore the more complex types of material behavior such as plasticity in the metals and viscoelasticity in the organic insulators. A certain amount of justification for this approach follows from the observation that the dominant state of loading in multilevel wiring structures occurs at low strain levels, which should give rise to a predominantly elastic response. Also, since the failure phenomena of interest are driven by stored elastic energy, assuming an elastic model will always give an upper bound on the true driving forces. Dissipative processes such as plasticity and viscoelasticity which operate in real structures will cause the strain energy available to be less than that calculated from an elastic model, thus giving rise to lower driving forces. For practical engineering purposes, linear elastic calculations can be used to set safe bounds on allowable loading conditions.

The above arguments should not be construed as saying that the situation is satisfactory. It should always be recognized that stress modeling in the absence of completely reliable constitutive data can lead to unwelcome surprises. If one is lucky, the assumption of linear elasticity will account for the largest part of the true material behaviour and the ensuing predictions will be reliable. At the very least, a linear elastic calculation gives a sound starting point for more elaborate investigations. Only experiment will tell whether the underlying assumptions were adequate.

This paper will first address the finite element stress analysis of copper–polyimide structures, which is followed by experimental confirmation of the finite element model.

2. FINITE ELEMENT STRESS ANALYSIS

Figure 1 shows an array of copper lines embedded in a polyimide insulator. A dense array of copper lines embedded in a polyimide insulator is idealized by an infinite array of parallel lines. Periodic boundary conditions can be applied to the infinite array so that only a small repeat unit needs to be considered. A thin layer of silicon nitride separates the top of the copper lines from the overlying polyimide. A similar layer of silicon nitride lies at the bottom of the trenches formed by the copper lines. The object is to anticipate what failure modes could arise from the thermal expansion mismatch between the copper and polyimide. If one is forced to consider the universe of all feasible mechanisms, this problem is for all practical purposes beyond our abilities. However, relying on past experience, the problem can be whittled down by making a number of appropriate assumptions:

(1) The failure is assumed to occur as a delamination.

(2) The dominant mode of loading comes from thermal expansion mismatch stresses.

(3) The primary response of all materials to the loading conditions is linear elastic.

(4) Elastic constants are estimated from values available in the literature and a small number of direct measurements made in our laboratory.

With these basic assumptions in place, a well-defined model of the copper line structure can now be constructed. Figure 2 shows a finite element model which was designed to explore possible delamination failure mechanisms.

The left-hand side of Fig. 2 shows a broad outline of the structure with each material type labeled. The right-hand side of the figure shows the actual element mesh. Note the high element density near the interface between the silicon nitride layer and the underlying copper and polyimide.

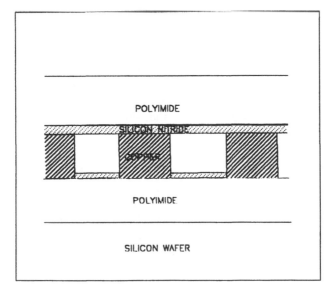

Figure 1. Idealized array of parallel copper lines in polyimide.

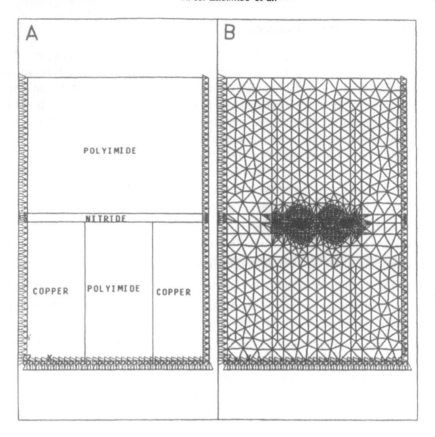

Figure 2. Finite element model for the array of copper lines. (A) structure; (B) finite element mesh plot.

2.1. Delamination of silicon nitride from copper and polyimide

The finite element mesh in Fig. 2 was designed explicitly to explore the possibility of silicon nitride delamination from the underlying copper and polyimide. This is a two-dimensional plane strain model incorporating six noded triangles as the basic element type. Note the very high density of elements in the vicinity of the silicon nitride layer. This feature is provided so that the stress field in this region can be more accurately gauged.

A thermal expansion mismatch loading was applied to this model and Fig. 3 shows the resulting stress distribution near the upper right hand corner of the copper line.

This figure exhibits a vector plot of the maximum principal stress. Arrows point in the direction of the maximum principal stress. The length of the arrows is proportional to the magnitude of the stress field. The maximum principal stress is plotted since this is the largest component of the stress tensor which can exist anywhere in the structure. Like every principal component, the maximum principal stress changes direction at every spatial point. Thus, there is a need for vector plots in order to follow the behavior of this stress component. The most important property of the maximum

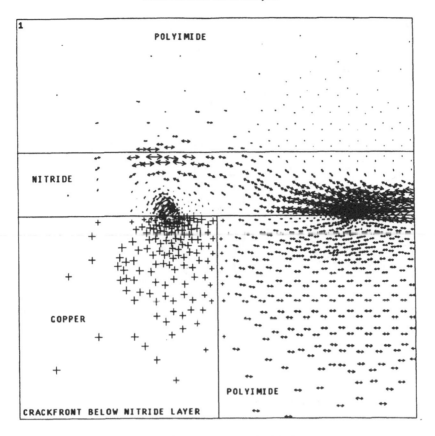

Figure 3. Vector plot of the maximum principal stress above the copper line.

principal stress for present purposes is the fact that any fracture or delamination process which might occur in the structure will tend to propagate in a direction perpendicular to the maximum principal stress vector.

Figure 3 shows two regions with a high density of arrows connected by a line which forms the boundary between the silicon nitride layer on top and the copper and polyimide on the bottom. This line represents a crack which has been deliberately imbedded in the model, and the regions of high arrow density are at the tips of the crackfront where there is a high element density. In fact, a special element type has been inserted at the cracktip in order to accurately reflect the possibly singular behavior of the stress field near such points. The level of resolution in Fig. 3 is too low to discern the nature of the stress field near the cracktips. Each region must be expanded separately to observe this behavior. Figure 4 shows an expanded view of the left-hand cracktip between the silicon nitride layer and the copper line. Note that stress contours as well as stress vectors for the maximum principal stress component are plotted. Above the cracktip, the silicon nitride material is an compression (arrowheads pointing inward). Below, in the copper, the stress vector points into the page. Neither of these stress conditions will tend to propagate a delamination along the silicon nitride–copper interface.

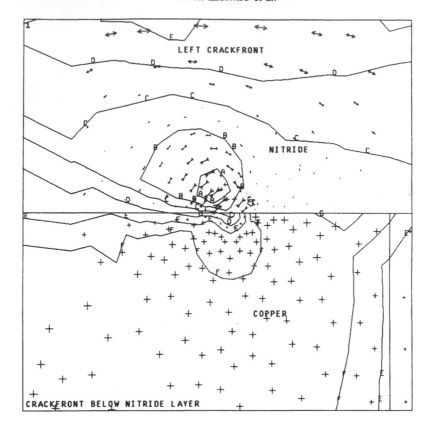

Figure 4. Expanded view of the left-hand cracktip region of Fig. 3.

The right-hand cracktip is examined next. An expanded stress plot is displayed in Fig. 5.

The stress state at this location is significantly different from the left-hand cracktip, but still the maximum principal stress vector is pointing in the wrong direction for propagating a delamination between the silicon nitride and the polyimide. Instead, we see large tensile stresses in the silicon nitride and polyimide which would tend to drive cracks in the vertical direction. Assuming that the materials are strong enough to resist cracking, they will not delaminate. Thus, as far as delamination is concerned, the investigation must be expanded to other parts of the structure.

2.2. Sidewall delamination of polyimide from copper

The calculations which have been completed up to this point indicate that delamination of the silicon nitride layer from the underlying copper or polyimide is not a likely mode of failure. However, at least one more delamination mechanism must be investigated. Notice that the maximum principal stress vector has shown a strong tendency to lie in the horizontal plane perpendicular to the copper line sidewalls (Figs 3 and 5). This strongly suggested that delamination of the polyimide from the sidewall of the copper line was a possible failure mode. Figure 6 presents a finite element model for

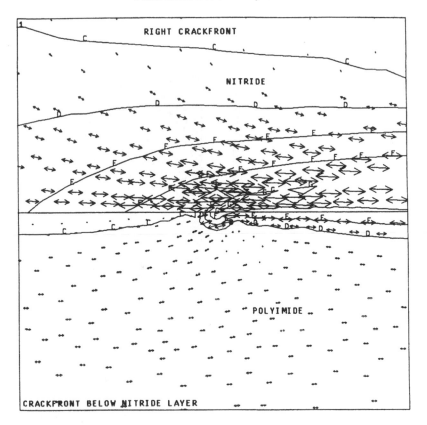

Figure 5. Expanded view of the right-hand delamination cracktip of Fig. 3.

investigating sidewall delamination. This figure is nearly identical to Fig. 2 except that an underlying layer of polyimide is now included as part of the model and the high element mesh density is now at the base of the copper line where failure is expected to occur. A virtual delamination crack has been inserted at the base of the left-hand copper line between the polyimide and the copper.

Figure 7 shows a highly expanded view of the maximum principal stress near the tip of a delamination crack which is propagating up the side of the copper line. Both stress contours and stress vectors are plotted. In contrast to the two previous cases, the maximum principal stress vector is perpendicular to the delamination crackfront between the copper and polyimide. Also note the high tensile stress field near the cracktip as indicated by the stress contours.

This large tensile stress at the cracktip and the fact that the maximum principal stress vector is now perpendicular to the delamination crackfront point out a crackfront with a strong potential for propagation. What remains to be done is to calculate the driving force for delamination. This is what is known as the strain energy release rate (G). The details of this type of calculation have been discussed in the literature [1]. For present purposes, note that the strain energy release rate gives a measure of how much elastic strain energy from the surrounding material can be unloaded into the process of creating a new surface area at the delamination cracktip. Thus, G is a quantitative

A   B

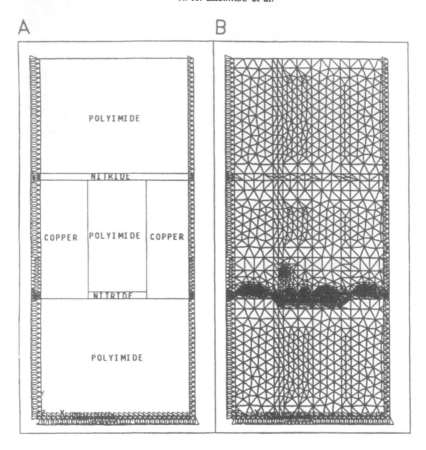

Figure 6. Finite element mesh for sidewall delamination. (A) structure; (B) finite element mesh plot.

measure of the driving force available for crack propagation. Figure 8 gives a plot of the strain energy release rate vs. the crack length for a pre-existing crack 0.03 μm long propagating up the copper sidewall to a length of 0.15 μm.

There are two major features to be noted in Fig. 8:

(1) The maximum driving force of 6 J/m^2 is a rather moderate value. In terms of the adhesion strength, as measured by a peel test for copper off polyimide, this would amount to about 300 J/m^2 nominal peel strength [2]. A good peel strength for the copper–polyimide system is considered to be of the order of 500 J/m^2 and can be as high as 1000 J/m^2.

(2) Even though the driving force is low, it increases with the crack length. The positive slope indicates an unstable crack growth. The crack has the potential to propagate in any environment which would degrade the sidewall adhesion. Thus, the main concern for the copper line structure is that if the sidewall adhesion were to degrade due to contamination or other processing irregularities, sidewall delaminations would be expected.

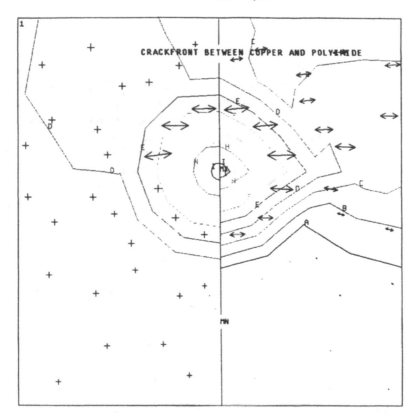

Figure 7. Stress countours near the sidewall delamination cracktip.

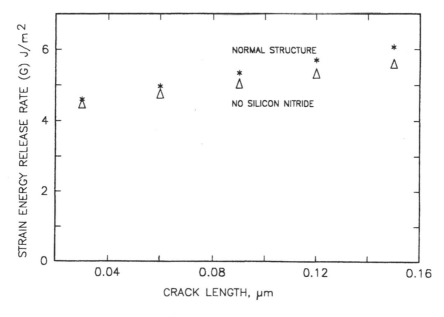

Figure 8. Strain energy release rate vs. crack length for the sidewall delaminationn shown in Fig. 7.

3. EXPERIMENTAL CONFIRMATION OF SIDEWALL DELAMINATION

The modeling suggests the possibility of delamination of the polyimide (PI) from the base of the Cu-line sidewalls. To avoid the expected failure at these locations, a strong interface is required. The interfacial strength needed should be greater than 6 J/m^2, which is about 300 J/m^2 in peel strength when peeling Cu off the PI surface [2].

The sidewall interfaces are Cu–Ta ($\approx$ 20–30 nm)–PI. Therefore the suggested delamination may occur between Cu and Ta, Ta and PI, or in any of the materials in the interfacial region. We chose to study Ta–PI and TaN–PI interfaces to assess the strength. The latter interface was chosen to see what effect a Ta compound might have on the peel strength and the locus of failure (LOF). The Cu–TaN/PI structure may fail between Cu and TaN instead of between TaN and PI, since a true metallic bond is not possible at the Cu–TaN interface.

3.1. Sample preparation and experimental procedure

3.1.1. Peel test. 3,3'4,4'-Biphenyl tetracarboxylic acid dianhydride-p-phenylene-diamine (BPDA-PDA) polyamic acid was spin-coated from a N-methylpyrrolidone (NMP) solution onto Si wafers with native oxide, which were cleaned in an oxygen plasma and treated with γ-aminopropyltriethoxysilane [APS, 0.1% solution in deionized (DI) water, freshly prepared]. The films were cured in a step-wise manner to 400°C. A 20 nm thick Cu release layer was deposited on one edge of the wafer for peel initiation. The PI surface was exposed to CF$_4$ reactive ion etching (RIE) and *in situ* Ar-sputtering. A 20–30 nm thick Ta film was deposited by sputtering in Ar, while TaN was sputter-deposited using a N$_2$–Ar mixture. The thin metal layers were backed with about 10 μm thick Cu for peel analysis. The conditions and the effects on surface chemistry and roughness have been described elsewhere [3].

An interface preparation, the metal film was diced into 2 mm wide peel strips using a Disco wafer dicer with a diamond dicing wheel and DI water coolant. The wafers were dried with a N$_2$ gun and stored in N$_2$-flushed ambient. The peel strength was measured with an Instron tester using 90° peel angle and a 5 mm/min peel rate in N$_2$-flushed ambient.

The peel analysis was carried out right after interface preparation (T–0) and after 10 thermal cycles (10 TC, room temperature to 400°C in forming gas or N$_2$) since the stresses created in the structures, as described earlier, are primarily due to thermal coefficient of expansion mismatch. The locus of failure (LOF) was determined using X-ray photoelectron spectroscopy (XPS) after each interface stressing step. The analysis was performed with a Surface Science Laboratories small spot XPS unit. The details of the method have been described elsewhere [4].

3.1.2. Transmission electron microscopy (TEM). In order to study the microstructure and identify the phases deposited by sputtering a Ta target in a N$_2$–Ar mixture, a 50 nm film was deposited onto an oxidized silicon substrate for observation by TEM. In order to image the electron-transparent 50 nm thick Ta film, the underlying Si-substrate was removed by mechanically grinding and polishing it from the

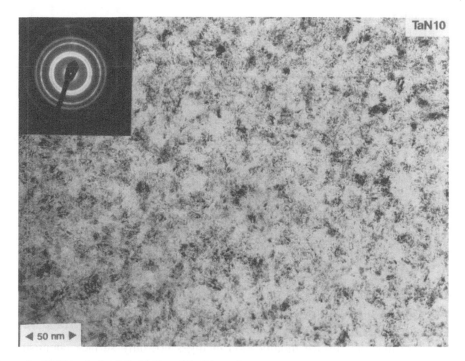

Figure 9. TEM analysis of the TaN used in this study.

backside. Then the remaining substrate and Si oxide were removed by Ar ion-milling. The film was observed in a JEOL 200CX microscope operating at 200 kV. The reactive deposition resulted in a cubic TaN phase with a grain size of about 20 nm as shown in Fig. 9.

A single-level patterned structure was also cross-sectioned for TEM observation. A BPDA-PDA polyimide film was spun and cured on a silicon nitride film on a 5″ Si wafer. A dense via pattern was etched into the polyimide film. A 50 nm Ta layer was sputter-deposited in Ar, and the vias were filled with copper. Part of the wafer was annealed in hydrogen at 400°C for 1 h. TEM samples of both annealed and unannealed structures were prepared by the Bravman–Sinclair method [5]. Each sample was bonded together with a blank substrate into a stack using epoxy. The stack was then sawed into cross-sections. The sections were mechanically ground and polished to 5 μm thickness, and the samples were ionmilled on both sides until they were electron-transparent. Both samples were observed in a JEOL 4000FX microscope operating at 400 kV.

3.2. Results and discusion

3.2.1. Surface analysis of BPDA-PDA. The surface elemental composition of the BPDA-PDA after CF_4-RIE and Ar-sputter surface treatment is given in Table 1. A more detailed analysis can be found in ref. 3.

Surface roughness has been shown to have an effect on adhesion [6, 7]. To determine the surface roughness change after the vacuum treatments, atomic force microscopy

Table 1.

BPDA-PDA surface elemental composition after modification: analysis samples taken at the center region of a 5″ wafer. The copper detected on the surface is sputtered off the Cu-release layer which takes about 10% of the wafer area

Exposure	Pallet coat	C%	O%	N%	F%	Cr%	Ta%	Cu%
CF$_4$ + Ar	Ta	66	23	3	1	–	4	3
Virgin	–	77	16	8	–	–	–	–

Table 2.

AFM surface roughness data in nm

Surface treatment	Roughness (6.25 μm^2 area)
Virgin	12 ± 2
CF$_4$ + Ar	17 ± 1

(AFM) data obtained on the exposed BPDA-PDA surface are shown in Table 2 compared with the virgin surface [3].

It is clear that the surface modification changes not only surface chemistry, but also the surface roughness. The increase in surface roughness is about 40% after the surface treatment. The primary differences in the surface chemistry of the modified PI as compared with virgin PI are the reduction of C=O species and introduction of metal contamination from the sputter tool and peel initiation layer. The oxygen found on the surface is primarily (50–60%) bonded to inorganic species with some C=O (30–40%) and C−O−(5–10%) type moieties. N 1s data suggest nitrogen bonding to inorganic species as well as the presence of imide and amine-type structures.

3.2.2. Effect of annealing on Ta/PI and TaN/PI interface integrity. Figure 10 summarizes the peel adhesion data for Ta/PI and TaN/PI interfaces at T–0, and 10 TC in forming gas (FG) or N$_2$.

Adhesion of Ta to the CF$_4$-RIE and Ar-sputter-treated BPDA-PDA surface is significantly better than that of TaN at T–0. Ten thermal cycles in FG cause about a 50% drop in the peel adhesion for the Ta/PI interface, while TaN/PI suffers a much smaller degree of peel adhesion drop after the same treatment. However, if the TC is done in N$_2$, the peel adhesion degradation is ≈ 35% for Ta/PI, while TaN/PI suffers ≈ 80% drop in peel adhesion.

3.2.3. Locus-of-failure (LOF) analysis of Ta/PI and TaN/PI samples. In order to gain more insight into the Ta/PI and TaN/PI peel adhesion results, (LOF) analysis with XPS was undertaken. The LOF results for the data presented in Fig. 10 are shown in Tables 3 and 4.

The drop in the peel adhesion from T–0 to 10 TC values (Fig. 10) is presumably due to physico-chemical processes which are not fully understood.

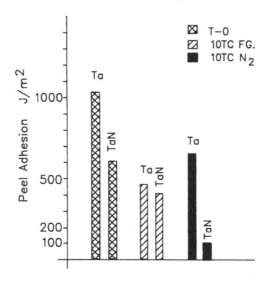

Figure 10. Ta/PI and TaN/PI peel adhesion vs. interface stressing.

The LOF at T–0 in Ta/PI and TaN/PI is in the 'virgin' PI* well below the modified PI surface and the metal/PI interface. The detection of the metal beneath the relatively thick PI is due to cracking of the residual polymeric material on the metal peel strip side [8]. The LOF after 10 TC in FG or N_2 is in the modified PI for the Ta/PI case, as evidenced by the presence of Ta on the PI interface surface.

TaN/PI interfaces after 10 TC fail between TaN and Cu, not between TaN and PI. This is shown in Tables 3 and 4 by the lack of N $1s$ peaks and the high concentration of copper on the TaN interface side, while the PI interface surface shows a significant concentration of Ta. The nitrogen observed on the PI interface surfaces is at 396.6 and 397.6 eV, which suggests inorganic nitrogen bonding.

The LOF in the Cu–Ta/PI structure can be moved to the inorganic interface by introducing N_2 into the Ta-sputtering gas, forming a TaN compound to which the *in situ* sputter-deposited Cu would have to bond. The data suggest that the Cu–Ta interface is sensitive to changes in the interfacial chemistry in a manner which would inhibit the metal–metal bond formation necessary for good adhesion.

The relatively better adhesion seen in the Cu–TaN/PI system when exposed to FG annealing may, in part, be due to a reduction occurring at the Cu–TaN interface, thereby allowing a higher-degree of metal–metal type bond formation as compared with the N_2 ambient annealing. No direct evidence is available at this time to show the presence of elemental Ta in the interfacial region of the FG cycled samples, but it is known that H_2 preferentially segregates to the interfaces [9].

---

*'Virgin' PI denotes PI material which is suggested to be chemically unmodified by XPS analysis. However, based on ref. 8, it is suggested that the mechanical properties (such as fracture toughness) are changed in this 'virgin' PI due to radiation-caused changes during plasma processing (such as cross-linking, increasing the molecular weight).

Table 3.

Elemental composition of Ta/PI and TaN/PI structures. Metal peel interface: all PI surfaces received CF_4-RIE and Ar-sputtering prior to interface preparation

Interface	Stress	C%	O%	N%	F%	Ta%	Cu%
Ta/PI	T–0	80	14	8	–	1	< 1
Ta/PI	10 TC FG	76	14	7	–	1	2
Ta/PI	10 TC N_2	69	19	7	–	1	4
TaN/PI	T–0	78	13	8	–	1	Yes
TaN/PI	10 TC FG	23	27	–	–	< 1	50
TaN/PI	10 TC N_2	22	29	–	–	–	49

Table 4.

Elemental composition of Ta/PI and TaN/PI structures. PI peel interface: all surfaces received CF_4-RIE and Ar-sputtering treatment prior to Ta deposition

Interface	Stress	C%	O%	N%	F%	Ta%	Cu%
Ta/PI	T–0	78	14	8	–	–	–
Ta/PI	10 TC FG	75	16	7	< 1	1	–
Ta/PI	10 TC N_2	67	21	9	–	3	–
TaN/PI	T–0	75	16	9	–	–	–
TaN/PI	10 TC FG	23	46	11	–	19	1
TaN/PI	10 TC N_2	23	47	10	–	18	2

Figure 11 shows a cross-section TEM micrograph of the single-level Cu–Ta/PI structure which had been annealed in hydrogen for 1 h at 400°C. Sidewall delamination between Cu and the Ta liner is documented in this sample. Cu/Ta adhesion has probably failed over the whole length of the sidewall, although only one part of it shows clear delamination. From the peel adhesion and modeling results, the Ta/PI interface would be expected to withstand the stresses created when the structure is annealed at elevated temperatures, while the Cu/Ta interface would be weak if metal–metal bonds had been affected. Additional stresses created in the TEM sample preparation may have exacerbated the delamination at the interface in question. Hydrogen in the annealing ambient also has an effect. At 400°C, hydrogen is known to occupy random interstitial sites in solid solution in α-Ta [9]. Upon cooling below 61°C, several TaH phases have been shown to form. If hydride formation has occurred, a volume increase, and thereby stress increase, in the interfacial region would result as well as degradation of the metal–metal bond. It has also been shown that Ta takes up a large amount of H_2 when annealed in forming gas [9]. Therefore,

Figure 11. TEM analysis of the hydrogen-annealed Cu–Ta/PI structure.

hydride formation is a likely culprit in the degradation of the interfacial integrity of the Cu–TaN/PI structure in strongly reducing environments.

4. CONCLUSION

Finite element analysis has shown the potential for delamination at the Cu–Ta/PI metal line sidewalls after thermal excursions to 400 °C and back to room temperature. Cross-sectional TEM analysis of a Cu–Ta/PI structure annealed in H_2 ambient shows metal line sidewall delamination between Cu and Ta. Peel adhesion measurements of Cu–TaN/PI structures reveal that the weak link in the system after themal stressing is between the Cu and TaN, suggesting that metal–metal bond disruption at the Cu–Ta interface results in poor adhesion after annealing.

Acknowledgements

We are grateful to the BEOL line at T. J. Watson Research Center for help in sample preparation, P. Gillen for the peel test, and H. Clearfield for making his XPS unit available for this work, E. Baran for TaN deposition, and S. Reynolds for supplying the single level structure and carrying out hydrogen annealing.

REFERENCES

1. M. F. Kanninen and C. H. Popelar, *Advanced Fracture Mechanics*. Oxford University Press, New York (1985).
2. The peel test measurement must be normalized to take into account the strain energy lost in deforming the peel strip (and the substrate). Details on how this type of normalization is accomplished have been covered in the literature: K. S. Kim and N. Aravas, *Int. J. Solids Struct.* **24**, 417 (1988); Y. H. Kim, J. Kim, G. F. Walker, C. Feger and S. P. Kowalczyk, *J. Adhesion Sci. Tech.* **2**, 95 (1988); and Y. H. Kim, G. F. Walker, J. Kim and J. Park, *J. Adhesion Sci. Technol.* **1**, 331 (1987). A peel strength of about 300 J/m^2 is close to an estimated adhesion strength of about 6 J/m^2 (though it should be recognized that the peel test analysis does not take into account stick-slip behavior or cracking of the residual polyimide on the metal peel interface surface).
3. L. P. Buchwalter and R. Saraf, *J. Adhesion Sci. Technol.* (in press).
4. L. P. Buchwalter, B. D. Silverman, L. Witt and A. R. Rossi, *J. Vac. Sci. Technol.* **A5**, 226 (1987).
5. J. Bravman and R. Sinclair, *J. Electron Microsc. Tech.* **1**, 53 (1984).
6. D. E. Packman, in: *Adhesion Aspects of Polymeric Coatings*, K. L. Mittal (Ed.), p. 19. Plenum Press, New York (1983).
7. B. R. Karas and D. F. Foust, in: *Metallized Plastics 3: Fundamental and Applied Aspects*, K. L. Mittal (Ed.), p. 319. Plenum Press, New York (1992).
8. L. P. Buchwalter, *J. Adhesion Sci. Technol.* (in press).
9. A. Sugerman and A. Marwick, IBM East Fishkill, unpublished results (1992).
10. K.-S. Kim and J. Kim, *J. Eng. Mater. Technol. ASME Trans.* **110**, 206 (1988).

Adhesion Measurement of Films and Coatings, pp. 299–321
K. L. Mittal (Ed.)
© VSP 1995.

Measurement of the practical adhesion of paint coatings to metallic sheets by the pull-off and three-point flexure tests

A. A. ROCHE,* P. DOLE and M. BOUZZIRI

Département de Chimie Appliquée (CNRS, URA 417), Université Claude Bernard, Lyon I, F-69622, Villeurbanne Cedex, France

Revised version received 14 September 1993

Abstract—In this paper, we compare the practical adhesion measurements of alkyd, vinyl, epoxy, and polyurethane paints on cold-rolled steel using the pull-off and three-point flexure tests. It has been shown that

(a) in the pull-off test, the ultimate load and ultimate strain values depend on the stud area. During trimming, cracks are created within the system which are responsible for a large decrease in the parameters measured. On the other hand, it is impossible to differentiate between the failure initiation and the failure propagation zones;

(b) in the three-point flexure test, the area (W) subtended by the load/displacement curves corresponding to samples with and without stiffener does not depend on the bonded width or on the substrate compliance. After carrying out such a mechanical test, it is also possible to discriminate between failure initiation and failure propagation using appropriate tools for observation, such as scanning electron microscopy, and for analysis, such as electron microprobe analysis and Fourier transform infrared spectroscopy. W is shown to be representative of the failure initiation energy.

To give a clear indication of the formation of the paint/substrate interphasial zone, we have studied the practical adhesion of different types of acrylic paint (thermoplastic acrylic binder filled with TiO_2, $BaSO_4$, or $CaCO_3$, with or without crosslinked polyurethane) applied onto aluminium substrates (treated in a sulpho-chromic acid bath). Five different types of failure initiation were observed, each type corresponding to quite different energies (W). Failure initiation takes place within the interphasial layer for thin coatings (thickness $\leqslant 70~\mu$m). For thicker paint layers, cohesive failures were observed. A model showing the composition of the paint/metal interphase is proposed.

Keywords: Practical adhesion measurement; pull-off test; flexure test; paint coatings onto metals; locus of failure; paint/metal interphase.

1. INTRODUCTION

Paint coatings are used for many purposes in diverse applications. Whatever their objective may be, i.e. functional, decorative, or protective, the coated system properties and performance all depend on the adhesion between the substrate and the paint coating. Unfortunately, it is well known that during ageing failure of painted systems may occur within the paint/adherend interphasial region, which is governed by adherend surface treatment and paint formulation. To achieve good adhesion, one needs knowledge of

---

*To whom correspondence should be addressed.

the interactions that occur within the interphase so created. Obviously, there is great practical importance in carrying out practical adhesion measurement and in determining the locus of the failure. Taking into account the most pertinent experimental data from previous works [1, 2], the aim of the present study was to point out the formation of an interphase which is created between the metallic substrate and the organic paint.

2. EXPERIMENTAL

To compare the information obtained by the pull-off test (AFNOR, * NFT 30 062) with that obtained by the three-point flexure test (AFNOR, NFT 30 010) [1–4], industrial paints (alkyd, vinyl, epoxy, and polyurethane binders) were applied onto cold-rolled steel. The locus of failure (initiation and propagation) occurring during mechanical tests was investigated using scanning electron microscopy (SEM), electron microprobe analysis (EMPA), and FTIR microspectrometry [5–9].

To show evidence of interphase formation between the metal and the paint, we used as the 'model substrate' a pure aluminium (1050 A from Péchiney) sheet and as the 'model paint' either a thermoplastic acrylic binder (B1) or a hydroxylated acrylic binder (B2) which can be crosslinked by polyurethane. The first binder (B1) was filled with TiO_2 at different pigment volume concentrations (PVCs). The second one (B2) was filled with TiO_2, $BaSO_4$, and $CaCO_3$ at different ratios. Two different surface treatments were used for the Al substrate: a chromic-sulphuric acid bath (250 g/l sulphuric acid, 50 g/l chromic acid, 87 g/l aluminium sulphate octadecahydrate; 30 min, 80°C), and 1 N NaOH (5 min, 20°C). The paint was applied with a spray gun in order to vary the thickness of the paint film. The samples were dried for 2 days at 20°C.

2.1. Pull-off test (AFNOR, NFT 30 062)

This test was performed with a tensile machine (DY25, Adamel Lhomargy, Instruments S.A., Ivry/Seine, France) according to the NFT 30 062 standard at a displacement speed of 10 mm/min (Fig. 1). Various aluminium studs of different diameters (from 20 to 50 mm) were used. After surface treatment of the studs in a sulphochromic acid

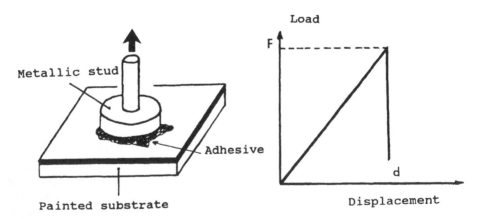

Figure 1. Principle of the pull-off test.

---

*AFNOR: Association Française de Normalisation, Tour Europe Cedex 7, 92049 Paris La Défense.

bath [10], a solvent-free epoxy adhesive (CIBA AW 134/HY 994) was applied. The studs were then adhesively bonded to the painted substrate (die-stamped to 70×150 mm^2) and a 4.9 N load was applied to each stud. The adhesive curing conditions were 2 h at 80°C. After curing, the samples were left in an air-conditioned room [22 ± 2°C, $55 \pm 5\%$ relative humidity (RH)] for 24 h. Mechanical tests were run in the same room. Before testing, the excess of adhesive around the stud due to flow was removed by trimming with a special cutting tool (co-axial cylinder with a toothed end). A set of six samples was prepared for each series. The average of the ultimate load (F) and ultimate stress ($C = F/S$, S being the stud area) values and their respective standard deviations were reported. After testing, the failure zone was studied visually.

2.2. Three-point flexure test (AFNOR, NFT 30010)

This test was performed with a tensile machine (FLEX3, Techmétal, Maizières-Les-Metz, France) according to the NFT 30010 standard at a displacement speed of 0.50 mm/min (Fig. 2). On the painted substrates (die-stamped to 10×50 mm^2), a solvent-free epoxy adhesive (CIBA AW 134/HY 994) was applied as a stiffener. In order to avoid any damage to the paint film, the curing conditions of the stiffener were 24 h at 40°C in the case of thermoplastic binders (without crosslinker) and 2 h at 80°C in the case of crosslinked binders. After curing, the samples were left in an air-conditioned room (22 ± 2°C, $55 \pm 5\%$ RH) for 24 h. Mechanical tests were carried out in the same room. A set of six samples was prepared to each series. The average values of the slope of the load/displacement curve (P/d), the ultimate load (F), the ultimate displacement (d), and the subtended area (W) and their respective standard deviations were determined. The parameter W corresponding to the area subtended by the load/displacement curves for samples with and without stiffener is shown in Fig. 2. After testing, the failure zone was studied visually. On one specimen from each series, SEM, EMPA, and FTIR were carried out in order to differentiate the failure initiation from the failure propagation zone.

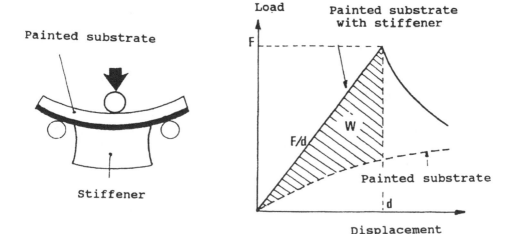

Figure 2. Principle of the three-point flexure test.

The influence of some experimental parameters such as the bonded area and the adhesive and substrate thicknesses on the P/d, F, d, and W values was studied on 0.8 mm thick titanium sheets without paint. Prior to bonding, titanium adherends were degreased with acetone, immersed in NH_4F, HF (10 g/l) acid solution at room temperature for 2 min, rinsed in deionized water, and dried with water-absorbent paper. This treatment was suitable to ensure failure initiation within the interphasial layer. A two-component adhesive (3525 from 3M) cured at 80°C for 2 h was used according to the AFNOR T 76-143 standard for structural bonding.

2.3. Surface observation and analyses

To differentiate the failure initiation from the failure propagation zone, a scanning electron microscope (Philips 515) fitted with an electron microprobe analysis accessory (Edax 9800) was used. The failed surface was neither coated with gold nor coated with carbon. The accelerating electron voltage was 10 kV; the spot for micro-analyses was 200 nm; and the tilt angle used was 15°. FTIR spectra were recorded on a Nicolet 710 spectrometer. As the FTIR spectrometer is a one-beam device, each sample spectrum has to be referred to a background, which was collected on a gold-coated substrate. Micro-spectrometry was performed with a Nic-Plan device, fitted to the 710 bench, and equipped with both conventional (40°, ×15) and grazing incidence (84°, ×32) objectives.

To locate, within the paint thickness, the different types of failure initiation area, EMPA analysis of bevel-edged systems was also carried out. The bevelled samples were obtained by (a) embedding the sample in an epoxy resin; (b) curing of the epoxy (20°C, 24 h); and (c) bevelling initially with fine sandpaper and the polishing with alumina powder. The higher the bevelling angle (close to 85°), the larger the surface that can be analysed.

3. RESULTS AND DISCUSSION

3.1. Influence of the test method and the sample geometry

3.1.1. Pull-off test. (a) *Influence of the stud area.* Irrespective of the paint formulation, an increase in the ultimate load (F) values and a decrease in the ultimate stress (C) values were observed when the stud area increased from 0.78 to 19.6 cm^2. For example, Fig. 3 shows the variation of F and C versus the stud area in the case of an epoxy paint applied on a cold-rolled steel degreased by immersion in trichloroethylene for 1 min. These data were obtained after trimming around the stud. In light of the stud cylindrical geometry, it is not possible to differentiate failure initiation from the propagation zone. Since ultimate parameters (F and C) are used, they are necessarily associated with failure initiation and not at all with flaw propagation. This comment points out the most important drawback of this mechanical test.

(b) *Influence of trimming.* The area of the studs used was 78 mm^2. F and C were compared for both trimmed and untrimmed samples. A steep decrease in the ultimate parameter values (80–200%) was observed for all the systems tested. For samples tested without trimming, an appropriate apparatus was used to prevent adhesive flow. As an

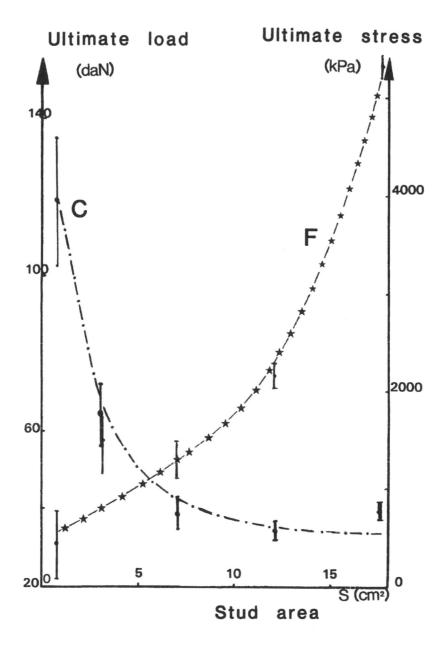

Figure 3. Variation of the ultimate load (*F*) and ultimate stress (*C*) values with the stud area.

example, Fig. 4 shows the variation of *F* versus the paint thickness in the case of an epoxy paint applied to a cold-rolled steel degreased by immersion in trichloroethylene (inhibited) for 1 min. These results suggest that failure initiation takes place during trimming. Under these conditions, the information obtained during the pull-off test for trimmed specimens is characteristic of flaw propagation (which is inappropriate to the use of ultimate parameters such as *F*). The *F* value increasing for trimmed samples with the paint film thickness was then characteristic of the increasing energy dissipation within the paint layer and/or of the paint cohesion.

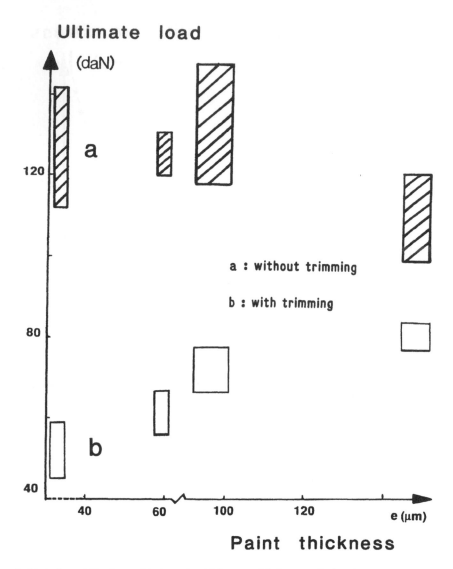

Figure 4. Variation of F values with the paint thickness and influence of trimming.

3.1.2. Three-point flexure test. Among the parameters which can be used, generally the ultimate load (F), displacement (d), and the subtended area (W) are chosen to characterize practical adhesion (Fig. 2), whereas the slope of the load/displacement curve (P/d) is used to characterize sample compliance. In previous work [11, 12], the dependence of P/d, F, and d on experimental parameters was studied. We present here the influence of some experimental parameters such as the bonded area and the adhesive and substrate thicknesses on the P/d, F, d, and W values. This part deals only with adhesive/uncoated titanium systems.

(*a*) *Influence of bonded width.* Table 1 shows the variation of the slope (P/d) of the load/displacement curves and the variation of the ultimate load (F), the ultimate displacement (d), and the subtended area (W) with the bonded width for adhesive/titanium

Table 1.
Variation of P/d, F, d, and W versus the bonded width for adhesive/titanium systems, as obtained in the flexure test

Width (mm)	P/d (N/mm)	F (N)	d (mm)	W (mJ)
3	142 ± 6%	41 ± 11%	0.29 ± 8%	3.8 ± 19%
4	157 ± 15%	41 ± 22%	0.27 ± 19%	3.7 ± 38%
5	199 ± 5%	46 ± 11%	0.24 ± 11%	4.2 ± 19%
6	209 ± 2%	46 ± 7%	0.22 ± 7%	3.9 ± 11%

systems. The P/d value, related to the system compliance or rigidity, increases with the bonded width. A slight decrease of d is observed when the bonded width increases, whereas F and W remain quite constant. SEM micrographs of the failed substrate surface for one of the tested samples (0.5 mm bonded width) are shown in Fig. 5. The distinction between the failure initiation and propagation zones is clearly pointed out by the convergence of the failure propagation lines discernible on the remaining adhesive which is left on the failed substrate surface. It was possible to observe (compared with the unbonded substrate surface) in the SEM micrograph of the failure initiation area (at the highest magnification) a thin layer of polymer material on the metallic substrate. This was observed for all the samples tested. EMPA and FTIR analyses (not represented here) showed that the failure initiation occurred within the interphasial region. This observation seems to indicate that the parameters measured (F and W) with the three-point flexure test are representative of failure initiation and not of failure propagation.

(b) Influence of the adhesive thickness or volume. To obtain samples with various adhesive thicknesses, different adhesive volumes were applied to titanium adherend sheets. Table 2 depicts the variation of P/d, F, d, and W with respect to the adhesive thickness. The P/d value, related to the system rigidity, increases with the adhesive thickness, whereas F remains quite constant. A drastic decrease of d and W was observed when the adhesive volume increased from 0.30 to 0.40 ml. For adhesive volumes higher than 0.45 ml, a slight decrease of the d and W values was noted. For all the specimens tested, failure initiation occurred within the interphasial region.

Table 2.
Variation of P/d, F, d, and W versus the adhesive volume for adhesive/titanium systems, as obtained in the flexure test

Volume (ml)	P/d (N/mm)	F (N)	d (mm)	W (mJ)
0.30	140 ± 3%	163 ± 15%	1.87 ± 36%	156 ± 53%
0.35	163 ± 4%	158 ± 13%	1.17 ± 17%	80 ± 32%
0.40	190 ± 8%	176 ± 7%	1.09 ± 11%	90 ± 17%
0.45	217 ± 3%	155 ± 13%	0.77 ± 14%	54 ± 26%
0.50	241 ± 2%	150 ± 2%	0.66 ± 15%	46 ± 28%
0.55	264 ± 1%	152 ± 11%	0.60 ± 13%	42 ± 23%
0.60	292 ± 2%	140 ± 24%	0.49 ± 26%	35 ± 52%
0.65	290 ± 3%	130 ± 16%	0.46 ± 17%	28 ± 32%
0.70	331 ± 2%	140 ± 15%	0.43 ± 17%	29 ± 29%

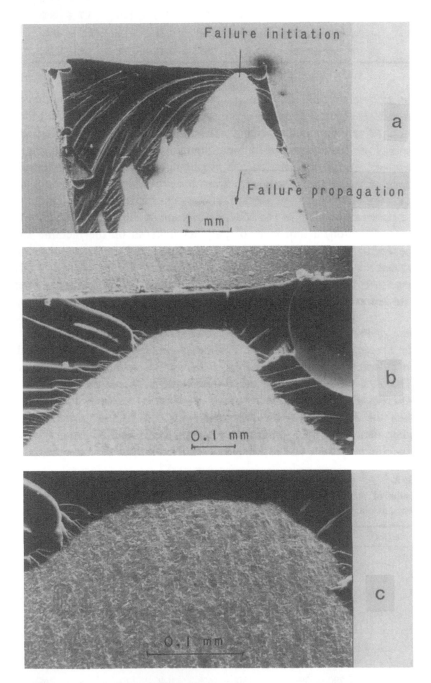

Figure 5. SEM micrographs of the failed substrate surface for an adhesive/titanium system with 0.5 mm bonded width. (a) Overall picture; (b, c) failure initiation area at two different magnifications.

Table 3.

Variation of P/d, F, d, and W versus the adherend thickness for adhesive/titanium systems, as obtained in the flexure test

Thickness (mm)	P/d (N/mm)	F (N)	d (mm)	W (mJ)
0.8	247 ± 4%	125 ± 6%	0.52 ± 10%	26 ± 15%
1.6	704 ± 2%	327 ± 15%	0.49 ± 16%	43 ± 28%
2.4	1406 ± 2%	794 ± 7%	0.68 ± 13%	35 ± 28%
3.2	2750 ± 5%	1378 ± 6%	0.56 ± 12%	47 ± 8%

Such a behaviour can be explained by the fact that stress concentration in the overall mechanical system is predominantly within the interphasial zone [13], inducing failure initiation within this region. However, in light of the F and W dependence on the adhesive volume, 0.5 ml was used in the following work.

(c) *Influence of the substrate thickness.* Table 3 shows the variation of P/d, F, d, and W versus the titanium adherend thickness. The titanium substrate thickness was increased by bonding several 0.8 mm thick titanium sheets. This method enables us to keep the substrate/adhesive interphase-related properties constant. The variations of the P/d and F values with the adherend thickness are explained by the increase in the bonded system rigidity, whereas d and W remain quite constant, showing the independence of these parameters on the mechanical behaviour of the substrate. Failure initiation occurred within the interphasial zone.

Tables 1, 2, and 3 show that for a given adhesive volume, the parameter W is independent of the experimental parameters used and is thus characteristic of the failure initiation energy. Taking into account this observation, practical adhesion in the following work was measured using W, even if the standard deviation for this parameter was always higher than for the other parameters.

3.2. Study of the paint/metal interphase

The paint/metal interphase was studied using the three-point flexure test. Since the ultimate parameters such as F, d and W are used, only failure initiation needs to be considered. One of the main advantages of the three-point flexure test is its ability, using visual observation for all tested specimens and using for one specimen from each series appropriate observation or analytical tools (SEM, EMPA, XPS, ISS, AES, FTIR, etc.), to differentiate failure initiation from flaw propagation. In the following work, only the loci of the failure initiation are reported using visual observation or EMPA and FTIR analyses. According to Mittal [14], three types of failure initiation can be observed: type a: interfacial failure; type ac: interphasial failure; and type c: cohesive (or bulk) failure.

After visual observation, the following notations are used: *interfacial failure* (type a) when the failed substrate surface is observed to be metallic in nature; *interphasial failure* (type ac) when a thin organic layer is observed on the failed substrate surface (the remaining colour is different from that of the bulk paint); and *cohesive failure* (type c) when the remaining colour of the failed surface is the same as that of the bulk paint.

After analysis (using EMPA and FTIR), the following notations are used: *interfacial failure* (type a) when it is impossible to detect any polymer material left on the failed substrate surface; *interphasial failure* (type ac) when EMPA and FTIR spectra of the remaining organic layer on the failed substrate surface are different from those of the bulk paint; and *cohesive failure* (type c) when EMPA and FTIR spectra of the remaining organic layer on the failed substrate surface are identical to those of the bulk paint.

It is interesting to note that the locus of failure initiation from visual observation may be quite different from that determined using EMPA or FTIR analytical tools. For example, all visually interfacial failures were found, after analysis, to be as a matter of fact interphasial failures. To understand the formation of the paint/metal interphase, 'model' paints were applied onto 'model' metallic substrates.

3.2.1. Three-point flexure test. In each case, a series of six samples was studied in order to obtain statistical reproducibility. In each figure providing practical adhesion data, results for each series are represented by a rectangle whose height more or less indicates the standard deviation and whose central line indicates the average value. Individual experimental points are also plotted using open circles.

(a) Influence of the paint thickness. Two different dry paint thicknesses of the binder B1, filled with 5% TiO_2, were applied to sulphochromic acid-treated aluminium samples. The variation and the standard deviation of W and the locus of failure initiation, by visual observation, versus the paint thickness are reported in Fig. 6. In

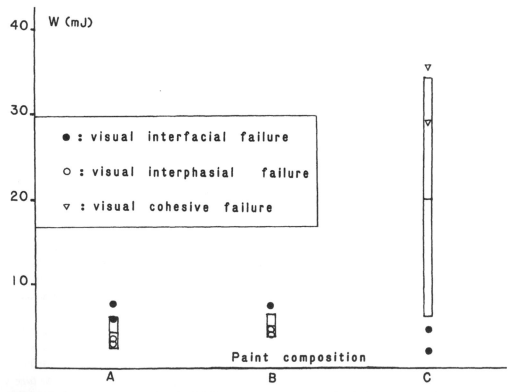

Figure 6. Variation of W and loci of failure for various paint compositions. (A) 5% TiO_2, 95% B1, thickness 50 μm; (B) 5% TiO_2, 95% B1, thickness 50 μm; (C) 5% TiO_2, 95% B1, thickness 120 μm.

the case of thin paint coatings (20–30 μm), visual interfacial failure (type a) was observed before the mechanical test (delamination of the samples occurred after cooling down of the epoxy stiffener). Series A and B correspond to a paint application (50 μm) under the same conditions, but series B was prepared 2 months later than series A. No significant variations in the W value and failure locus (type a or ac) were observed for series A and B, which clearly illustrates the reproducibility of the three-point flexure test. For thicker paint (120 μm) layers (series C), two different types (a and c) of failure were observed. For this binder (B1), a significant scatter in the W values is reported which corresponds to different loci of failure. However, visual interphasial failure (type ac) and cohesive failure (type c) correspond to quite different energies: W is less than 10 mJ for interphasial failures (type ac), while W is greater than 10 mJ in the case of cohesive failures (type c). This observation indicates that the composition of the interphase is different from that of the bulk paint.

(b) *Influence of the pigment volume concentration (PVC)*. The binder B1, filled with 5 (A), 7.5 (B), 10 (C), and 15% TiO$_2$ (D), was applied to sulphochromic acid-treated aluminium samples to the same dry paint thickness (50 μm). The variation and the standard deviation in W and the locus of failure initiation, by visual observation, versus the PVC are reported in Fig. 7. When the PVC value was higher than 7.5%, no significant variation in W was observed. In the case of visual interphasial failures (type ac), no

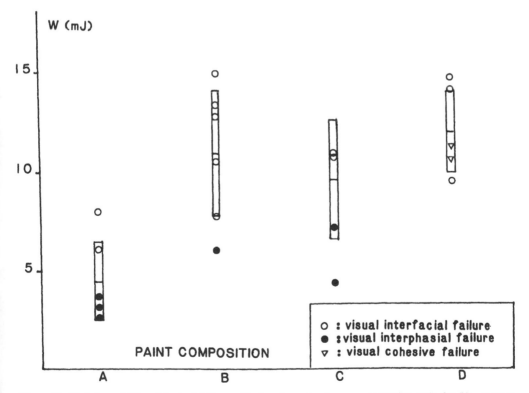

Figure 7. Variation of W and loci of failure with the pigment volume concentration ratio for 50 μm thick paint layers. (A) 5% TiO$_2$, 95% B1; (B) 7.5% TiO$_2$, 92.5% B1; (C) 10% TiO$_2$, 90% B1; (D) 15% TiO$_2$, 85% B1.

effect of the PVC value (ranging from 5 to 15%) was seen. This observation suggests that the composition of the interphase does not depend on the composition of the bulk paint, i.e. the PVC value.

(c) *Influence of application and drying modes.* To determine the influence of the paint viscosity, the binder B1, filled with TiO_2, was applied either with a spray gun (low viscosity) or with a hand-coater (high viscosity) to sulphochromic acid-treated aluminium samples to the same dry paint thickness (50 μm). For the two different paint applications, paint drying was carried out with the painted side up for one set and down for a second set in order to observe the influence of gravity or filler sedimentation. The relevant data are not plotted in the figure. In all cases, no significant variation in the W values was obtained ($W \approx 10$ mJ). Visual interphasial failures (type ac) were observed. These observations point out that the composition of the interphase depends neither on the filler sedimentation nor on the paint application process.

(d) *Influence of crosslinking.* The binder B2, with or without polyurethane crosslinker, filled with TiO_2, $BaSO_4$, or $CaCO_3$ at various ratios and PVC was applied to sulphochromic acid-treated aluminium samples to the same dry paint thickness (50 μm). The variation and the standard deviation in W and the locus of failure initiation, by visual observation, versus the paint formulation are reported in Fig. 8. One can see that the W values increase from uncrosslinked to crosslinked series. Visual cohesive failure was

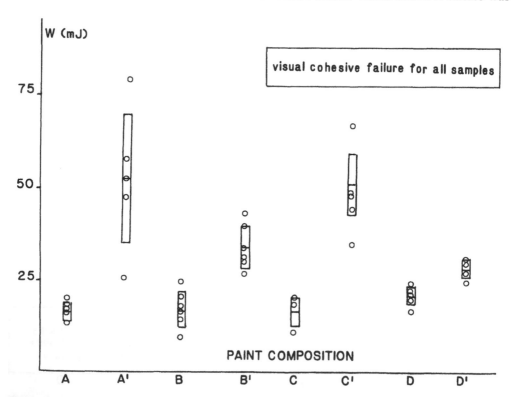

Figure 8. Variation of W and loci of failure for 50 μm thick paint layers: A, B, C, and D without crosslinker; and homologous series A′, B′, C′, and D′ with crosslinker (binder B2; PVC from 5 to 20%; fillers: TiO_2, $BaSO_4$, $CaCO_3$ in various ratios).

always observed. It is important to note that the locus of failure determined by EMPA was in all cases within the interphase region, in a polymer-enriched layer (see Section 3.2.2). These observations can explain the increase in the W value for crosslinked paints irrespective of the nature of the filler.

(*e*) *Influence of the filler nature.* The binder B2 with polyurethane crosslinker, filled with TiO_2, $BaSO_4$, or $CaCO_3$ at various ratios, was applied to sulphochromic acid-treated aluminium samples to the same dry paint thickness (50 μm). The filler size as measured by SEM was 0.5 μm for TiO_2 and 1 μm for $BaSO_4$, and $CaCO_3$. The variation and the standard deviation in W and the locus of failure initiation, by visual observation, versus the filler nature are reported in Fig. 9. Good filler dispersion within the dry paint layer was observed by SEM in all systems. The filler nature has a considerable influence on the W values, the best mechanical performance being obtained with the TiO_2 filler. $BaSO_4$-filled samples gave better results than $CaCO_3$-filled samples. No significant influence due to the filler size was observed. Visual cohesive failure was always observed. Here, too, it is important to note that the locus of the failure determined by EMPA was in all cases within the interphase region, in a polymer enriched layer (see Section 3.2.2). Consequently, the filler nature has an effect on the constitution of the polymer-enriched layer.

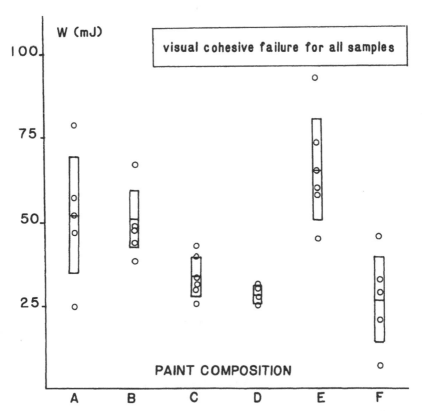

Figure 9. Variation of W and loci of failure for 50 μm thick paint layers. (A) 20% TiO_2, 80% B2; (B) 16% TiO_2, 4% $BaSO_4$, 80% B2; (C) 20% $BaSO_4$, 80% B2; (D) 20% $CaCO_3$, 80% B2; (E) 7.5% TiO_2, 5% $BaSO_4$, 2.5% $CaCO_3$, 85% B2; (F) 7.5% TiO_2, 7.5% $CaCO_3$, 85% B2.

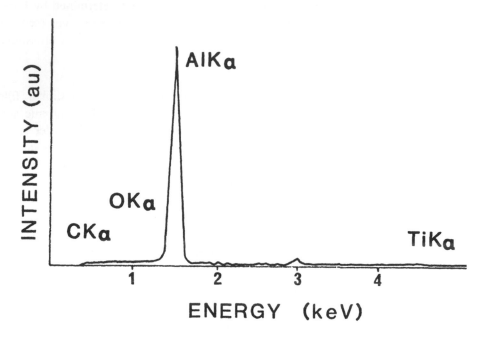

Figure 10. EMPA spectrum of the interfacial failure initiation area obtained for a 20–30 μm thick paint coating (binder B1; 5% TiO_2 filler).

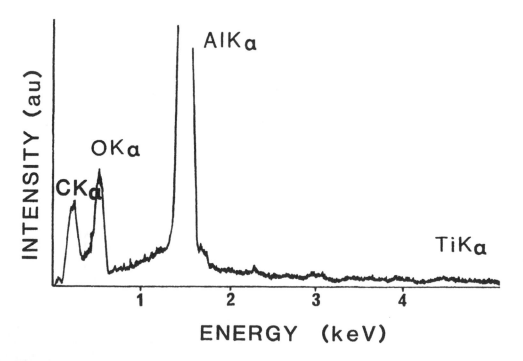

Figure 11. EMPA spectrum of the interphasial failure initiation area (type 1) obtained for a 50 μm thick paint coating (binder B1; 5% TiO_2 filler).

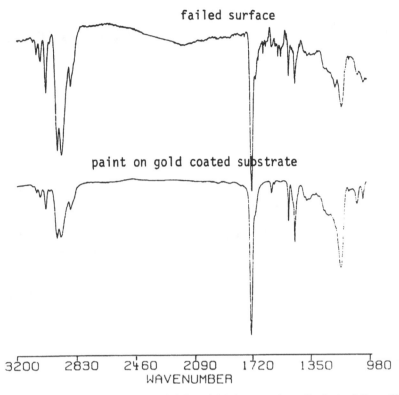

Figure 12. FTIR spectrum of the interphasial failure initiation area (type 1) obtained for a 50 μm thick paint coating (binder B1; 5% TiO$_2$ filler).

3.2.2. Locus of the failure initiation investigated by EMPA and FTIR. For each series tested in Section 3.2.1, the locus of failure was determined using EMPA and FTIR analyses. Analyses show five different types of failure initiation:

• *interfacial failure*, i.e. only the aluminium is detected by EMPA (Fig. 10);

• *interphasial failure*, three types of interphasial failures have been observed by EMPA and FTIR:

— type 1: failure leaving a very thin organic layer (Figs 11 and 12) on the metallic substrate;

— type 2: failure in a filler-enriched layer (Figs 13 and 14);*

---

*The EMPA peak intensity of characteristic X-ray radiation depends, among other parameters, on the incident electron energy (E)/radiation excitation potential (E_0) ratio. The intensity increases when E/E_0 increases to the maximum value obtained, i.e. $\approx$ 5–7, and then decreases. The excitation potentials of CK_α, OK_α, AlK_α, CaK_α, and TiK_α are 0.283, 0.531, 1.559, 4.038, and 4.985 keV, respectively. In order to obtain information about light elements (C, O), the electron energy used for EMPA analysis was 10 keV. This energy does not allow a good sensitivity for the CaK_α and TiK_α radiations. Moreover, for a 5% TiO$_2$ filler, the titanium weight percent in the paint is only 2%. These comments point out why titanium was not detected on a 5% TiO$_2$-filler paint bulk. However, the titanium or calcium concentration observed on the failed surface was then highly significant.

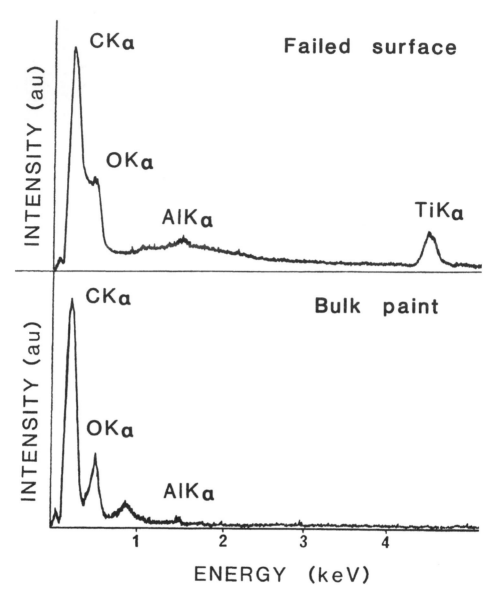

Figure 13. EMPA spectra of the interphasial failure initiation area (type 2) obtained for a 50 μm thick paint coating (binder B1; 5% TiO_2 filler) dried up side down, and of the bulk paint.

— type 3: failure in a polymer-enriched layer on top of the filler-enriched layer. The thickness of this layer is quite important, because the filler-enriched layer is not detected unless the electron energy of the EMPA beam is higher than 10 keV (Fig. 15);

• *cohesive failure*, i.e. the failure initiation area and the bulk paint have exactly the same compositions.

3.2.3. SEM observation and EMPA analysis of the paint layer. To locate within the paint thickness the different types of failure initiation area, SEM observations and

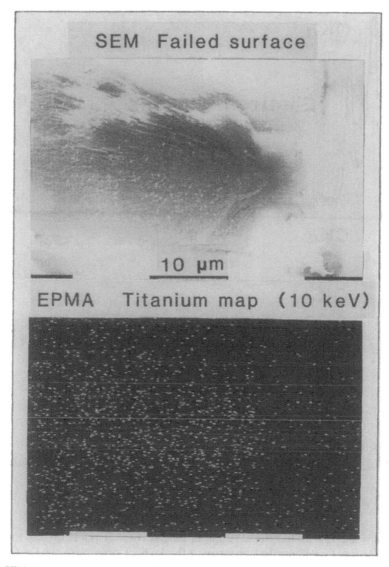

Figure 14. SEM photomicrograph (top) and X-ray titanium map (bottom) of the interphasial failure initiation area (type 2) obtained for a 50 μm thick paint coating (binder B1; 10% TiO_2 filler).

EMPA analyses of different bevelled systems (Fig. 16) were carried out. Polishing a section for subsequent surface analysis is fraught with artifacts such as contamination with the grinding medium, selective polishing and/or loss of components (especially filler particles), smearing of soft material (binder) over fillers, etc. However, the EMPA analysis of the bevelled systems never revealed SiC (grinding paper) or Al_2O_3 (polishing powder) contamination. In light of Fig. 17, the loss of filler particles or the smearing of soft resin over fillers is possible, but there is no way (using EMPA or FTIR) to show conclusively the loss of filler particles or binder smearing. For a 50 μm thick paint layer, an apparent paint thickness close to 500 μm was observed by SEM; this is

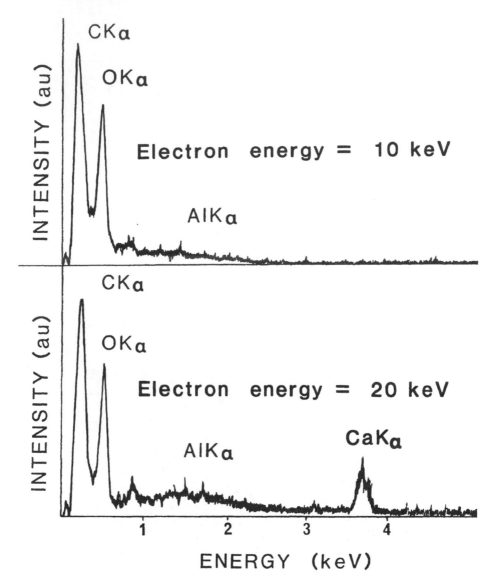

Figure 15. EMPA spectra of the interphasial failure initiation area (type 3) with 10 keV and 20 keV electron beam energies obtained for a 50 μm thick paint coating (binder B2 + crosslinker; 20% $CaCO_3$ filler).

due to the bevelled cut of the sample. EMPA spectra were recorded at different points along a line normal to the substrate surface using an electron spot size of 200 nm. The intensities (after background correction) of the different characteristic X-ray peaks were determined at each point. EMPA results are reported in Fig. 17 (carbon being the marker of the binder; the C/Ti, C/S, and C/Ca peak intensity ratios represent the binder/filler ratio of TiO_2, $BaSO_4$, or $CaCO_3$ filled systems). The average values obtained for the entire paint are shown as a dashed line. It can be observed that the C/Ti, C/S, and C/Ca ratios are at a maximum close to the polymer/metal 'interface', decrease to a minimum value, and after that, increase to a local maximum value and then decrease to reach the paint bulk ratio value. An interphasial layer is created when paints are applied onto metallic sheets and can be represented as shown in Fig. 18. The three different

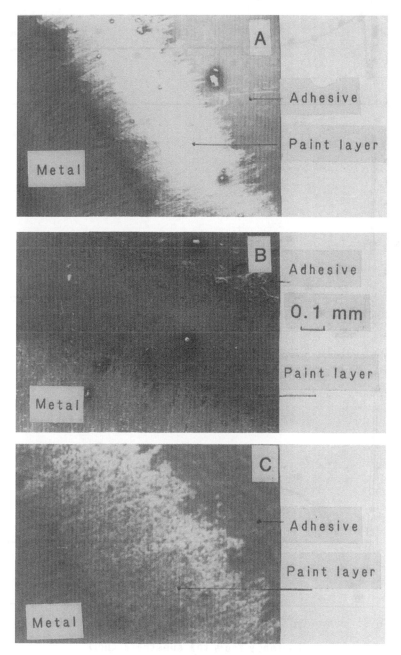

Figure 16. SEM micrographs of three different bevelled systems (50 μm thick paint (binder B1)). Paints were filled with 20% TiO_2 (A), 20% $BaSO_4$ (B), and 20% $CaCO_3$ (C).

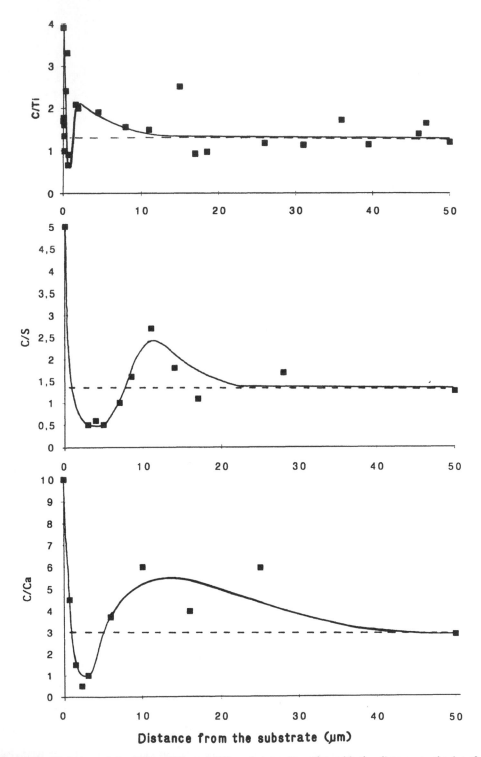

Figure 17. Variation of the C/Ca, C/Ti, and C/S peak intensity ratios with the distance to the interface in the case of (A) 20% $CaCO_3$ + 80% + B2 + crosslinker, (B) 20% TiO_2 + 80% B2 + crosslinker, and (C) 20% $BaSO_4$ + 80% B2.

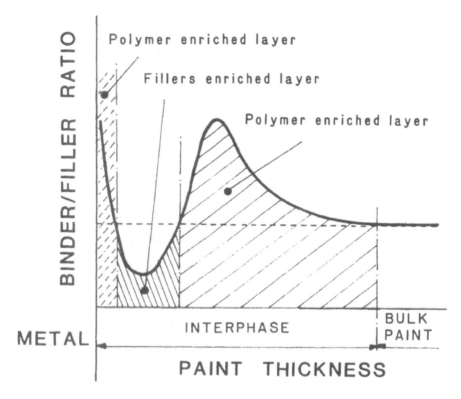

Figure 18. Paint/metal interphase 'model'.

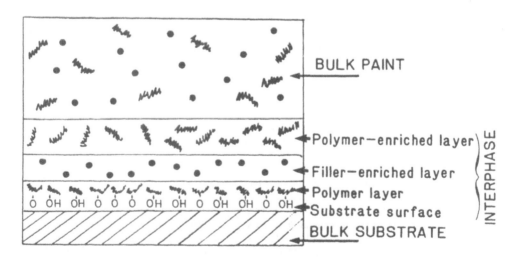

Figure 19. Schematic representation of the interphase formation.

types of interphasial failure as observed in Section 3.2.2 are then clearly shown. The paint/aluminium interphase thickness and the relative thickness of the different zones depend on the paint composition, i.e. the nature of the filler. The interphase formation may be explained, as shown in Fig. 19, by:

(a) the migration of low molecular weight material (LMWM) towards the paint/aluminium interface and/or migration of organic molecules exhibiting the greatest affinity (OMGA) to the metallic surface [15, 16];

(b) then a specific orientation of the binder molecules (LMWM and/or OMGA) reacting with the metallic surface;

(c) this specific orientation of binder molecules should lead to a preferential interaction with fillers. This could explain the filler migration onto the polymer layer to create the filler-enriched layer; and

(d) subsequently, the filler migration leads to the formation of the depleted-filler layer, i.e. the formation of the polymer-enriched layer.

5. CONCLUSION

The three-point flexure test provides a parameter which characterizes the paint/metal system practical adhesion and an easily recognizable failure initiation area. Using appropriate analytical tools, five different types of failure initiation area were observed, and a complementary study of the paint composition variation from the metallic surface towards the paint bulk allowed us to propose a model for the paint/aluminium interphase. The influence of the parameters studied, mainly the paint thickness, on the failure initiation locus tends to validate the proposed model. However, the results obtained do not agree with the classical approach, which tends to consider paints as perfectly homogeneous systems.

Acknowledgements

We wish to express our gratitude to Dr J. C. Laout, Ceripec S.A., Lyon, France for performing the pull-off test, and to Dr G. W. Scherer, E.I. Du Pont de Nemours & Co, Central R&D Department, Wilmington, USA for helpful discussions. Appreciation is also expressed to the referees for providing valuable comments on the original manuscript.

REFERENCES

1. J. C. Laout and A. Roche, in: *Proc. EUROCOAT 1991*, Nice, France, pp. 93–107. Erec, Puteaux (1991).
2. A. Roche, J. C. Laout, F. Gaillard and M. Romand, in: *Proc. EURADH '92*, Karlsruhe, Germany, pp. 568–573. Dechema, Frankfurt (1992).
3. A. A. Roche, A. K. Behme, Jr and J. S. Solomon, *Int. J. Adhesion Adhesives* **2**, 249 (1982).
4. A. A. Roche, M. J. Romand and F. Sidoroff, in: *Adhesive Joints: Formation, Characteristics and Testing*, K. L. Mittal (Ed.), pp. 19–30. Plenum Press, New York (1984).
5. F. Gaillard, A. A. Roche and M. J. Romand, in: *Adhesive Joints: Formation, Characteristics and Testing*, K. L. Mittal (Ed.), pp. 85–102. Plenum Press, New York (1984).
6. A. Roche, N. Psychoyos and D. Mazella, in: *Proc. XVIth Int. Conf. Org. Coatings Sci. Technol.*, Athens, Grece, pp. 391–412. A.C.S., Polymeric Materials: Science and Engineering Division, York (1990).
7. J. C. Laout and A. A. Roche, *Vide Couches Minces* **251**, 36 (1991).
8. D. Mazella and A. A. Roche, *Vide Couches Minces* **251**, 75 (1991).
9. F. Gaillard and A. Legros, *Vide Couches Minces* **251**, 78 (1991).
10. J. M. Cuntz, Diplôme d'Ingénieur CNAM. Paris (1986).
11. F. Gaillard, A. Roche and M. Romand, in: *Proc. ADHECOM '86*, Bordeaux, France, pp. 439–450. Adhecom, Bordeaux (1986).
12. A. Roche, F. Gaillard, M. Romand and M. von Fahnestock, *J. Adhesion Sci. Technol.* **1**, 145 (1987).

13. D. Benazet, A. Roche, F. Gaillard, S. Aivazzadeh and F. Zidani, in: *Proc. ADHECOM '89*, Bordeaux, France, pp. 359–366. Adhecom, Bordeaux (1989).
14. K. L. Mittal (Ed.), *Adhesion Measurement of Thin Films, Thick Films and Bulk Coatings*, pp. 5–17. ASTM, Philadelphia (1978).
15. J. W. Holubka, R. A. Dickie and J. C. Cassatta, *J. Adhesion Sci. Technol.* **6**, 243 (1992).
16. M. W. Urban, *J. Adhesion Sci. Technol.* **7**, 1 (1993).

Adhesion Measurement of Films and Coatings, pp. 323–330
K. L. Mittal (Ed.)
© VSP 1995.

Analysis of pull tests for determining the effects of ion implantation on the adhesion of iron films to sapphire substrates

J. E. PAWEL[1,*] and C. J. McHARGUE[2]

[1] *Oak Ridge National Laboratory, Bldg. 5500, MS 6376, PO Box 2008, Oak Ridge, TN 37831-6376, USA*
[2] *Center for Materials Processing, 121 Perkins Hall, The University of Tennessee, Knoxville, TN 37996-2000, USA*

Revised version received 19 March 1993

Abstract—Various chemical species were implanted at the interface between evaporated iron films and their sapphire substrates. From interfacial free energy considerations, it was postulated that implantation of Cr ions should increase the strength of the bond between iron and sapphire, whereas implantation of Ni ions should weaken the bonds. Implantation of Fe ions should cause changes due to irradiation-produced defects but should exhibit no chemical effect. The pull test was used to evaluate the bond strength of unimplanted and implanted specimens. Since there was large scatter in the data, it was necessary to conduct a statistical analysis. It was assumed that the separation of the film from the substrate was nucleated by interfacial defects in the bond. Because the Weibull analysis was developed to describe the failure probability due to a population of flaw-induced cracks, this technique was used to interpret the data.

Keywords: Ion implantation; pull test; thin films; adhesion; Weibull analysis.

1. INTRODUCTION

Ion implantation has been used for many years to improve the adhesion of thin films to their substrates (e.g. [1–8]). The research reported here focuses on the role that the ion species, implanted at the interface, plays in the adhesion enhancement of a metal–ceramic system. Small amounts of impurities at interfaces have been found to affect the adhesion of many systems [9–11]. For example, Hondros [9] showed that some elements, present at the interface between a metal and a ceramic, could lower the interfacial energy and thereby increase the work of adhesion. This possible adhesion mechanism, however, has not been fully utilized in ion implantation experiments. In this project, chromium, iron, or nickel ions were implanted at the interface of an

---

*To whom correspondence should be addressed.

iron film–sapphire substrate system. This system was chosen because an interfacial Cr impurity was found to enhance the adhesion of an Fe–Al_2O_3 system while Ni had the opposites effect [9]. The implantation of Fe ions produces defects but exhibits no chemical effect.

The adhesion change as the result of the implantation was investigated using the pull test. In this test, pins are attached to the film surface with an epoxy and then pulled off with a normal force. The force required to remove the film from the substrate is referred to as the pull strength. While apparently straightforward, this test is an indirect measure of the adhesion bond strength and the results can be difficult to interpret due to large scatter in the individual strength measurements [4, 12, 13]. The film failure is nucleated at interfacial defects and the propagation of the cracks is controlled by the adhesion bond strength. For a system with a low bond strength or work of adhesion, a small defect can initiate failure of the film. Because Weibull statistics were developed to describe the failure probability due to a population of flaw-initiated cracks [14, 15], the Weibull distribution was chosen to analyze the data.

2. EXPERIMENTAL

High-purity Al_2O_3 single crystals with an ⟨0001⟩ surface normal and an optical grade polish were annealed at 1350°C for 120 h in flowing air to remove residual polishing defects and surface contaminants. This technique has been found to leave an extremely clean Al_2O_3 surface [12, 16, 17]. Iron films were then electron beam-evaporated onto the substrate pieces under a base pressure of less than 5.3×10^{-5} Pa and at a rate of 1 nm s^{-1}. After film deposition to a thickness of 100 nm, the specimens were implanted with chromium (300 keV), iron (320 keV), or nickel (340 keV) ions. The fluence in each case was 1×10^{15} ions cm^{-2} and the implantations were performed at room temperature. One half of each specimen was masked during implantation to preserve an unmodified (as-deposited) region for adhesion test comparisons. The implantation energy required to place the peak of the ion concentration curve at the interface for each species was calculated using the computer codes TRIM [18] and E-DEP-1 [19]. The maximum ion concentration given by these calculations was approximately 0.1 at.%.

Both pull tests and scratch tests were used to measure the adhesion of the films. The pull tests were performed with a Sebastian I Adherence Tester. For this test, bonding pins with a 3 mm diameter were attached to the film surface with epoxy. The tensile stress required to remove the pin (with the film attached) was then measured. The maximum tensile strength of the epoxy was 70 MPa. In the scratch test, a rounded steel stylus (1.6 mm tip radius) was drawn across the film surface at a constant speed of 0.067 mm s^{-1}. A series of scratches under normal forces ranging from 0.098 to 14.7 N were made across both the as-deposited and the implanted regions of the specimens. Inspection in an optical microscope clearly showed the segments of the scratch from which the film had been removed, as well as the position of the mask such that the implanted and as-deposited regions could be distinguished. This latter test was used as a qualitative confirmation of the pull test results.

3. INTERFACIAL ANALYSES

Because the chromium, iron, and nickel ions have nearly the same mass, the slight adjustments in the accelerating voltage required for identical ion concentration profiles also resulted in cascade damage profiles and recoil distributions that were the same for each species. Several experiments were performed to confirm that the only implantation variable was the ion species. Nuclear damage and the extent of interfacial mixing were among the parameters investigated. RBS channeling (2 MeV He^+) experiments were conducted to confirm that the three ion species (chromium, iron, and nickel) caused similar nuclear damage cascades in the specimens. In these experiments, the aligned (channeling) spectra taken from the implanted region of each specimen were compared with the random spectra. The ratios of the aligned to the random yield were essentially the same for each species, implying that the damage levels in the crystals were the same. The depth range of this damage in each case was about 100 nm, and corresponded well to the damage range predicted by TRIM (i.e. the depth to which the ions still cause vacancies in Al_2O_3). This agreement lent confidence to the other computer calculations, such as ion concentration profiles.

Auger electron spectroscopy (AES) experiments also gave information about the interfacial mixing. The specimen was continually sputtered with 3 keV argon ions at a chamber pressure of 6.7×10^{-3} Pa. Simultaneous with the sputtering, the Auger signal from the specimen was monitored for the presence of O, Cr, Fe, Ni, and Al. By sputtering through the interface, and knowing the film thickness from surface profilometry, the sputter rate of the iron film was calculated and the time scale replaced with a depth scale for the element concentration profiles. The depth scale was then applied across the interface profiles for an approximation of the depth of mixing. At several points during this process, the sputter gun was

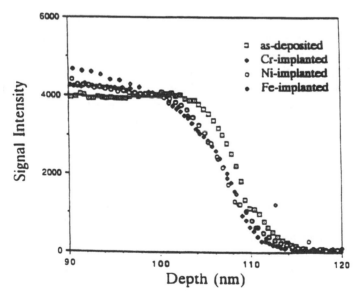

Figure 1. Iron signals from AES spectra taken from an as-deposited film superimposed over those from implanted films. The 100 nm depth marks the beginning of the interfacial region.

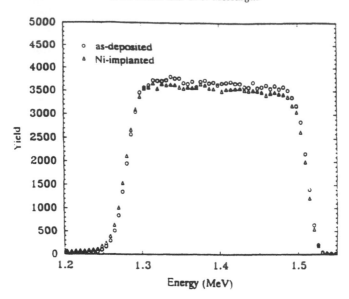

Figure 2. RBS spectra showing the iron film signals taken from an as-deposited and a Ni-implanted film at glancing angle incidence.

turned off and a survey spectrum was taken to identify all the elements present in that layer. Figure 1 shows a typical AES sputter depth profile of the iron concentration generated from an as-deposited area superimposed over those from implanted regions. Because no significant differences exist in the extent of the apparent mixed layer in the four types of specimen, it can be inferred that the transition layer seen (approximately 15 nm wide) is largely the result of the sputtering necessary to obtain the AES depth profile. Any mixing due to implantation must be 'hidden' in this regime and therefore the upper limit on the extent of a mixed layer was determined to be 15 nm. An RBS analysis performed at glancing angle incidence implied that the thickness of any mixed layer must be less than 10 nm (the resolution of the technique). Figure 2 shows RBS spectra from a Ni-implanted specimen and compares the as-deposited and implanted iron film signals. Had the implantation resulted in a mixed layer greater than 10 nm, the slope of the low-energy edge from the implanted signal would have been less than that for the as-deposited case. These spectra are typical of those taken for each implantation species.

The presence of a mixed layer was detected by energy dispersive X-ray (EDX) spectroscopy. In this experiment, the specimens were first etched in a 2% Nital solution (98% ethanol, 2% nitric acid). This effectively removed the iron film but did not attack the substrate. After complete removal of the surface film, the exposed Al_2O_3 was examined for evidence of iron, chromium, and nickel. The EDX experiments showed that some iron had been mixed into the Al_2O_3 by the implantation because it escaped chemical etching: a distinct iron signal remained in the implanted Fe/Al_2O_3 region after etching but there was no such signal in the as-deposited Fe/Al_2O_3 region. Thus, a mixed layer exists but it is very thin (less than 10 nm).

4. WEIBULL ANALYSIS

The Weibull distribution gives the probability of failure as a function of the failure stress and applies to a wide range of problems [20]. This type of analysis can be used to evaluate engineering changes, such as ion implantation on adhesion strength, because a shift in the Weibull distribution indicates a change in the probability of failure. Weibull analysis can offer several advantages over other methods; for instance, it works well even with small sample sizes and there are ways to handle various kinds of inadequacies in the data.

The Weibull distribution is defined as

$$F(t) = 1 - \exp{-\left[(t - t_0)/A\right]}^B,\tag{1}$$

where $F(t)$ is the fraction failing, t is the failure stress, t_0 is the origin of the distribution, A is the scale parameter with the dimension of stress, and B is a dimensionless shape parameter. Assuming that $t_0 = 0$ [20], this function can be rearranged to

$$\ln \ln \left[1/[1 - F(t)]\right] = B \ln t - B \ln A.\tag{2}$$

By choosing $\ln \ln [1/[1 - F(t)]]$ as the scale for the ordinate, and $\ln(t)$ as the scale for the abscissa the cumulative Weibull distribution can be represented as a straight line with slope B.

The analysis of the pull test results involves the comparison of individual test results. In some tests, however, the adhesion strength of the film exceeded the maximum strength of the epoxy (or the epoxy failed prematurely at a lower load) resulting in epoxy, rather than film, failure. In this event, the pull strength value was censored from the rest of the data set using a standard technique that is described elsewhere [20–22]. The pull test value was also censored if there was only partial film failure. Because these latter types of failure occurred by a different failure mode than the one of interest (i.e. by epoxy rather than film failure), they do not belong to the same failure distribution [20]. These values cannot be ignored when establishing the Weibull plot, however, because if they had undergone complete film failure, they would have influenced the distribution. The potential influence of these data is accounted for in the censoring technique [22].

The Weibull parameters A and B in equation (2) were found by the maximization of the log-likelihood method, which consists of finding the values of A and B which maximize the 'likelihood' of obtaining the observed data t and $F(t)$ [22]. The quantiles are values of the random variable t (failure stress) that divide the area under the probability distribution in such a way that a given proportion of the area lies to the left of the dividing line. Thus, the 10% quantile denotes the stress value for which 10% of the population has failed. The 50% quantile denotes the median of the distribution. The quantiles, Q_p, are also found from the log-likelihood equation.

5. RESULTS AND DISCUSSION

Each of the individual specimens was pull-tested (often several times) in both the as-deposited area (protected under the mask during implantation) and the ion-implanted region. The tests indicated an increase in adhesion as a result of the implantation of chromium and no significant increase in the adhesion strength as a result of nickel implantation. The iron-implanted specimens also showed an increase in the adhesion strength, although not as pronounced as that after Cr implantation.

The Weibull parameters and quantiles for the Cr- and Ni-implanted specimens are given in Table 1. The pull test results from the implanted and as-deposited regions of each specimen set were compared to determine whether the data were part of the same distribution or came from two separate distributions, one from each region. The Cr results show a clear separation, with the implanted regions having greater adhesion strengths. The data from the two regions on the Ni-implanted specimens overlap: this implies that the data were part of the same failure distribution and that the Ni implantation did not change the adhesion strength. For example, while 50% of the as-deposited samples had failed by 32 MPa, the median of the Cr-implanted distribution was 81 MPa. The median decreased slightly after Ni implantation. The Weibull plots, generated from the *A* and *B* estimates of Table 1, are shown in Fig. 3. Implantation of chromium shifted the as-deposited distribution to higher failure stresses but the implantation of nickel did not significantly shift the distribution.

Table 1.

Weibull pull test summary

Type of specimen (No. of individual measurements)	Parameter			
	A	*B*	10% quantile	50% quantile
Cr: As-deposited (23)	50.5	0.77	2.7	31.4
Cr: Implanted (13)	89.8	3.73	49.1	81.4
Ni: As-deposited (16)	45.5	1.00	4.7	31.5
Ni: Implanted (17)	31.6	1.16	4.5	23.0

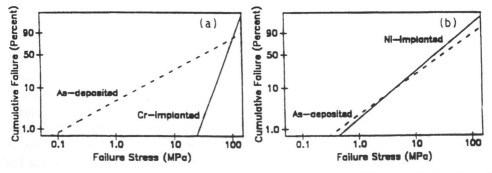

Figure 3. Weibull plots generated from the *A* and *B* estimates given in Table 1. (a) Implantation of Cr shifts the Weibull distribution to higher failure stresses with respect to the as-deposited distribution. (b) Ni implantation does not significantly shift the failure distribution.

The scratch test was used to qualitatively confirm the pull test results [12]. For the Cr- and Fe-implanted specimens, stylus loads that completely removed the film in the as-deposited region did not cause de-adhesion in the implanted area. For the Ni-implanted specimens, however, each load that removed the as-deposited film also removed the implanted film, implying no increase in adhesion strength.

The implanted ions can change the interfacial energy, and thereby the work of adhesion, in two ways: by the presence of the new species at the interface, and/or by creating damage that changes the configuration of the interface. Because the ion concentration profiles, cascade damage profiles, and recoil distributions are all very similar for the implantations in this study, adhesion processes dependent on these mechanisms should be the same for the three ion species. The only implantation variable is the ion species. The differences in adhesion between specimens implanted with the different species may be due to variations in the interfacial energy resulting from the presence of the ion species at the interface [12]. It is thought that each of the three species causes sufficient damage to promote the adhesion of the system: this is made apparent by the beneficial effect of the iron ions, which is not a new element at the $Fe-Al_2O_3$ interface. The chromium ions, however, have the additional effect of decreasing the interfacial energy for this system [9] and so these specimens have a more pronounced increase in adhesion. The nickel ions, on the other hand, have the opposite effect on the interfacial energy [9], and to a large enough extent that no increase in adhesion is seen.

6. CONCLUSIONS

Chromium, iron, and nickel ions were implanted at the interface between an iron film and an Al_2O_3 substrate. The adhesion of the implanted films was measured quantitatively by pull testing and qualitatively by scratch testing. In every case, the adhesion was compared with that of as-deposited films. There is an adhesion increase after either chromium or iron implantation but not after nickel implantation. The adhesion effect as a result of ion implantation is proposed to be due to changes in the interfacial energy due to damage and interfacial mixing, and the presence of the ion species at the interface. Thus, while ion implantation can be used to improve the adhesion of many film–substrate combinations, not all ion species are equally effective. Extensive interfacial mixing is not necessary for adhesion enhancement. In this study, the mixed layer was less than 10 nm thick. Weibull statistics can be a useful and effective method to represent and analyze pull test data.

Acknowledgements

This research was sponsored by the US Department of Energy, Assitant Secretary for Conservation and Renewable Energy, Office of Transportation Technologies, as part of the Ceramic Technology Project of the Materials Development Program, and by the Division of Materials Sciences, under contract number DE-AC05-84OR21400 with Martin Marietta Energy Systems, Inc. Additional financial support was received under a National Science Foundation Graduate Fellowship (J. E. Pawel). This research was conducted while J. E. Pawel was a guest of Oak Ridge Associated Universities under contract number DE-AC05-76OR00033 between the US Department of Energy and Oak Ridge Associated Universities and assigned to Oak Ridge National Laboratory. We are grateful to Dr G. R. Rao and Ms S. G. Winslow for critical review of the manuscript.

REFERENCES

1. L. E. Collins, J. G. Perkins and P. T. Stroud, *Thin Solid Films* **4**, 41 (1969).
2. G. Auner, Y. F. Hsieh, K. R. Padmanabhan, J. Chevallier and G. Sorensen, *Thin Solid Films* **107**, 191 (1983).
3. S. Noda, H. Doi and O. Kamigaito, *Radiat. Phys. Chem.* **30**, 253 (1987).
4. J. E. Pawel and C. J. McHargue, *J. Adhesion Sci. Technol.* **2**, 369 (1988).
5. J. E. E. Baglin and G. J. Clark, *Nucl. Instr. Meth. Phys. Res.* **B7/8**, 881 (1985).
6. L. Romana, P. Thevenard, G. Massouras, G. Fuchs, R. Brenier and B. Canut, in: *Structure–Property Relationships in Surface-Modified Ceramics*, C. J. McHargue, R. Kossowsky and W. O. Hofer (Eds), p. 181. Kluwer, Dordrecht (1989).
7. D. K. Sood, W. M. Skinner and J. S. Williams, *Nucl. Instr. Meth. Phys. Res.* **B7/8**, 893 (1985).
8. T. A. Tombrello, in: *Thin Films and Interfaces II*, J. E. E. Baglin, D. R. Campbell and W. K. Chu (Eds), p. 173. North-Holland, New York (1984).
9. E. D. Hondros, in: *Science of Hard Materials*, E. A. Almond, C. A. Brookes and R. Warren (Eds), p. 121. Adam Hilger, Bristol (1986).
10. A. G. Evans and M. Rühle, *MRS Bull.* 46 (Oct. 1990).
11. J. E. E. Baglin, in: *Ion Beam Modification of Insulators*, P. Mazzoldi and G. W. Arnold (Eds), p. 585. Elsevier, Amsterdam (1987).
12. J. E. Pawel, PhD Dissertation, Vanderbilt University, Nashville, TN (1991).
13. E. Abonneau, G. Fuchs, M. Treilleux and A. Perez, *Nucl. Inst. Mech. Phys. Res.* **B46**, 111 (1990).
14. M. R. Lin, J. E. Ritter, L. Rosenfeld and T. J. Lardner, *J. Mater. Res.* **5**, 1110 (1990).
15. D. G. S. Davies, *Proc. Br. Ceramic Soc.* **22**, 429 (1973).
16. E. Abonneau, A. Perez, G. Fuchs and M. Treilleux, *Nucl. Instr. Meth. Phys. Res.* **B59/60**, 1183 (1991).
17. E. Abonneau, PhD Dissertation, Université Claude Bernard, Lyon, France (1990).
18. J. F. Ziegler, J. P. Biersack and U. Littmark, *The Stopping and Range of Ions in Solids*. Pergamon Press, New York (1985).
19. C. M. Davisson and I. Manning, *New Version of the Energy-Deposition Code E-DEP-1: Better Stopping Powers*. Naval Research Laboratory Report 8859 (1986).
20. R. B. Abernethy, J. E. Breneman, C. H. Medlin and G. L. Reinman, *Weibull Analysis Handbook*. Aero Propulsion Laboratory, Air Force Wright Aeronautical Laboratories, Wright Patterson Air Force Base, Ohio (1983).
21. J. F. Lawless, *Statistical Models and Methods for Lifetime Data*. John Wiley, New York (1982).
22. J. E. Pawel, W. E. Lever, D. J. Downing, C. J. McHargue, L. J. Romana and J. J. Wert, in: *Thin Films: Stresses and Mechanical Properties III*, W. D. Nix, J. C. Bravman, E. Artzt and L. B. Freund (Eds), p. 541. Materials Research Society, Pittsburgh, PA (1992).

Adhesion Measurement of Films and Coatings, pp. 331–344
K. L. Mittal (Ed.)
© VSP 1995.

Measurement of the adhesion of diamond films on tungsten and correlations with processing parameters

M. ALAM,[1,]* D. E. PEEBLES and J. A. OHLHAUSEN[2]

[1]*Materials Engineering, Jones Hall 114, New Mexico Institute of Mining and Technology, Socorro, NM 87801, USA*
[2]*Sandia National Laboratories, Albuquerque, NM 87185, USA*

Revised version received 24 June 1993

Abstract—Adhesion between diamond films and tungsten substrates is reported as a function of the deposition processing parameters. Diamond films were grown by a hot filament method as a function of seven different processing parameters: substrate scratching prior to diamond deposition, substrate temperature, methane content of the input gas mixture, filament temperature, filament–substrate distance, system pressure, and total gas flow rate. Adhesion was measured by using a Sebastian Five A tensile pull tester. Testing was complicated by the non-uniformity of the film thickness, diamond quality, film cohesion, and surface preparation across the full substrate surface area. Various types of film failure mode were observed, which did not correlate with the film processing parameters. The measured adhesion values showed larger variations from point to point across the sample surface and from identically prepared samples than variations as a function of the film processing parameters. Weak correlations of adhesion with the processing parameters were found using statistical analysis of the results from multiple pulls on a large number of samples. The statistical results suggest that substrate preparation, gas flow rate, and gas pressure are the most important processing parameters affecting the film adhesion, while the temperature of the hot filament has little or no effect on the adhesion of the film. However, improvements in film processing and adhesion testing need to be made before true quantitative adhesion testing of high-quality diamond films can be accomplished.

Keywords: Diamond films; tungsten; adhesion; tensile pull testing; film failure modes.

1. INTRODUCTION

Chemically vapor-deposited (CVD) polycrystalline diamond films are being considered for use as protective films on high-wear surfaces. One of the critical issues in this regard is the adhesion between the diamond film and the underlying substrate material. An understanding of the diamond film–substrate bond strength and how it can be improved is therefore of considerable importance. Very few studies dealing with the measurement of adhesion between CVD diamond films and substrate materials have been published and all of them are very recent. Adhesion measurements of diamond films are complicated by the very properties of diamond films that are desirable—high hardness, well-defined grain structure, and resistance to mechanical damage. Several research groups have presented quantitative data on adhesion based on the cutting life of diamond-coated tool inserts [1–3], the wear of diamond film surfaces as measured by the pin-on-disk method [4], and the resistance of films to erosion by impact with sand particles [5]. Other groups have used more direct methods for

---

*To whom correspondence should be addressed.

the quantitative determination of diamond film–substrate pair adhesion, such as scratch testing [6], tensile pull testing [7, 8], and indentation testing [9–11]. Since the adhesion measured by the above-mentioned tests is dependent on a wide variety of variables, the measured value is called the practical adhesion [12]. It is not a fundamental quantity, but it can be useful for comparison purposes. Although considerable progress has been made, systematic studies dealing with the quantitative measurement of the adhesion of diamond films to various substrate materials and correlations with deposition variables have yet to be reported. Such studies are extremely important in arriving at some general guidelines for synthesizing diamond films which exhibit optimum adhesion to the substrate. In this paper, we present adhesion data obtained by tensile pull testing for CVD diamond films on tungsten. The diamond films were prepared by a hot filament-assisted process for a range of critical deposition parameters. The objective of the study was to determine how the adhesion of diamond films on tungsten varied with the processing parameters, with the ultimate goal of establishing the optimum set of conditions for growing diamond films on tungsten with good adhesion.

2. EXPERIMENTAL

Diamond films were deposited on $25 \times 25 \times 1$ mm polycrystalline tungsten substrates (less than 500 ppm total impurities) on a 380 mm^2 area (22 mm diameter) from methane–hydrogen gas mixtures by a hot filament-assisted CVD method in a bell jar system, using tungsten ribbon filaments (0.762 mm wide $\times$ 0.0762 mm thick). Prior to deposition, the substrates were polished with alumina powder of 5 μm size, ultrasonically cleaned with methanol for 12 min, and seeded by scratching the surface with natural diamond powder. The films were prepared as a function of seven different process variables (controllable factors): substrate scratching prior to deposition, substrate temperature, methane content of the input gas mixture, filament temperature, filament–substrate distance, system pressure, and total gas flow rate. For each of the seven series of experiments, the range of the variables studied and the values of other parameters kept constant during that series of experiments are given in Table 1. As an example, the second row of the table indicates that diamond films were prepared as a function of the substrate temperature, which was varied from 1173 to 1323 K. During this series of depositions, other deposition variables were kept at the values listed in this row. Two samples were prepared at each condition and a total of 50 samples were made.

Adhesion testing was performed by the tensile pull method, using a Sebastian Five A adhesion tester manufactured by Quad Group of Spokane, WA. For each sample, an epoxy-coated aluminum pull stud with a 5.73 mm^2 head (2.7 mm diameter) was mounted on the sample by clamping the stud normal to the substrate using a mounting clip, as shown in Fig. 1a. The assembly was placed in an oven on an aluminum block preheated to 433 K for 1 h to cure the epoxy. The assembly was then removed from the oven and allowed to cool to room temperature for 30 min. The mounting clip was then removed and the study was clamped into the tester. Assembly of the pull stud to the sample to be tested is illustrated schematically in Fig. 1b. The tester applies a constantly increasing pull

Table 1.
Diamond film deposition parameters

For the series varying:	S (μm)	T_s (K)	V_{CH_4} (vol%)	T_f (K)	d (mm)	P (Pa)	Q (sccm)
Substrate scratching (S, μm)[a]	0.5–1.50	1223	0.5	2213	5	4000	100
Substrate temperature (T_s, K)	43	1173–1323	0.5	2213	5	4000	100
Gas composition (V_{CH_4}, vol %)	43	1223	0.1–2.0	2213	5	4000	100
Filament temperature (T_f, K)	43	1223	0.5	2053–2423	5	4000	100
Filament-substrate distance (d, mm)	43	1223	0.5	2213	5–15	4000	100
System pressure (P, Pa)	43	1223	0.5	2213	5	1333–6665	100
Total gas flow rate (Q, sccm)	43	1223	0.5	2213	5	4000	40–600

[a]This represents the size of the natural diamond particles used to seed the tungsten surface by scratching.

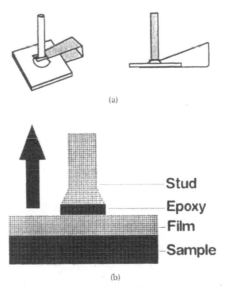

(a)

(b)

Figure 1. (a) Schematic diagram of the pull stud clamped to the substrate by the use of the mounting clip. (b) Schematic diagram of the assembly of the pull stud to the sample to be tested.

force to the stud, normal to the substrate surface, causing the stud to pull from the substrate. The maximum force applied to the stud (in N) and the force per unit area (in Pa) were determined by the tester. Multiple pulls (6–10) were performed on each sample to try to determine an estimate of the average adhesion across the sample surface. Typically, the range of adhesion values across the surface was up to ±50% of the average adhesion value for the sample. All of the individual pull values were averaged to obtain a single adhesion value for each sample.

3. RESULTS AND DISCUSSION

3.1. Adhesion failure modes

In order to get a better understanding of the pull test results, the pulled areas on the samples were examined by optical microscopy. The load at which the stud is pulled from the sample is representative of the film adhesion only if the film is pulled cleanly from the entire head area of the stud. The desired pull mode of film failure (good, clean pull) is shown in Fig. 2. When a sufficient number of pulls could be obtained from a sample, only the data from good, clean pulls such as that depicted in Fig. 2 were used. On some occasions partial film failure occurred, as shown in Fig. 3. These results occurred when the epoxy bonding was uneven across the surface as a result of inhomogeneity in the epoxy or the film. Such tests provide good film removal from only a portion of the stud area, producing an area

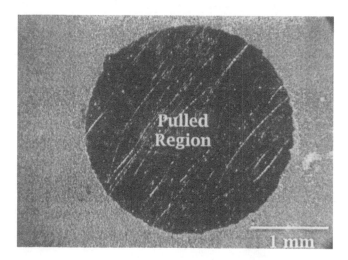

Figure 2. Photograph of a sample showing a good, clean pull.

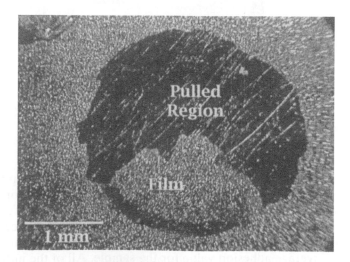

Figure 3. Photograph of a sample showing partial film failure.

of film pulled that is not equal to the area of the head of the stud. In such cases, the measured value was corrected for the actual area of film pulled from the sample to obtain a representative corrected adhesion value. Where only a few good, clean pulls were obtained from a sample, corrected adhesion data from partial pulls were incorporated to provide better statistics.

Besides the failure modes mentioned above, there are other possible failure modes which do not produce usable adhesion data. One possibility is failure at the film–epoxy interface as a result of poor epoxy–film bonding, due to surface contamination or poor epoxy adhesion, as shown in Fig. 4. Data from these types of tests were not used because no information was gained about the actual value of the film adhesion. Occasionally, adhesion of the film to the tungsten substrate exceeded the cohesive strength of the diamond film itself, producing a failure within the film rather than at the film–substrate interface, as shown in Fig. 5. Data

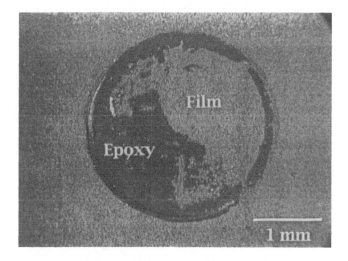

Figure 4. Photograph of a sample showing failure at the film–epoxy interface.

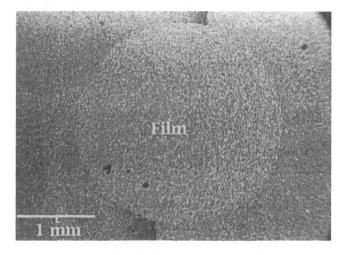

Figure 5. Photograph of a sample showing cohesive failure within the film.

from tests resulting in such a failure mode were also discarded because they do not provide an accurate value of the adhesion, but rather a minimum estimate of the adhesion strength. Some tests resulted in film spallation, as shown in Fig. 6. In these tests, the film surrounding the pulled area was also debonded from the substrate for an extensive distance. These tests provided artificially large adhesion values, since additional work was performed other than just removing the film from under the stud. Measured values from such tests were therefore not used in any further data processing. Finally, some tests resulted in total delamination of the film from the sample, with the resulting fracture of the film into many small pieces, as shown in Fig. 7. Adhesion values obtained from such tests are clearly not directly related to the stud area and were not used.

All the tensile pulls (even the good, clean pulls) include some degree of cohesive failure within the film as well. The film is removed not only from the

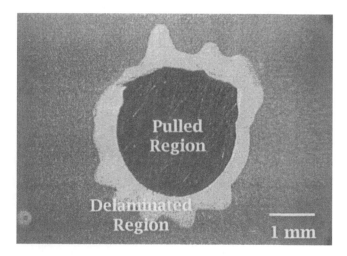

Figure 6. Photograph of a sample showing film spallation.

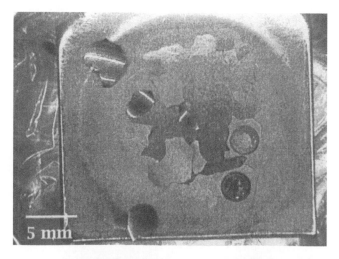

Figure 7. Photograph of a sample showing total delamination of the film into small fragments.

substrate, but also from the remainder of the film itself, requiring cohesive rupture around the perimeter of the pulled area. To test the relative contribution of this cohesive component, several samples were prepared by depositing diamond films through masks so that only small areas of film having diameters very close to that of the pull stud were obtained. On these samples, no statistical differences in adhesion values were found relative to those on identically prepared large area samples, suggesting that the cohesive component in these films is relatively small. The various film failure modes observed throughout this study did not exhibit any correlation with the deposition parameters.

3.2. *Adhesion* vs. *processing parameters*

Variations in the measured adhesion values vs. the processing parameters are presented in Figs 8–14. Each plot shows the effect of one processing variable; the other variables were kept constant at the values presented in Table 1. The effect of the size of the diamond grit used to scratch the tungsten surface prior to film deposition (substrate scratching) is shown in Fig. 8. Scratching the substrate surface with diamond grit of a larger size should promote adhesion due to the increased surface area and mechanical interlocking between the film and the substrate. Excessive roughness due to scratching with an even larger grit, on the other hand, could result in film defects leading to poor adhesion. The figure does not show any strong correlation between adhesion and the diamond grit size used to scratch the surface.

The other film processing parameters should cause changes in the film structure, growth mechanisms, and composition, which could influence the film adhesion values. Figure 9 is a plot of adhesion vs. the substrate temperature during deposition. Figure 10 shows the correlation of adhesion with the filament temperature, while Fig. 11 shows the correlation with the filament–substrate

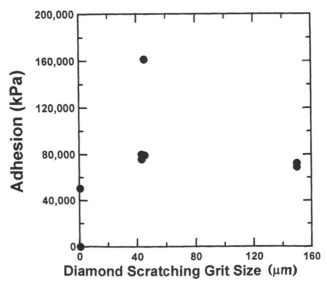

Figure 8. Adhesion values obtained with the Sebastian Five A vs. the diamond grit size used for scratching during substrate preparation.

M. Alam et al.

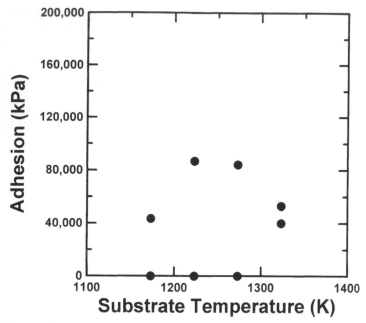

Figure 9. Adhesion values obtained with the Sebastian Five A vs. the substrate temperature during deposition.

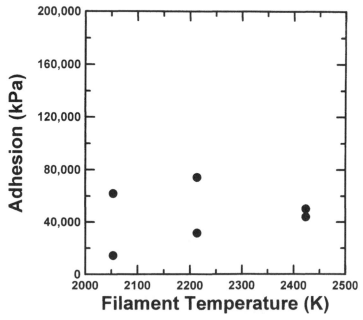

Figure 10. Adhesion values obtained with the Sebastian Five A vs. the filament temperature during deposition.

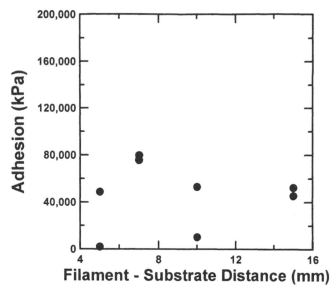

Figure 11. Adhesion values obtained with the Sebastian Five A vs. the filament–substrate separation distance during deposition.

separation distance. The remaining figures plot the adhesion vs. processing gas parameters—total gas pressure in Fig. 12, gas flow rate in Fig. 13, and methane content in Fig. 14. None of these plots shows any strong correlation of adhesion with any processing parameter. In general, the reproducibility of adhesion values is very poor. Changes in the measured adhesion values as a function of any processing variable are no larger than the variations in adhesion values measured on samples prepared under identical conditions, or even from different locations on the same sample.

3.3. Statistical analysis

The measured adhesion values are widely scattered. This wide scatter in the adhesion data obtained by the tensile pull method has been reported in other systems as well. Pawel *et al.* [13] assumed that the scatter was due to a random distribution of interfacial flaw sizes controlling the failure nucleation. Based on the above assumption, they used Weibull statistics, which has been developed to describe the failure probability due to a population of flaw-initiated cracks, to analyze their data [14]. In the present investigation, no assumption is made with regard to the origin of scatter in the data. Statistical analysis was performed by utilizing a software program called Strategy (Experiment Strategies Foundation, Seattle, WA; version 4.91). After the data are analyzed and entered into a results matrix, a model is chosen for the interaction of the different variables (process parameters) and the response (adhesion). For the data set at hand, a 'Main Effects Only' model could be explored, which allows each process parameter to influence adhesion, but not to affect other process parameters, and does not allow for any synergistic effects between variable factors. The model chosen selects the type of equations used to fit the data. The 'Main Effects Only' model uses the most

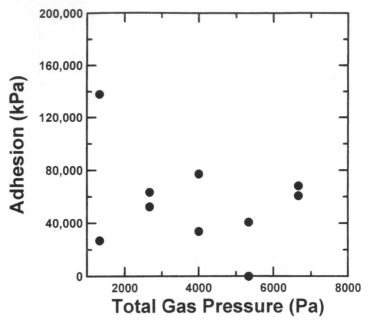

Figure 12. Adhesion values obtained with the Sebastian Five A vs. the total gas pressure during deposition.

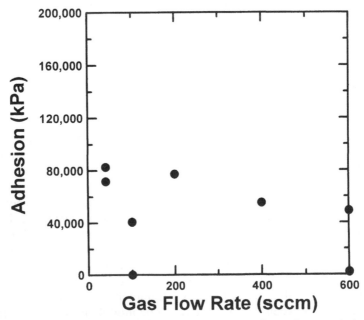

Figure 13. Adhesion values obtained with the Sebastian Five A vs. the gas flow rate during deposition.

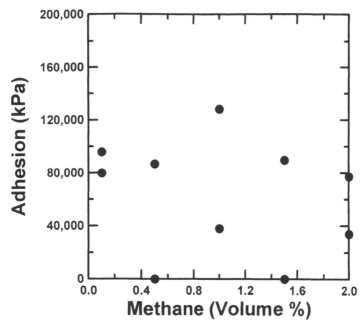

Figure 14. Adhesion values obtained with the Sebastian Five A vs. the methane content of the feed gas during deposition.

simplistic behavior of the process parameters and measured responses (first-order linear responses). The advantage of the statistical fitting is that adhesion values from all 50 samples are used to help determine the effects of each processing parameter, not just the few samples that are a part of any single series.

All the data were fitted to first-order linear equations, using a least-squares curve-fitting method. Least-squares fitting of the eight variables (seven process parameters and one measured response, adhesion) for all 50 samples yielded equation (1), describing the correlation of adhesion to each of the process parameters:

$$A = 161\,760 + 225\,S - 86.9\,T_s + 15.3\,T_f - 1740\,d - 5.53\,P - 59.2\,Q - 5700\,V, \quad (1)$$

where A is the adhesion (in kPa), S is the substrate scratching (diamond grit size, in μm, for scratching) T_s is the substrate temperature (in K), T_f is the filament temperature (in K), d is the filament–substrate distance (in mm), P is the gas pressure (in Pa), Q is the gas flow rate (in sccm), and V is the methane content of the feed gas (in vol%). These results are presented in the form of normalized fitting coefficients in Table 2. The normalized fitting coefficients are obtained by allowing each variable to range from -1 for the minimum value to $+1$ for the maximum value during the fitting process. Consequently, the normalized fitting coefficients describe the relative importance of each of the process parameters on adhesion. Comparison of the normalized fitting coefficients for each of the process parameters suggests certain conclusions, in spite of the large scatter. The most important process parameters affecting the film adhesion are the substrate scratching, gas flow rate, and gas pressure. Adhesion is affected less strongly by the substrate temperature, filament–substrate distance, and composition of the

Table 2.
Normalized fitting coefficients for the first-order
equation representing the effects of process
parameters on adhesion (see the text for a
discussion of the interpretation and use of these
coefficients)

Constant	5.86
Substrate scratching	2.45
Substrate temperature	−0.94
Methane content	−0.79
Filament temperature	0.41
Filament–substrate distance	−1.26
Pressure	−2.13
Flow rate	−2.40

Table 3.
Experimental deposition parameters (as determined by strategy) for growing diamond films of
desired adhesion values

adhesion = 91.7 kPa	
Substrate scratching (μm)	150
Substrate temperature (K)	1173
Methane content (vol%)	1.05
Filament temperature (K)	2423
Filament–substrate distance (mm)	5
Pressure (Pa)	4000
Flow rate (sccm)	40

feed gas. The temperature of the hot filament has little or no effect on film
adhesion. The fitting coefficients and equation (1) can also be used to predict the
adhesion between tungsten and diamond films (prepared in the particular
system), for a given set of process parameters. A sample prediction result is
presented in Table 3, which should produce the high adhesion value noted at the
top when the value of the process parameters listed in the table are used.

4. CONCLUSIONS

Adhesion between tungsten substrates and diamond films deposited on them by a
hot-filament chemical vapor deposition method as a function of seven different
processing parameters was measured by Sebastian tensile pull testing. Various
types of film failure mode were observed, including good pulls, partial film
failures, failure at the film–epoxy interface, cohesive failure within the film, film
spallation, and film delamination. The mode of film failure does not correlate with
the film processing parameters. A significant amount of scatter was observed in
the measured adhesion values across the film surface, perhaps due to non-
uniformity of the film thickness, diamond quality, film cohesion, and surface
preparation across the full substrate surface. In fact, variations in the measured
adhesion values from spot to spot across a sample surface or between identically
prepared samples exceeded any variations observed as a function of the film

processing parameters. Consequently, no clear correlations of adhesion with the processing parameters were observed. Although no strong correlations were found, statistical methods were used to determine any weak correlations between the measured adhesion values and the diamond film processing parameters. The statistics suggest that substrate scratching, gas flow rate, and gas pressure are the most important factors affecting the film adhesion, while the temperature of the hot filament has little or no effect on film adhesion. The lack of any strong correlation of adhesion with the film processing parameters suggests that one of two situations applies: (1) the parameter which has the greatest control over film adhesion has not been identified and studied; or (2) the variations in the adhesion-controlling parameter across the sample surface and from sample to sample are so large that a sample with a 'representative' control value cannot be obtained. Improvements in the uniformity of film deposition and properties are required before true quantitative adhesion results can be obtained. In addition, alternative methods of adhesion testing still need to be explored to get a reliable and clear quantitative analysis of the adhesion of diamond films. However, these efforts are hampered by the very nature of the diamond films being tested: their high hampered by the very nature of the diamond films being tested: their high hardness, well-resolved crystallinity, and facets. Prior to settling on tensile pull-failed, due to the destruction of the testing stylus tip by the hard, cutting facets contained within the diamond films. Although great effort went into these studies, with careful control of the processing parameters and extensive analysis of the resulting films, it is clear that quantitative adhesion testing of high-quality diamond films is still not readily available.

Acknowledgements

Part of this work was performed at Sandia National Laboratories supported by the US Department of Energy (DOE) under contract No. DE-AC04-76-DP00789. We gratefully acknowledge the efforts of Susan F. Hunt, who completed most of the statistical analysis of the results.

REFERENCES

1. M. Murakawa, S. Takeuchi, Y. Hirose, K. Komaki and M. Yanagisawa, in: *New Diamond Science and Technology*, R. Messier, J. T. Glass, J. E. Butler and R. Roy (Eds), pp. 815–820. Materials Research Society, Pittsburgh, PA (1990).
2. M. Murakawa and S. Takeuchi, in: *Diamond and Diamond-like Films and Coatings*, R. E. Clausing, L. L. Horton, J. C. Angus and P. Koidl (Eds), pp. 757–764. Plenum Press, New York (1990).
3. K. Saijo, K. Uno, M. Yagi, K. Shibuki and S. Takatsu, in: *Applications of Diamond Films and Related Materials*, Y. Tzeng, M. Yoshikawa, M. Murakawa and A. Feldman (Eds), pp. 69–76. Elsevier, Amsterdan (1991).
4. M. N. Gardos, B. L. Soriano, D. L. Patterson, R. H. Hauge and J. L. Margrave, in: *Proceedings of the 2nd Interational Symposium on Diamond Materials*, A. J. Purdes, B. M. Meyerson, J. C. Angus, K. E. Spear, R. F. Davis and M. Yoder (Eds), pp. 365–373. The Electrochemical Society, Pennington, NJ (1991).
5. F. Davanloo, T. J. Lee, D. R. Jander, J. H. You, H. Park and C. B. Collins, *Thin Solid Films* **212**, 216–219 (1992).
6. D. E. Peebles and L. E. Pope, *J. Mater. Res.* **11**, 2589–2598 (1990).
7. M. Alam, D. E. Peebles and D. R. Tallant, in: ref. 4, pp. 348–356.
8. R. Ramesham, T. Roppel, R. W. Johnson and J. M. Chang, *Thin Solid Films* **212**, 96–103 (1992).

9. C. T. Kuo, T. Y. Yen, T. H. Huang and S. E. Hsu, *J. Mater. Res.* **11**, 2515–2523 (1990).
10. M. E. O'Hern and C. J. McHargue, in: ref. 2, pp. 715–721.
11. M. D. Drory and M. G. Peters, in: ref. 4, pp. 340–347.
12. K. L. Mittal (Ed.), *Adhesion Measurement of Thin Films, Thick Films and Bulk Coatings*, K. L. Mittal pp. 5–17. ASTM, Philadelphia, PA (1978).
13. J. E. Pawel, W. A. Lever, D. J. Downing, C. J. McHargue, L. J. Romana and J. J. Wert, *Mater. Res. Soc. Symp. Proc.* **239**, 541–546 (1992).
14. M. R. Lin, J. E. Ritter, L. Rosenfeld and T. J. Lardner, *J. Mater. Res.* **5**, 1110–1117 (1990).

Adhesion Measurement of Films and Coatings, pp. 345–356
K. L. Mittal (Ed.)
© VSP 1995.

Adhesion measurements of non-crystalline diamond films prepared by a laser plasma discharge source

F. DAVANLOO,* T. J. LEE, H. PARK, J. H. YOU and C. B. COLLINS

Center for Quantum Electronics, University of Texas at Dallas, P.O. Box 830688, Richardson, TX 75083-0688, USA

Revised version received 30 June 1993

Abstract—Films of amorphic diamond can be deposited with a laser plasma discharge source of multiply-charged carbon ions in an ultrahigh vacuum (UHV) environment without the use of any catalyst in the growth mechanism. The beam from a pulsed Nd–YAG laser is focused at very high power densities onto a graphite feedstock and the resulting plasma ejects carbon ions carrying keV energies through a discharge space to the substrates to be coated. The high energies of condensation produce interfacial layers between the films and substrate materials, resulting in levels of adhesion which can support the protection of fragile and sensitive substrates subjected to harsh environmental conditions. Coatings of 2–5 μm thicknesses have extended the lifetimes of substrate materials such as Si, Ti, Ge, ZnS, and stainless steel against the erosive wear from high-speed particles and droplets by factors of tens to thousands. In this paper, we give details of the adhesion and mechanical properties of amorphic diamond films. Emphasis is placed on the quantitative methods to assess and measure the adhesion of films to a variety of substrates. Parameters such as the bonding of the films are examined by Rutherford backscattering spectrometry and nanoindentation techniques. Resistances to wear and debonding are estimated with a modified sand blaster which simulates erosive environments.

Keywords: Amorphic diamond; laser ablation; adhesion; interfacial layer.

1. INTRODUCTION

Amorphic diamond films have been distinguished by their mechnical strength from other diamond-like films prepared by a variety of techniques [1–7]. They have been produced by accelerating and quenching an intense layer plasma of C^{3+} and C^{4+} onto a cold substrate [8]. Recent microstructural studies of amorphic diamond films have shown them to be composed of hard dense nodules with grain sizes of the order of 10–100 nm. The diamond characteristics of this material have been evaluated by several analytical methods. Different measurements agree in supporting sp^3 contents of better than 75% [3–5]. No other source has been reported for these materials and, to date, the quenching of C^+ ion beams alone onto a substrate has been shown to produce only i-C or defected graphite [4, 7]. The importance of this amorphic diamond material has been suggested by recent reports of its unique mechanical properties [5, 6]. It was shown that a combination of low internal stress and high bonding strength produced coatings with exceptional resistance to wear and erosion.

In those studies, a beam bending method was used to measure the internal

---

*To whom correspondence should be addressed.

stress of nominally produced amorphic diamond films and a relatively low value of compressive stress was found. The dependence of stress on the laser intensities at the graphite ablation target indicated that the stress was largely due to the graphic contents and that it could be reduced further by improving the sp^3 fraction of the films. Analyses of the interfaces of amorphic diamond films on several substrates showed significant interfacial layers [5, 6]. This was caused by the highly localized but intense levels of energy density created by impacts of C^{3+} and C^{4+} ions. The mechanical properties, namely, hardness, Young's modulus, and stiffness, were obtained by a nanoindentation measurement and it was shown that coatings of laser plasma diamond could protect substrates and increase the lifetime againt abrasive wear from impacts of particles or rain droplets [5, 9].

Recently, with critical control of the point of ablation, films have been produced that are as hard as natural diamond [10]. These unique properties, together with the room-temperature growth environment, make this material suitable for abrasion and corrosion protection of important substrates such as medical implants, prostheses, computer disks, and fragile infrared (IR) optics.

2. DEPOSITION METHOD AND STRUCTURAL REVIEW

The deposition of amorphic diamond by a laser plasma discharge has been realized with a Q-switched Nd–YAG laser at the University of Texas at Dallas. As described previously [1–7], the laser delivers 250–1400 mJ to a graphic feedstock in a UHV system at a repetition rate of 10 Hz. The beam is focused to a diameter chosen to keep the intensity on the target near 5×10^{11} W cm^{-2} and the graphite is moved so that each ablation occurs from a new surface. A high current discharge confined to the path of the laser-ignited plasma is used to heat and process the ion flux further. Discharge current densities typically reach 10^5–10^6 A cm^{-2} through the area of the laser focus, but the laser power alone is sufficient to insure that the resulting plasma is fully ionized [8]. A planetary drive system for rotating substrates within the core of the plasma where they are exposed only to ions insures the simultaneous deposition of uniform layers of amorphic diamond over several substrate disks 30 mm in diameter [1–7]. Despite the relatively high ion fluxes, bulk temperatures of the substrates monitored by a thermocouple do not exceed 35°C over deposition periods of several hours.

Modeling studies have shown that at laser intensities around 5×10^{11} W cm^{-2} the plasma is composed of multiple-charged carbon ions with kinetic energies of the order of 1 keV [8]. The impact of the laser plasma upon a substrate is equivalent to an irradiation with a very high fluence ion beam. Quenching of such energetic ions yields diamond, while the condensation of neutral carbon produces only graphite [3, 4].

The initial energy density in the ablation plume has been shown to be the dominant factor in determining the diamond content of the material subsequently deposited. Marquardt *et al.* [11] determined a critical threshold laser intensity of 5×10^{10} W cm^{-2} on the carbon feedstock above which diamond-like films were condensed from the fully-ionized carbon plasmas and below which only soft, graphite layers were deposited that resembled those produced by the thermal evaporation of carbon. As a result of these findings, amorphic diamond films are grown with significant fluxes of C^{3+} and C^{4+} with kinetic energies around 1 keV, arriving from the core of an intense ablation plume. A combined process of breaking sp^2 bonds and encouraging sp^3 bonding by an overconstraint during implanting caused by the impact of these high-energy ions grows high-quality film. In contrast, very dark graphitic layers are deposited by carbon materials grazing only through the cooler periphery of the plume [4].

Recently, different types of amorphic diamond have been identified and their properties have been found to depend on the energies of the ions condensed from the laser plasma [4, 7]. Comparative microstructural studies of films condensed from ions passing through the core and periphery of a laser plasma have been performed extensively by scanning tunneling microscopy (STM) [4].

The unique nodular structure of amorphic diamond films has also been reported [4, 7] from examinations of films with STM. Typically, the nodule size ranges from 10 to 100 nm in diameter and imparts the properties of diamond found in the finished films [5, 6, 9]. The microstructure of the film grown at the periphery of the laser plasma looks very graphitic and shows none of the hardened nodules associated with sp^3 bonding [4]. For convenience, Fig. 1 reproduces [6] a typical appearance of an amorphic diamond obtained by transmission electron microscopy (TEM). A nodular structure is clearly seen in this figure.

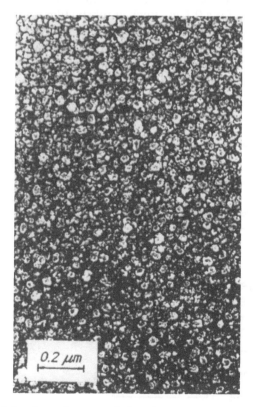

Figure 1. Diamond nodules shown by transmission electron microscopy (TEM) of a gold-coated replica of a film of amorphic diamond.

3. INTERFACE ANALYSES BY RUTHERFORD BACKSCATTERING SPECTROMETRY

With our laser plasma deposition system, we have produced over 3000 different films of thicknesses varying from 0.1 to 5 μm on substrates including Si, fused silica, glass, Ag, Cu, Mo, Ni, Ge, W, SiC, ZnS, stainless steel and polyimide material [1–7, 9]. The fact that the nominally produced films do not wrinkle, buckle, crack, or debond is indicative of good adhesion and the deep penetration of the coating into the substrate materials. The impact of C^{3+} and C^{4+} ions carrying keV energies from the laser plasma leads to interfacial layers of disturbed compositions of considerable thicknesses on a variety of materials. Rutherford backscattering spectrometry (RBS) is an important diagnostic technique for the adhesion of such films and a profile of elemental composition with depth below the original surface can be obtained with this technique.

Generally, RBS spectra can be used to identify elements or analyze compositions of materials. By projecting He^+ ions traveling at a few MeV onto the target materials in a vacuum environment and detecting backscattered ions, quantitative information about layered structures or bulks can be obtained. A schematic representation of the RBS apparatus is shown in Fig. 2.

Analysis of the interfaces of amorphic diamond films on different substrates by RBS have confirmed the formation of interfacial layers which are considered to be the prevalent cause of the good adhesion experienced with our diamond films [5, 6]. In this study, RBS spectra were obtained for a nominal 0.1 μm thick amorphic diamond and for a 0.1 μm thick graphitic film prepared at the periphery of the laser plasma for comparative purposes. A 2.0 MeV He $^+$ ion beam with fluence and current of 20 μC and 20 nA, respectively, was impinged upon these films which had been deposited on Si substrates. Backscattered ions were collected at a scattering angle of 165° through a detector into a multichannel analyzer. The relative numbers of ions in each channel were plotted as data points

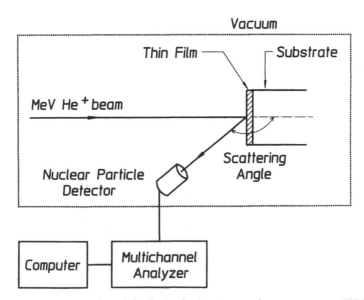

Figure 2. Schematic representation of the Rutherford backscattering spectrometry (RBS) setup used in this work.

against the channel numbers and their corresponding energies, as shown in Figs 3 and 4, and identified by the normalized yields.

The composition of the interfacial layers can be determined by a conventional numerical simulation which allows the extraction of the concentrations and depth distributions of the composite materials [12]. The simulated spectrum is assumed to be composed of the superimposed contributions from each element in each sublayer. The critical part of the computation is the evaluation of ion energy loss through the sublayers. The solid curve fits in Figs 3 and 4 are the results of such simulations by assuming that the carbon and silicon concentrations in each interfacial sublayer are indicated in these figures. The only natural unit for RBS is the area density, i.e. atoms cm^{-2}, which specifies the interlayer thicknesses in Figs 3 and 4. Conversion between the areal density and distance unit requires

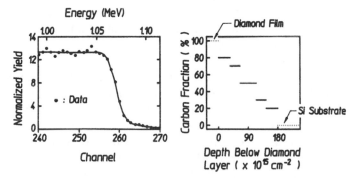

Figure 3. Left: Rutherford backscattering spectrometry (RBS) measurements of the amorphic diamond film on Si substrates. The solid line plots the computer-generated fit obtained by considering interfacial SiC layers. Right: Plot of the concentration of the C atom component as a function of depth below the diamond film that was used to generate the solid curve fit to the RBS data. Diamond film and Si substrate are distinguished from the interlayers by dotted lines. (Measurements courtesy of J. C. Pivin, CSNSM Laboratory, Orsay.)

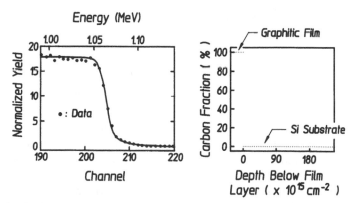

Figure 4. Left: Rutherford backscattering spectrometry (RBS) measurements of the graphitic comparison film on Si substrates. The solid line plots the computer-generated fit obtained by assuming no interfacial layer. Right: plot of the concentration of the C atom component as a function of depth below the graphitic film that was used to generate the solid curve fit to the RBS data. Film and Si substrate are shown by dotted lines. (Measurements courtesy of J. C. Pivin, CSNSM Laboratory, Orsay).

knowledge of the atomic density of the compounds in the interlayer. However, due to interdiffusion of carbon atoms into the silicon substrate, the exact atomic density of the amorphous Si–C compound in the interfacial layer is not known and only approximate thicknesses can be obtained by using the SiC atomic density. In order to obtain an accurate amount of interlayer between the films and the Si substrates, we used the natural unit of areal density.

The best fit to the RBS spectra of Fig. 3 was obtained by assuming that an interfacial layer was composed of six sublayers with variable concentrations of carbon and silicon atoms. The carbon fraction decreases from 80% in the top sublayer below the diamond film to 20% in the bottom sublayer right above the Si substrate. An effective interlayer of about 1.8×10^{17} cm^{-2} with a mean stoichiometry of SiC can be seen from this figure. No interfacial layer could be simulated for the RBS data of Fig. 4 obtained from the graphitic film. The solid curve in this figure presents the best fit to the data by considering two distinct layers of carbon and Si with no interlayer. The fits to RBS spectra such as those shown in Figs 3 and 4 are sensitive to the amount of interlayer assumed and only a 10% change results in simulation curves which cannot fit the data.

The presence of an interfacial layer between the amorphic diamond films and the substrates enhances film adhesion, as will be shown later by nanoindentation and sand erosion tests. In this work, only two examples of films with extreme adhesion properties, i.e. poor and excellent, are given. However, the RBS technique can also be used to assess comparatively the adhesion of amorphic diamond films prepared under different deposition process parameters. The thicknesses of the interfacial layers obtained can be correlated to the degree of adhesion, provided that films are deposited on similar substrates.

Examination of the interfaces between amorphic diamond and a variety of substrates with TEM has confirmed the formation of interfacial layers [5, 6, 13]. The electron diffraction patterns have also identified crystalline precipitates of carbide compounds formed at the interface of amorphic diamond and some substrates [5, 6]. The condensation of crystalline or amorphous interfacial layers of compounds or alloys is considered to be the cause of the good adhesion realized with amorphic diamond [5, 6, 13].

4. INTERFACE FAILURE ANALYSIS BY NANOINDENTATION TECHNIQUES

Recently, a comparative study of the hardnesses of amorphic and polycrystalline diamond films as indicated by indentation at the nanometric scale has been reported [10]. A Pollock-type [14] instrument was used to obtain the hardness, and both amorphic diamond and columnar polycrystalline diamond films displayed very comparable values that approached the hardness of natural diamond [10].

In concept, the determination of the hardness value from a depth-sensing test is straightforward, but the interpretation can be difficult for very elastic materials. The current measurement techniques [14–16] at the nanometer scale are distinguished in the details of implementation, but in all cases they depend on the continuous measurement of the depth of penetration of a pyramidal tip of diamond into a film under increasing loads followed by decreasing loads. This procedure gives load–displacement curves from which the hardness and elasticity can be extracted with some interpretations [10].

Many methods for adhesion measurements of thin films have been detailed [17]. The analytical technique of nanoindentation with a depth-sensing instrument can be used to obtain reliable information about the adhesion of thin films [18]. This is done by recording several load–displacement curves up to different maximum loads and maximum penetration depths [16]. An interface failure may occur at a critical value of the penetration depth and is evidenced by reproducible discontinuities in the variation of the indentation depth with the load. This procedure was applied to a low-density, $0.27\mu m$ thick amorphic diamond film [4] deposited on a Si substrate and also to a 0.4 μm thick graphitic film prepared for comparative purposes on a stainless steel substrate at the periphery of the laser plasma. A sharp triangular diamond indenter of angle 35° was used [18] in the Pollock-type Nanotest instrument manufactured by Micro Materials Limited, Clwyd, U.K., to obtain the load–displacement indentation curves as shown in Figs 5 and 6. Device setting conditions were kept the same as in these tests.

The critical depth was reached at around four times the thickness of amorphic diamond as shown in Fig. 5. The discontinuity in the indentation curve correlated with flaking of the film caused by a crack on the Si-substrate. For the low-quality graphitic film, however, the film began to fail along the interface as the indenter penetrated only 0.27 μm into the film. It is interesting to note that interfacial layer analysis with RBS showed a considerable amount of interlayer for the amorphic diamond, but no interlayer for the graphitic film. As expected, these observations were consistent with the nanoindentation test results, which indicated a superior adhesion quality for the amorphic diamond film.

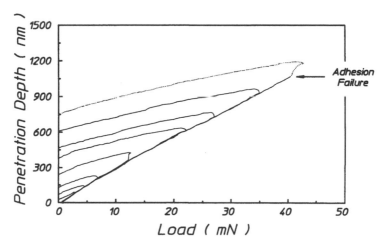

Figure 5. Penetration depth plotted as a function of the force applied to the diamond tip penetrating into a low-density amorphic diamond film having a thickness of 0.27 μm on a Si substrate. Eight separate loading and unloading curves are shown. The film began to fail during the loading process at the critical penetration depth and load values of 1050 nm and 41 mN respectively, as shown by the dotted line. (Measurements courtesy of J. C. Pivin, CSNSM Laboratory, Orsay).

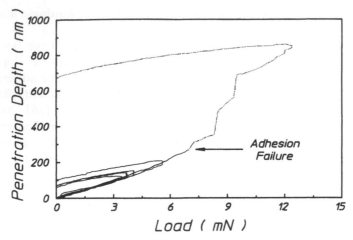

Figure 6. Penetration depth plotted as a function of the force applied to the diamond tip penetrating into a graphitic film having a thickness of 0.4 μm on a stainless steel substrate. Several separate loading and unloading curves are shown. The film began to fail during the loading process with the critical penetration depth and load values of 260 nm and 7 mN respectively, as shown by the dotted line. (Measurements courtesy of J. C. Pivin, CSNSM Laboratory, Orsay.)

5. ADHESION ASSESSMENT WITH THE PARTICULATE EROSION

Recent reports [5, 6, 9] have described the use of a convenient system to eliminate the increase in lifetime against the particulate erosion of substrates afforded by different thicknesses of amorphic diamond films. With this system, we have demonstrated considerable resistances to chipping by high-speed particles with diameters larger than the thicknesses of films deposited on Si, Ge, Ti, ZnS, and stainless steel substrates. To date, the results of these studies indicate that a coating thickness of less than 5.0 μm would give increased lifetimes by factors of thousands against dust erosion in all these cases [9]. In this paper, a brief review of this novel scheme is given. The technique is applied to the samples of amorphic diamond and graphitic film to demonstrate its usefulness in assessing the degree of adhesion of thin films.

In order to establish the general magnitude of improvements in the resistance of films to dust erosion, a quick-look test apparatus was constructed in our laboratory. A sand blaster was modified and incorporated into a test fixture for measuring the effect of surface erosion both on the diamond film coating and on the substrate material. To accomplish this, a jet of compressed gas was arranged to carry 20–40 μm diameter glass beads through a venturi to impact upon a 1 cm^2 target region. The pressure of the compressor could be altered to provide a range of impact momenta for the particles. In this work, glass beads were driven from the venturi at 40 psi.

It is difficult to quantify the resistance to damage imparted by a thin film coating. In this work, we used $\log[(I - I_0)/I_s]$ as a measure of damage, where I is the scattered laser intensity from the damaged sample, I_0 is the same quantity measured before damage, and I_s is the intensity from a standard scattering object for calibration. Of the possible angles of incidence with the surface, 20° was found to work best. At such nearly grazing incidence, the scattering from the rims of the pits created by impacts of the larger particles provides a better contrast between damaged and undamaged regions of film. Test samples and corresponding uncoated substrates were damaged by erosion under similar conditions. The light scattering was measured after each erosion period and $\log[(I - I_0)/I_s]$ was plotted as the damage data shown in Figs 7 and 8. The dotted curves in these figures were plotted to distinguish the damage data for the uncoated substrate and the test samples with different coating thicknesses and to guide the eye between the data.

It should be emphasized that the sizes of the impacting particles are about an order of magnitude larger than the film thicknesses and the resistance to the damage imparted by amorphic diamond is against chipping by these particles. Only the first onset of damage is fully sensitive to the attenuation of the shocks of impact in the coating because damage is cumulative. Once the coating is penetrated at a certain point on the sample surface, the resistance to further damage at that point is reduced by the exposure of the weaker substrate material. Since catastrophic failure occurs more rapidly once the coating is breached in a significant number of places, the curves tend to converge at the highest levels of scattering. At this level of damage, the coating is removed completely and the substrate is damaged to a level of light scattering similar to that of the uncoated substrate at the damage plateaus shown in Figs 7 and 8.

Differences in the details of the damage mechanisms between the uncoated substrate and amorphic diamond give different slopes and forms to the curves, making it difficult to extract a single number for the lifetime improvement of the substrate coated with a particular thickness of amorphic diamond. However, from Fig. 7, reproduced from previous work [5], it can be seen that at a modest level of damage of about 10% of the highest level of light scattering, a 1 μm coating of amorphic diamond extends the lifetime against chipping of a Si wafer

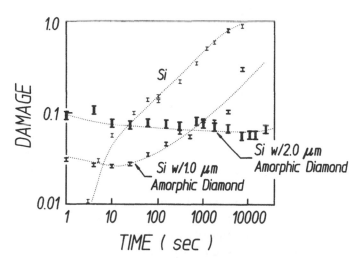

Figure 7. Plot of the damage observed on test samples as a function of the times for which they were exposed to the high velocity impact of a flux of glass beads 20–40 μm in diameter. Damage is plotted in units of the log ratio of the relative scattering of laser radiation incident at a 20° angle from the surfaces as described in the text. Data describe measurements on samples of 1.0 and 2.0 μm amorphic diamond chemically bonded to Si substrates in comparison with a similar uncoated substrate. The dotted curves guide the eye between data for the same coating thickness.

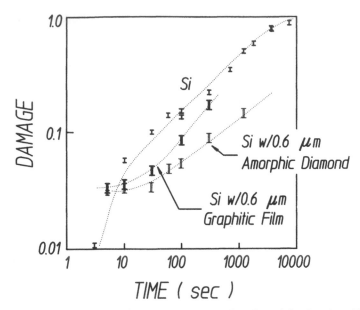

Figure 8. Plot of the damage observed on test samples as a function of the time for which they were exposed to the high velocity impact of a flux of glass beads 20–40 μm in diameter. Damage is plotted in units of the log ratio of the relative scattering of laser radiation incident at a 20° angle from the surface as described in the text. Data describe measurements on samples of 0.6 μm amorphic diamond and 0.6 μm of graphitic comparison film deposited on Si substrates together with data for a similar uncoated substrate. The dotted curves guide the eye between the data for the same coating thickness.

by a factor of 30–40. The ratio of erosion times at the 10% level of light scattering for the coated and uncoated samples gives this factor. Scanning electron microscopy (SEM) of the surface of a sample of Si coated with 1 μm thickness of amorphic diamond damaged at the level of 10% light scattering has been studied [19]. It has been found that there is quite an increase in the total amount of light scattered by the damage and the actual fraction of the surface that is pitted by particle impacts. Analysis of the frequency of pitting has shown that only 1.5% of the surface is pitted at the damage threshold of 10%. Thus, it seems reasonable to obtain the substrate lifetime improvement factor for a particular coating thickness at this level where the damage to the bulk substrate is minimal.

Most important are the data for 2 μm coating shown in Fig. 7. The initial level of scattering is higher because the intrinsic surface roughness is proportional to the thickness and the thickness is greater. However, the effect of the abrasive particle flux is to polish the diamond coating, lowering the roughness and the amount of light it scatters. A measureable damage level was not achieved within the test times available, meaning that a 2 μm coating of amorphic diamond extends the lifetime against chipping of a Si wafer by a factor of better than 1000.

This technique can be used to quantify comparatively the degree of adhesion strength of thin films, provided that the films are deposited on similar substrates and have the same thicknesses. Figure 8 shows one example of such a comparative study. The data describe the damage measurements on samples of amorphic diamond and graphitic films with thicknesses of 0.6 μm that were deposited on Si substrates together with data obtained with a similar, but uncoated substrate. It can be seen that the graphitic film fails to protect the Si substrate and film failure occurs more rapidly. In fact, at levels of damage around 10%, scatter in the data plotted for the graphitic film approaches that of the uncoated substrate. In contrast, protection of the substrate by amorphic diamond is relatively better and a longer erosion time is required to reach the levels of damage around 10% as shown in Fig. 8. This indicates the high adhesion strength of the amorphic diamond film which can be attributed in part to the significant amount of interlayer formed between the film and the Si substrates as confirmed by the RBS studies.

6. CONCLUSIONS

Amorphic diamond films are produced by impacts of C^{3+} and C^{4+} ions carrying keV energies from laser plasmas which encourage the formation of interfacial layers with considerable thicknesses on a variety of substrates. This natural interfacial layer formation is the cause of the good adhesion experienced by amorphic diamond. Information about the amount and composition of interfacial layers can be obtained by RBS examination, as described in this paper. The RBS technique can be extended to other thin film coatings that can form interfacial layers with substrates. Information obtained can then be correlated to the degree of film adhesion as described in this paper.

The interfacial strength, and thus the adhesion quality, of amorphic diamond films can also be comparatively studied by means of nanoindentation technique with a depth-sensing instrument. This method is most suitable for films with sub-micrometric thicknesses [18]. Reproducible discontinuities in the variation of the penetration depth with load can be obtained when adhesion failure occurs under indentations.

The resistance to the damage imparted by amorphic diamond in harsh erosive environments can be quantified by sand erosion and light scattering measurements as described in this work. This method, combined with RBS and nano-indentation studies, has given quantitative information about the adhesion quality of amorphic diamond. It is reasonable to expect that other thin film coatings may also find benefit in these analytical techniques.

Acknowledgements

We gratefully acknowledge the contribution of J. C. Pivin of the Centre de Spectrométrie Nucléaire et de Spectrométrie de Masse (CSNSM), Orsay, France for the RBS spectra and the nanoindentation data. We express our sincere appreciation to R. K. Krause for arranging the deposition system. This work was supported by the Texas Advanced Technology Program under Grant No. 9741–011.

REFERENCES

1. C. B. Collins, F. Davanloo, E. M. Juengerman, W. R. Osborn and D. R. Jander, *Appl. Phys. Lett.* **54**, 216–218 (1989).
2. F. Davanloo, E. M. Juengerman, D. R. Jander, T. J. Lee and C. B. Collins, *J. Appl. Phys.* **67**, 2081–2087 (1990).
3. F. Davanloo, E. J. Juengerman, D. R. Jander, T. J. Lee and C. B. Collins, *J. Mater. Res.* **5**, 2398–2404 (1990).

4. C. B. Collins, F. Davanloo, D. R. Jander, T. J. Lee, H. Park and J. H. You, *J. Appl. Phys.* **69**, 7862–7870 (1991).
5. F. Davanloo, T. J. Lee, D. R. Jander, H. Park, J. H. You and C. B. Collins, *J. Appl. Phys.* **71**, 1446–1453 (1992).
6. C. B. Collins, F. Davanloo, T. J. Lee, D. R. Jander, J. H. You, H. Park and J. C. Pivin, *J. Appl. Phys.* **71**, 3260–3265 (1992).
7. C. B. Collins, F. Davanloo, D. R. Jander, T. J. Lee, J. H. You, H. Park, J. C. Pivin, K. Glejbøl and A. R. Thölen, *J. Appl. Phys.* **72**, 239–245 (1992).
8. J. Stevefelt and C. B. Collins, *J. Phys.* **D24**, 2149–2153 (1991).
9. C. B. Collins, F. Davanloo, D. R. Jander, T. J. Lee, J. H. You and H. Park, *Diamond Films Technol.* **2**, 25–50 (1992).
10. C. B. Collins, F. Davanloo, T. J. Lee, J. H. You and H. Park, *Mater. Res. Soc. Symp. Proc.* **285** (in press).
11. C. L. Marquardt, R. T. Williams and D. J. Nagel *Mater. Res. Soc. Symp. Proc.* **38**, 325 (1985).
12. L. R. Doolittle, *Nucl. Instrum. Methods* **B9**, 344–351 (1985).
13. T. J. Lee, H. Park, J. H. You, F. Davanloo and C. B. Collins, *Surface Coat. Technol.* **54/55**, 581–585 (1992).
14. J. D. J. Ross, H. M. Pollock, J. C. Pivin and J. Takadoum, *Thin Solid Films* **148**, 171–180 (1987).
15. M. F. Doerner and W. D. Nix, *J. Mater. Res.* **1**, 601–609 (1986).
16. W. C. Oliver, *MRS Bull.* **XI**, 15–19 (1986).
17. K. L. Mittal, *J. Adhesion Sci. Technol.* **1**, 247 (1987).
18. J. C. Pivin, *Thin Solid Films* (in press).
19. C. B. Collins, F. Davanloo, T. J. Lee, J. H. You and H. Park, *Am. Ceram. Soc. Bull.* **71**, 1535–1542 (1992).

Adhesion Measurement of Films and Coatings, pp. 357–365
K. L. Mittal (Ed.)
© VSP 1995.

Nondestructive dynamic evaluation of thin NiTi film adhesion*

QUANMIN SU, SUSAN Z. HUA and MANFRED WUTTIG†

Department of Materials and Nuclear Engineering, University of Maryland, College Park, MD 20742-2115, USA

Revised version received 1 November 1993

Abstract—Thin layer materials may be considered as composites of the substrate and deposited film. The degree of stress transfer between the two components, i.e. the adhesion, will influence the dynamical response of the thin layer composite characterized by its resonance frequency. A semi-empirical adhesion parameter γ was defined which varies between 1, indicating perfect adhesion, and 0, indicating no adhesion. Resonance measurements were carried out on the $Ni_{50}Ti_{50}/SiO_2/Si$ system, employing a high-resolution mechanical vibration technique. To confirm that the parameter γ truly reflects the quality of the adhesion, NiTi films of different thicknesses were sputter-deposited on a Si substrate and $1 - \gamma$, reflecting the degree of deviation from pure elastic coupling through the interface, was measured. It was found that NiTi/Si yields γ values of about 0.8, reflecting partial adhesion. Annealing improves the adhesion of NiTi films on Si. This improvement of the adhesion was followed *in situ*.

Keywords: Adhesion; nondestructive; NiTi thin films; dynamic response; vibrating reed.

1. INTRODUCTION

The adhesion of a thin film to the substrate is an important issue controlling the reliability of the film's performance. This subject has been most recently reviewed by Mittal [1] and Valli [2]. The mechanics for the adhesion and development of measuring techniques were mainly focused on interface fracture or bonding failure [3, 4]. Our present work studies the adhesion at the interfaces from a different point of view, i.e. it considers the load transfer across the interface. Adhesion is then defined, instead as a detaching force, as the degree of load which can be elastically transferred from one side of the interface to the other side. Any partial or nonelastic transfer is considered as an imperfection of the adhesion, though it may not lead to any permanent damage such as crack formation or detachment. In fact, detachment may be considered to represent the final stage of interface deformation. As in bulk materials, information on the nonelastic response exists in the stress–strain curve long before the material fails. Most information on mechanical properties is extracted in the region immediately following elastic deformation. Analogously, there exist a wide range of mechanical responses between elastic film/substrate bending and complete film detachment in which

---

*Supported by the Office of Naval Research.

†To whom correspondence should be addressed.

information on partial adhesion may be obtained. The load applied to the film can be static with varying magnitudes or it can also be dynamic with varying frequencies. In this paper we will show and example where the response was detected dynamically when an extremely small load was applied. The nonelastic response at interfaces, which reflects adhesion, was determined.

2. DYNAMICAL RESPONSE OF THE FILM/SUBSTRATE COMPOSITE

The dynamic mechanical properties of thin film layered materials consisting of a substrate of thickness d_s and a perfectly adhering film of thickness d_f are usually determined by a cantilever method. Since such an oscillating cantilever comprises a parallel composite, its eigenfrequency f_c is given by [5]

$$\frac{f_c^2 - f_s^2}{f_s^2} = \left(\frac{3E_f}{E_s} - \frac{\rho_f}{\rho_s}\right)\left(\frac{d_f}{d_s}\right),$$ (1)

where the subscripts s and f denote substrate and film quantities, and E and ρ represent the elastic modulus and density, respectively.

Consider now the case of imperfect adhesion with the degree of adhesion characterized by a parameter γ varying between 1, indicating perfect adhesion, and 0, indicating no adhesion. As can be seen from Fig. 1, a partially adhering film causes a discontinuity of the displacement through interfacial sliding upon flexure of the composite. The sliding may occur locally and hence not be visible macroscopically. The lack of adhesion leads to a step-like drop in the displacement at the interface, resulting in an overall reduced strain in the film. This strain reduction, in turn, reduces the restoring force contributed by the film during flexural oscillations, resulting in an eigenfrequency lower than the one to be anticipated for perfect bonding. An analysis of this situation leads to the relationship [6]

$$\frac{f_c^2 - f_s^2}{f_s^2} = \left(\frac{3\gamma^2 E_f}{E_s} - \frac{\rho_f}{\rho_s}\right)\left(\frac{d_f}{d_s}\right).$$ (2)

Equation (2) shows that the eigenfrequency of the composite, f_c, is a function of the degree of adhesion and can, contrary to expectations, be *less* than the eigenfrequency of the substrate for weak bonding,

$$\gamma^2 < \frac{1}{3}\frac{\rho_f}{\rho_s}\left(\frac{E_f}{E_s}\right)^{-1},$$

as in this case the film contributes predominantly to the inertia.

3. EXPERIMENTAL

3.1. Film deposition and characterization

NiTi films of various thicknesses were deposited onto oxidized (100) Si substrates by DC magnetron sputtering using a Kurt Lesker super sputtering system II. The target material was a $Ni_{50}Ti_{50}$ alloy which has applications as actuator material. The commercially available (100) oxidized Si wafers (International Wafer Service) were cleaned by acetone

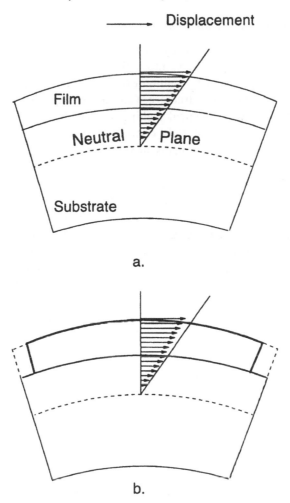

Figure 1. Schematic diagram of the displacement discontinuity resulting from non-perfect interface adhesion. (a) Perfect adhesion where displacement shows continuity at the interface; (b) partial adhesion which allows sliding of film relative to the substrate.

and methanol and dried in methanol vapor. The wafers were then placed in a high vacuum chamber and baked at 300°C for 1/2 h. The base pressure was in the range of $6 \times 10^{-8} - 2 \times 10^{-7}$ Torr and the films were sputtered at an Ar pressure of 10 m Torr. The films were deposited at a rate of about 0.1 nm/s, while the substrate was kept at room temperature. Each deposition allows for four specimens to be deposited simultaneously and one specimen was shadowed to produce a sharp edge to determine the thickness, using a Sloan Dektak II thickness scanner with a resolution of a few nanometers.

The films of different thickness and heat treatment conditions were analyzed by cross-sectional transmission electron microscopy (TEM) and X-ray diffraction. The details of these investigations have been reported elsewhere [7]. The important information for the present paper is that a sharp interface between the film and substrate was preserved throughout all heat treatment processes. Below the recrystallization temperature of the film, around 520°C, cross-sectional TEM showed a homogeneously amorphous film throughout its thickness. Rutherford back scattering also indicated an equally homogeneous compositional profile.

3.2. *Dynamic measurements of the film/substrate composites*

Although the trigger for the present dynamic adhesion measurements came from observations on recrystallized specimens, most films chosen for this adhesion study were
amorphous and carefully treated to maintain this structure such that films of different
thicknesses had an identical microstructure. Cantilever-shaped substrates were prepared
from SiO_2/Si wafers by wet chemical etching so that the cantilever part, which was
thinned to about 100 μm, was naturally connected to the thicker Si wafer. This configuration minimizes mounting problems arising in dynamic measurements [8]. The
cantilever specimen was electrostatically excited to vibrate at its resonance frequency.
The vibration was detected by applying a frequency modulation method with mechanical
Q^{-1} resolution as high as 10^{-8} [9]. As a result of the very low intrinsic damping of the
Si substrate and the fact that all measurements were performed in a high vacuum, the
resonance frequency of the film/substrate composite could be determined with a relative
resolution up to 10^{-6}. This accuracy translates into an uncertainty of the parameter γ
of approximately 2×10^{-4}. The temperature was measured by a thermocouple inserted
into the specimen holder and controlled by a furnace positioned outside the vacuum
chamber.

4. RESULTS

The first example of determining γ is shown in Fig. 2, in which the temperature dependence of the flexural stiffness of the substrate serves as a reference. The specimen was
a 2 μm thick NiTi film after grain growth at 660°C. It is clear that the square of the
eigenfrequency, f^2, is proportional to the flexure stiffness and, for the same specimen,
the proportional constant is the same. Therefore, in all figures, f^2, which can be experimentally measured, is used instead of the stiffness. After deposition, the film/substrate
composite was expected to be stiffened due to the positive difference on the right-hand
side of equation (1). The amount of stiffening calculated from equation (1) on the basis
of the data available in the literature is plotted as curve 3 in Fig. 2. By contrast, the results of the measurements indicate a pronounced softening (curve 2, Fig. 2). The slight
cusp around 50°C is due to modulus softening which occurs when the film transforms
from the martensitic B19 phase to the austenitic B2 phase.

It is worth noting that even the phase transformation-induced softening in the film
(curve 2, Fig. 2), which gives rise to a substantial variation of the bulk modulus [10],
is much less than the overall change of the composite stiffness in comparison with the
substrate stiffness. Any other modulus defects induced by imperfections inside the film
are expected to be less than that induced by the martensitic transformation. Therefore,
the observed drop in the composite modulus cannot be explained by relaxations inside
the film. It is thus proposed that the large difference is the result of imperfect interface
bonding. Using equation (2) to analyze this difference yields a value of $\gamma = 0.70$. The
analysis has been based on a literature value for the elastic constant of single crystalline Si, 130 GPa [11]. Young's modulus of NiTi, 129 GPa, was measured directly
using a sample made from the target material which had a resonance frequency similar
to the film/substrate composite. In order to obtain a reliable value for the modulus, the
strain amplitude used in the dynamic measurement was kept below 10^{-7}.

Although the deviation from equation (1) is obvious in Fig. 2, the simple model leading
to equation (2) applies strictly only to homogeneous microstructures. Since the cross-

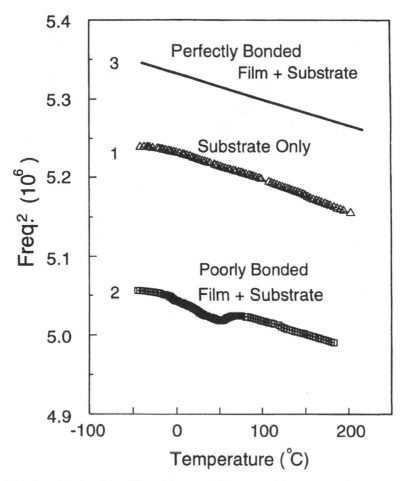

Figure 2. Variation of the bending stiffness in terms of the square of the resonance frequency as a function of temperature. Curve 1, center: reed consisting of SiO$_2$/Si substrate only, measured; curve 2, bottom: NiTi/SiO$_2$/Si composite reed annealed at 660°C, measured; curve 3, top: NiTi/SiO$_2$/Si composite reed as above, calculated from equation (1).

sectional TEM studies revealed inhomogeneous microstructures, a direct comparison of films with different thicknesses but of the same microstructure is required.

A systematic series of experiments was performed in which the as-deposited amorphous homogeneous microstructure of NiTi films was maintained throughout. Knowing empirically that the NiTi film adhesion degrades when the film becomes thicker, as evidenced from the observation that very thick films can be taped off while thin films adhere well, the thickness of the film served as a control factor while the other processing conditions were kept constant. In order to further ensure a reliable comparison, the same vibrating reed was used for all measurements. The reed was measured before deposition of the film to establish the reference curve 1 shown in Fig. 3. A 0.5 μm film was subsequently deposited on his reed and measured, the results of which are shown as curve 2 in Fig. 3. This 0.5 μm film was then chemically removed, after which the reed reverted to its 'bare' resonance frequency. Subsequently a thicker film was deposited onto the same reed. Curves 3 and 4 were obtained in this manner.

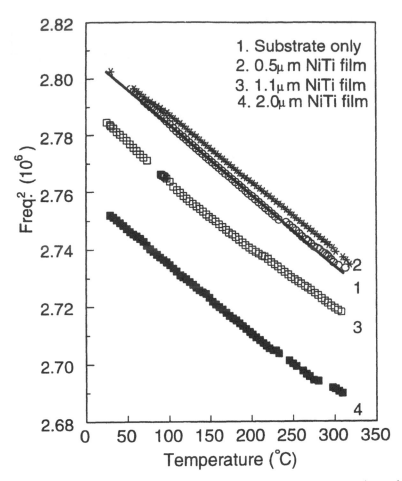

Figure 3. Dependence of the bending stiffness of a NiTi/SiO$_2$/Si film/substrate composite on the NiTi film thickness as a function of temperature. Curve 1: SiO$_2$/Si substrate only; curves 2–4: 0.5 μm, 1.1 μm, and 2 μm thick NiTi films deposited individually onto the same substrate (see text).

Curve 2 in Fig. 3 indicates that the bending stiffness, proportional to f^2 in the figure, of the film/substrate composite increased as compared with the substrate. This implies that the bending stiffness increment overrides the inertia. It may thus be anticipated from equation (1) that a thicker film should increase the bending stiffness even further. However, the experimental results indicate the opposite, as shown by curves 3 and 4. Equation (2), which allows for imperfect adhesion, is therefore applicable to this situation.

Figure 4 shows how annealing affects the composite stiffness and thus the adhesion properties. Compared with the substrate reference, curve 3 in Fig. 4, the as-deposited film has a noticeably lower bending stiffness. After annealing for 2 h at 300°C, the bending stiffness was enhance (see curve 2 in Fig. 4). Since the 300°C annealing brought about not change in the microstructure, as confirmed by X-ray diffraction and cross-sectional TEM studies, it is reasonable to assume that the enhancement of the bending stiffness is due to an improvement in the interface bonding properties. Such an improvement results in a better load transfer across the interface. The same behavior was reported by George *et al.* for the Au/Si and Au/SiO$_2$ systems [12].

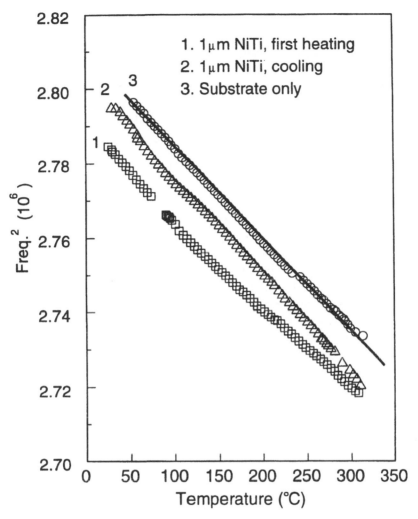

Figure 4. Variation of the bending stiffness of the film/substrate composite upon annealing. Curves 1 and 2: 1 μm NiTi/SiO₂/Si before and after 300°C annealing; curve 3: SiO₂/Si substrate only, for reference.

The adhesion coefficient, γ, was calculated from equation (2) for differently heat-treated films. The results are summarized in Table 1. For the 0.5 μm film annealed at 300°C, the value of γ is very close to 1, but it falls significantly below 1 for both the 1 μm and 2 μm films. Annealing results in a general improvement in the adhesion and brings γ closer to 1. Heat treatment at 660°C results in a lower value of γ, however. This change may be related to the transformation of the microstructure.

5. DISCUSSION

One major concern about the dynamic evaluation of the adhesion is the physical significance of the parameter γ. This parameter characterizes the load transfer across an interface and manifests in an apparent decrease of the modulus. Decreases of Young's modulus of this kind are known in the field of anelasticity of solids containing grain boundaries, which has been thoroughly studied [13]. It is known that the nonelastic

Table 1.

Variation of the adhesion parameter γ with thickness and heat treatment of amorphous NiTi films. Treatment 1: the specimen was heated to 300°C and the adhesion parameter γ was taken immediately upon arriving 300°C. Treatment 2: the specimen was annealed at 300°C for 2 h and the adhesion parameter γ was taken after the annealing

Thickness	γ		
	As deposited 25°C	Treatment 1 300°C	Treatment 2 300°C
0.5 μm	0.90	0.90	0.97
1.1 μm	0.79	0.70	0.74
2 μm	0.77	0.75	0.77
2 μm[a]	0.70 measured at room temperature		

[a]after recrystallization and grain growth for 0.5 hour at 660°C.

load transfer across a grain boundary will result in a decrease of the modulus, known as a modulus defect. Such a modulus defect corresponds to an intense damping peak, as first found by Kê [14]. The important feature of the grain boundary damping peak is that it is a dynamic response and high damping appears only in a limited frequency range. Below this frequency range, a strong modulus defect, or reduced stiffness in our language, remains while no damping occurs. Above this frequency range, neither a modulus defect nor damping can be observed. In the course of the present experiments, we did not observe pronounced internal friction related to the composite stiffness. This observation can be explained by assuming that our frequency of measurement is below the high damping region. The microstructural mechanism responsible for the nonelastic load transfer cannot be inferred from the present data, although many possibilities exist. More detailed studies at different frequencies which combine microstructural and mechanical measurements will be desirable.

6. CONCLUSION

We have considered adhesion through a load transfer parameter γ of a film/substrate composite. Dynamic measurements were performed to study this parameter. It was found that the measured values of γ vary in parallel with the expected change of the film/substrate adhesion. Dynamic mechanical measurements may thus serve as a tool for the nondestructive study of adhesion.

REFERENCES

1. K. L. Mittal (Ed.), *Adhesion Measurement of Thin Films, Thick Films and Bulk Coatings*, p. 5. ASTM, Philadelphia, PA (1978).
2. J. Valli, *J. Vac. Sci. Technol. A* **4**, 3007 (1986).
3. M. D. Thouless, *Thin Solid Films* **181**, 397 (1989).
4. R. J. Good, in: *Adhesion Measurement of Thin Films, Thick Films and Bulk Coatings*, K. L. Mittal (Ed.), p. 18. ASTM, Philadelphia, PA (1978).
5. B. S. Berry, in: *Diffusion Phenomena in Thin Films and Microelectronic Materials*, D. Gupta and P. S. Ho (Eds), p. 73. Noyes, Park Ridge, NJ (1988).
6. M. Wuttig and C. M. Su, in: *Dampings of Multiphase Inorganic Materials*, R. B. Bhagat (Ed.), p. 159. ASM International (1993).

7. S. Z. Hua, C. M. Su and M. Wuttig, in: *Thin Films — Stress and Mechanical Properties.* Mater. Res. Soc. Proceedings, Vol. 308, p. 525. MRS Press, Pittsburgh (1993).

8. F. Vollkommer, PhD Thesis, Forschungszentrum KFA, Germany (1989).

9. H. G. Bohn, F. Vollkommer and K.-H. Robrock, in: *Internal Friction and Ultrasonics Attenuation in Solids,* T. S. Kê (Ed.), p. 587. Pergamon Press, Oxford (1990).

10. A. S. Nowick and B. S. Berry, *Anelastic Relaxation in Crystalline Solids.* Academic Press, New York (1972).

11. W. A. Brantley, *J. Appl. Phys.* **44**, 534 (1973).

12. M. A. George, Q. C. Bao, I. W. Sorensen, W. S. Glaunsinger and T. Thundat, *J. Vac. Sci. Technol. A* **8**, 1491 (1990).

13. H. Gleiter and B. Chalmers, *Progr. Mater. Sci.* **16**, 1 (1972).

14. T. S. Kê, *Phys. Rev.* **71**, 533 (1947).

Adhesion Measurement of Films and Coatings, pp. 367–401
K. L. Mittal (Ed.)
© VSP 1995.

Recent developments in the laser spallation technique to measure the interface strength and its relationship to interface toughness with applications to metal/ceramic, ceramic/ceramic and ceramic/polymer interfaces

VIJAY GUPTA,* JUN YUAN and ALEXANDER PRONIN

Thayer School of Engineering, Dartmouth College, Hanover, NH 03755, USA

Revised version received 12 January 1994

Abstract—This paper summarizes recent developments in the laser spallation technique for measuring the tensile strength of planar thin film interfaces. In this technique, a laser-produced compressive stress pulse in the substrate, reflecting from the coating's free surface, pulls the interface in tension and leads to its failure if the tensile amplitude is high enough. Earlier, the critical stress amplitude that accomplishes the removal of the coating was determined through a computer simulation of the process. Recently, the technique was modified so that the interface stress can be determined directly by recording the coating or substrate free-surface velocities using a Doppler interferometer. The recorded surface velocity is related to the interface stress via an elastic wave mechanics simulation. Interface strengths of several metal/ceramic, ceramic/ceramic and ceramic/polymer systems are summarized from our recent efforts. In addition, two developments, the first a novel interferometer to record velocities from rough surfaces, and the second a technique to produce subnanosecond rise-time stress pulses with no asymptotic post-peak decay, are discussed which further allows the technique to be applied to rough thermal spray coatings and also to films as thin as 0.1 μm. Finally, it is shown how the tensile strengths obtained from the laser spallation experiment can be related to the interface fracture energies through a Griffith-type relationship, which, in turn, is derived by using the concepts of universal bonding correlations. It is shown that the estimated values obtained through this relationship are in good agreement with experimentally obtained values if the interfaces are free of defects, suggesting that the laser spallation experiment measures a fundamental strength value that is intrinsic to the material system and makes the measured value a suitable parameter for characterizing the interface.

Keywords: Interface strength; laser spallation experiment; interface toughness; metal/ceramic interfaces; Doppler interferometer.

1. INTRODUCTION

Mechanical properties of interfaces between dissimilar or similar materials (e.g. grain boundaries) have become the focal point of research in several fields, including composite materials (metal, ceramic and intermetallic matrix composites), tribology and in the solid state device area. This is not surprising because the interfaces between dissimilar materials are sites for mechanical stress concentrations and often become the nucleus of the overall failure process. Interfaces of interest in composite materials exist between fibers and their diffusion barrier coatings or between the matrix and

---

*To whom all correspondence should be addressed.

continuous, discontinuous or particulate reinforcements. In the field of tribology, interfaces exist between various types of functional (magnetic, conducting, optical and electrical), protective (thermal barrier, corrosion, wear resistant) or decorative coatings and their underlying substrates, and finally, metal/ceramic interfaces are of interest in multilayer devices and in magnetic discs and head technology. In all of the above applications, mechanical properties of the interface (tensile and shear strength, toughness, etc.) often control the overall functionality of the coated part. Therefore, improving the mechanical properties of the interface for prolonged life of the coated part is of fundamental interest. However, in ceramic and metal matrix composites, where the fiber/matrix interface is used to deflect impinging cracks from the matrix [1–5], it is often desirable to impair the strength of the interface. To accomplish either of the above goals, the first step is to measure reliably the fundamental mechanical properties of the interface (free from artifacts generated by the experimental setup). Among the various mechanical interface properties, the tensile strength tends to dominate the behavior and thereby determines the mode of interface failure and the resulting toughness or ductility, according to Varias *et al.* [6]. Accordingly, the measurement and control of interface strength through control of interface chemistry and structure is important.

Several techniques [7–12] exist in the literature to determine some average mechanical properties of the interface. The reader is referred to references 13 and 14 for a critical evaluation of these methods, and bibliography on adhesion measurement, respectively. The thrust of this article, however, is to provide a comprehensive review of recent developments in characterizing the tensile strengths of thin film interfaces by using a laser spallation technique. In this technique [13, 15–19] a high-energy laser pulse of 3 ns duration from a Nd:YAG laser is converted into a pressure pulse of critical amplitude and width, and sent through the substrate toward the free surface of the coating. The reflected tensile pulse from the coating's free surface pries-off the coating. The feasibility of the spallation process was shown initially by Vossen [20] but no quantification of the phenomenon was provided. This was first attempted by Gupta *et al.* [13, 15], where the critical stress amplitude that accomplishes the removal of the coating was determined through a computer simulation of the process, which, in turn, was verified by means of a piezo-electric crystal probe that was capable of mapping out the profile of the stress wave generated by the laser pulse. This is briefly discussed in Section 2 below. More recently, the spallation experiment was modified [16–19] and the interface strength was determined more accurately by recording the transient displacement and velocities of the coating free-surface by a Doppler interferometer, and related to the interface stress using wave mechanics, albeit by assuming the density and the elastic modulus of the coating. Since with the new setup the stress pulse is recorded directly, the strength of interfaces involving ductile components can be determined as long as the wave mechanics simulation includes the elastic–plastic constitutive law for the ductile component. However, for all materials considered here the measured strengths were within the Hugoniot elastic limit, thereby permitting an elastic analysis to quantify the free-surface velocities. A discussion on the modified setup is also provided in Section 2. The material presented here is drawn heavily from references [15–19, 21, 22].

Two additional developments were recently made by our group which increased the versatility of the technique, allowing coatings with rough surfaces (e.g. thermal spray coatings) and those with thicknesses less than 0.1 μm to be tested. The first invention was that of producing sub-nanosecond rise-times stress pulses with amplitudes in excess of 3 GPa in the substrate [21]. In contrast to pulse profiles assuming an asymptotic tail at about 5% to 10% of the peak stress, the new pulses show hitherto unreported, much sharper post-peak decays resulting in a zero stress at about 17 ns. Since in the laser spallation experiment, the tensile stress at the interface results after reflection of the parent compressive stress pulse from the coating surface, a sharper rise-time and a sharper post-peak decay enhance the interface tensile stress in thin film interfaces, and as shown below, films 0.1 μm in thickness could be separated from the substrate. The setup for producing such short stress pulses is given in Section 3 below.

The second development was that of an optical interferometer that allowed us to record the transient velocities of rough surfaces [22], which otherwise destroy the geometrical coherence of the reflected beam and make the regular interferometer unusable. This setup is briefly discussed in Section 4.

Using the modified laser spallation assembly in conjunction with the above two ideas, we determined the tensile strengths of various metal/ceramic, ceramic/ceramic and ceramic/polymer interfaces, which are of interest to the composites, thermal spray and microelectronic device industries. These results are presented in Section 5. Further, we show how the interface strength can be varied over a wide range by depositing controlled thicknesses of Cr and Sb innerlayers: this we show for Nb/alumina and Nb/sapphire interfaces. By taking an example of the Nb/sapphire interface in Section 6 we show how the measured interface strength can be related to the interface atomic structure and chemistry.

Since in the modified laser spallation experiment the interface is loaded at a strain rate of almost 10^7/sec, the inelastic deformation that usually accompanies the decohesion process is largely avoided. Thus, the measured tensile strength can be related to the interface fracture energy. In Section 7 we derive such a relationship by using the universal bonding correlations discovered by Ferrante and Smith [23], and we examine if the tensile strength measurements of various interfaces in Section 5 can be related to the intrinsic fracture toughness by this relationship. A similar relationship derived by Griffith [24], albeit on continuum arguments and for homogeneous materials, is different by a significant factor of 3.70.

The paper is organized in the following way. In Section 2 we describe both the initial and the modified versions of the laser spallation experiment. Sections 3 and 4, respectively, deal with the experimental strategy to produce short stress pulses and an optical interferometer that can record transient velocities from non-specular surfaces. Section 5 provides the measurements made to date on various metal/ceramic, ceramic/ceramic and ceramic/polymer interfaces. Section 6 provides the strength-structure-chemistry relationship for the Nb/sapphire interfaces. In Section 7, a strength-toughness relationship is derived as suggested by the universal bonding correlations, and is validated by relating it to the measured interface strength values. Finally, we end by discussing a relevant mechanical parameter to characterize the interface.

2. MEASUREMENT OF INTERFACE STRENGTH BY THE LASER SPALLATION EXPERIMENT

2.1. Previous version

The experimental approach utilizes a planar arrangement of a substrate and coating combination, as shown in Fig. 1. Details of the sample holder are given in Gupta *et al.* [13, 15]. The collimated laser pulse is made to impinge on a 1-μm thick gold film that is sandwiched between the back surface of a substrate of interest and a fused quartz confining plate, transparent to the wavelength of the laser (Nd:YAG laser operated at 1.06 μm). Gold was chosen as the laser absorbing film. Absorption of the laser energy in the confined gold leads to a sudden expansion of the film which, due to the axial constraints of the assembly, leads to the generation of a compressive shock wave directed toward the test coating/substrate interface. A part of the compressive pulse is transmitted into the coating as the compression pulse strikes the interface. It is the reflection, from the free surface of the coating, of this compressive pulse into a tension pulse that leads to the removal of the coating, given a sufficiently high pulse amplitude. A key element of the experiment was to determine the amplitude of the interface tensile stress at the threshold laser fluence causing interface failure.

Previously, the approach consisted of a three-part strategy. The first part was the development of a finite element computer simulation of the conversion of the laser light pulse into a pressure pulse, and of the resulting history of tensile stress which develops at the interface as the wave is reflected from the free surface of the coating. In the second part of the strategy, the pressure pulses were measured in a microelectronic, piezoelectric device in which the conditions of the computer simulation were experimentally achieved. This permitted verifying and fine tuning of the computer simulation. Finally, in the third part, actual spallation experiments were carried out for several thin coating/substrate interfaces (e.g. SiC/C and SiC/Si interfaces). The laser fluence necessary for the

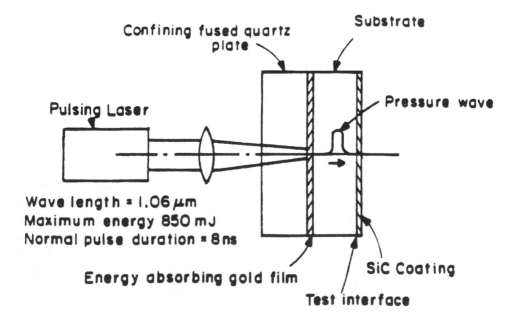

Figure 1. Schematic of the old version of the laser spallation experiment.

removal of the probed portion of the coating at the interface was recorded, and the tensile stress across the interface that accomplished this was determined from the computer program. The analysis of the post-spalled samples through in-depth Auger spectroscopy showed the failure to occur at the interface in the SiC coated pyrolytic graphite and Si wafer discs. The SiC/Si interface was varied by changing the deposition conditions of the SiC coating, and accordingly, interface strength values ranging from 3.7 to 10.5 GPa were determined. An average value of 7.2 GPa was determined for the SiC/PG interface.

2.2. Modified version

A schematic of the modified setup is shown in Fig. 2. The novel setup [16–19] involves producing stress pulses using a 0.5 μm thick Al laser-absorbing film sandwiched between a 5 μm thick layer of solid water glass and the substrate of interest. The solid water glass layer was obtained by applying a thin layer of liquid water glass ($H_2SiO_3 : (H_2O)_x$) to the laser absorbing Al film; after several minutes' exposure to the air, water evaporates from the water glass leaving behind a layer of SiO_2 film that is transparent to the laser wavelength. The use of solid water glass in producing sub-nanosecond rise-time stress pulse with no tail is discussed in Section 3 below.

When the stress pulse is reflected from the free surface of the coating or the substrate, the free surface experiences a transient velocity which is proportional to the transient profile of the striking stress pulse. In the modified version, two strategies are used to quantify the threshold laser fluence. The transient velocity is measured either at the coating's (first strategy) or substrate's (in a separate experiment in the second strategy) free surface directly by a a laser Doppler velocity interferometer, and also more accurately, by recording the displacement history of the surfaces by a displacement interferometer. The peak interface tensile stress generated at each level of the laser

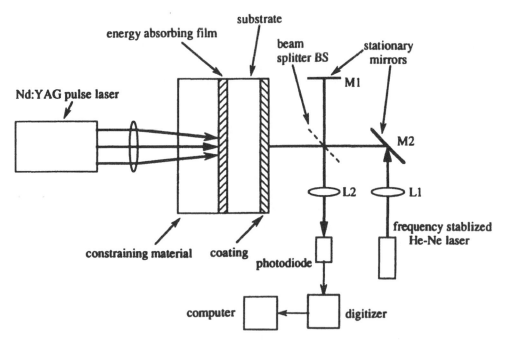

Figure 2. Schematic of the modified version of the laser spallation experiment along with the displacement interferometer.

fluence is related to the maximum compressive stress at the free surface (coating's or substrate's) through a computer simulation. In this program, a stress pulse measured at the substrate's free surface (measured in separate experiments) is made to impinge at the interface from the substrate side, and the resulting peak tensile amplitudes of the stresses at the interface and the coating's free surface are determined. We simply take the ratios of these calculated amplitudes and use them as transfer coefficients to convert the measured peak stress at the coating or the substrate free surface to the interface peak stress. Since the generated stress profile is similar for different laser fluence levels [18], the simple approach of using transfer coefficients is very accurate provided the coating remains elastic. The above procedure results in a plot of laser fluence vs. the interface stress. This information is then used to convert the threshold laser fluence causing coating spallation to the interface strength. In cases where the substrate's free surface can be polished to a mirror-finish, we use it to characterize the laser fluence, otherwise we use the velocity measurements on the coating's free-surface for characterization of the laser fluence. Unfortunately, for the diamond/Al_2O_3 interfaces pursued here, the 3-μm thick coating assumes the roughness of the substrate and makes both strategies ineligible for quantification of the interface tensile strength. To tackle this problem, we developed a new displacement interferometer which is also capable of measuring the free surface velocities of the rough surfaces, as discussed in Section 4.

For specular surfaces, it is possible to measure both the free surface displacement and the velocity by using the standard laser Doppler displacement and velocity interferometers [25–27] which are used widely in plate impact research. Since in our experiment, the stress pulses were sharp (rise-time < 5 ns) and produced over a small area of 7 mm^2, recording of sub-nanosecond displacement and velocity fringes posed a significant challenge.

In a displacement interferometer (see Fig. 2), a beam splitter (BS) divides the He–Ne laser beam into two equal beams. The beam which is reflected from the stationary mirror (M1) is the reference beam, while the one reflected from the substrate's rear surface is the signal beam. When the stress wave arrives at the rear surface, the movement of the surface results in a change in the signal beam frequency through the Doppler effect. If the rear surface velocity $v(t)$ is much less than the velocity of light c, the Doppler shift in frequency $\nu(t)$ is given by [25]

$$\nu(t) = \frac{2v(t)}{\lambda_0},\tag{1}$$

where λ_0 is the wavelength (632.8 nm) of the He–Ne laser. When the reference beam and the signal beam are remixed at the beam splitter, the beam incident onto the photodiode is, therefore, amplitude modulated at the Doppler shift frequency (i.e. the fringe frequency). Equation (1) can be integrated to relate the surface displacement $u(t)$ to the fringe count $f(t)$ as [25]:

$$u(t) = \frac{\lambda_0 f(t)}{2}.\tag{2}$$

The photodiode output voltage amplitude $A(t)$ is recorded on a transient digitizer and related to the fringe count as:

$$A(t) = \frac{(A_{max} + A_{min})}{2} + \frac{(A_{max} - A_{min})}{2} \sin\left(2\pi f(t) + \delta\right),\tag{3}$$

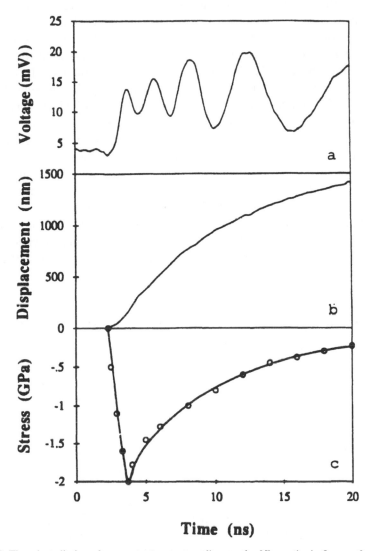

Figure 3. (a) The photodiode voltage output corresponding to the Nb coating's free surface obtained by the Doppler displacement interferometer, (b) the free-surface displacement obtained from the fringe record of (a), (c) the stress pulse profile generated by the pulsed laser obtained from (b).

where A_{max} and A_{min} are the maximum and minimum amplitudes, respectively, and δ is a phase angle. Once $f(t)$ is determined from equation (3), $u(t)$ can be obtained through equation (2). Finally the free surface velocity $v(t)$ is obtained by differentiating displacement $u(t)$ with respect to time. The displacement interferometer gives very precise data since one complete fringe shift corresponds to a $\lambda_0/2$ displacement of the free surface. In addition, the frequencies of the displacement fringes corresponding to the stress pulses with rise times shorter than 5 ns require 10–100 GHz transient digitizers. Since the useful duration of the stress pulses generated in the spallation experiment is about 20 ns, only about 10 displacement fringes are produced within this duration albeit with a minimum 0.5 ns rise time. The use of a Tektronix SCD 1000 digitizer with a transient rise time of 5 ps, and Newport's 877 ultra-high speed photodetector with a rise time of less than 200 ps allows us to record such fringes with a resolution of 0.2 ns. Furthermore, our setup is capable of recording velocities to about 800 m/s, which is three times higher than that in [25]. This resolution suffices for typical velocities obtained in our experiments. The process used for obtaining the stress pulse is presented in Fig. 3. Figure 3(a) shows the photodiode voltage output corresponding to the Nb coating's free surface obtained by the Doppler displacement interferometer. The coating thickness was 3 μm and deposited on a polycrystalline alumina substrate. Figure 3(b) corresponds to the free surface displacement obtained from the fringe record of Fig. 3(a), and finally, the stress pulse profile generated by the pulsed laser obtained from Fig. 3(b) is shown in Fig. 3(c). Because of the similarity in shape of the incident stress profile (in the substrate) and that at the coating free surface, the stress profile of Fig. 3(c) was put into the computer program, and the resultant transfer coefficient II (ratio of the interface stress to that striking the coating surface) for a 3-μm thick Nb coating was determined to be 0.31. Multiplying the recorded amplitude at the threshold laser fluence by the transfer coefficient yields a value of the interface strength. For the Nb/alumina system, an average value of 0.28 GPa was determined.

3. NANOSECOND RISE-TIME STRESS PULSES WITH NO ASYMPTOTIC DECAY

In this section, we discuss the mechanics of generating transient stress pulses with no asymptotic post-peak decay. Apart from demonstrating a short rise time of 1.14 ns, and an amplitude in excess of 3 GPa, this pulse shows a remarkably sharp, post-peak decay resulting in zero stress at about 17 ns. This observation contrasts with all other reported cases of laser-produced stress pulses [15, 20, 28–30], which show a gradual post-peak decay resulting in a long tail, assuming an asymptotic value at about 5% to 15% of the peak amplitude. As explained initially by Vossen et al. [20], and more quantitatively by Gupta et al. [13], the long tail results from the long penetration depth of the thermal pulse in the substrate. The use of a confining medium with higher thermal diffusivity than that of air helps decrease the penetration depth by providing an energy sink away from the direction of the substrate. Therefore, when air was used a constraining medium for the laser-absorbing film, a long gradual tail reaching an asymptotic value of about 10% of the peak amplitude was observed at 100 ns by Vossen et al. [20]. Using the old version discussed in Section 2.1, the stress pulse was produced by focusing a 2.5 ns long YAG laser pulse onto a 1.0 μm thick, laser-absorbing gold film, sandwiched between a piezoelectric quartz substrate and a fused quartz plate. Both quartz plates were 25.4 mm in diameter and 2 mm in thickness. The stress pulses obtained by recording the piezoelectric current signal showed a sharper decay resulting in an asymptotic value of 15% of the peak stress at about 45 ns.

Our new procedure [21] involves producing stress pulses by using a 0.5 μm thick Al laser-absorbing film sandwiched between a 5 μm thick layer of solid water glass, and a Si substrate (used here as an example) as described in Section 2.2. The stress pulse profiles were obtained by monitoring the transient free surface velocity of the Si surface by using a displacement interferometer. Figure 4 shows the stress pulse profile obtained for a laser fluence of 8.67×10^4 J/m^2, measured at the YAG source. Apart from an amplitude increase, and a pulse rise-time of 1.14 ns, this profile displays a sharper post-peak decay with the stress finally decaying to zero in 19 ns. Interestingly,

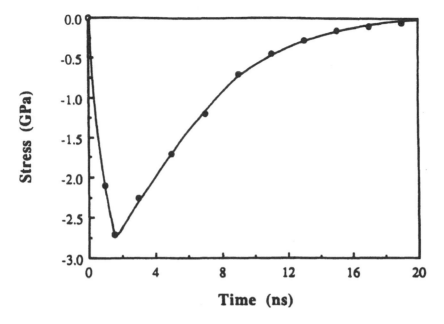

Figure 4. Nanosecond rise-time stress pulse with no post-peak tail.

the layer of solid water glass spalls off from the Si surface along with the 0.5 μm thick Al film during the microexplosion of the laser-absorbing Al film.

Although a quantitative proof of the mechanism leading to such a stress profile is not attempted, a phenomenological model presented in Gupta *et al.* [21] provided a possible mechanism. As shown by Gupta *et al.* [15], in the laser spallation setup, one compressive pulse is produced in each of the substrate and the constraining medium. It is the reflection of the compressive pulse into a tension pulse from the free surface of the solid water glass that leads to the removal of the Al film from the Al/Si interface. For a 5 μm-thick layer of solid water glass and an Al film of 0.5 μm thickness, the tension pulse reaches the Al/Si interface in about 1.27 ns (assuming a longitudinal wave velocity of 8380 m/sec in SiO_2 and 6400 m/sec in Al). Since the laser heating pulse is 3 ns long, such a spallation process within 1.27 ns removes the heat source, resulting in the absence of a thermal tail in the generated stress pulse. Using the computer simulation developed in Gupta *et al.* [15], the spatial temperature distribution within the first 1.27 ns was determined in the vicinity of the Al/SiO_2, and of the Al/Si interface. The maximum temperature at the Al/Si interface was measured to be only 700 K, which decays to the ambient temperature within 0.5 μm of the Si substrate. Although such a temperature gradient and amplitude remain within the Si upon the removal of the heat source entrapped within the Al film, these are insufficient to produce any comparable stresses. Therefore, the decay tail of the stress pulse shown in Fig. 4 is a mechanical tail produced by the sudden impact of the Si substrate in the first 1.27 ns! In other words, the Si surface sees only a transient pressure loading in the first 1.27 ns, resulting from a pressure pulse in the Al film, produced through the laser-Al interaction and transmitted through the Al/Si interface. Figure 5 shows the stress history at the Al/Si interface in the first 1.4 ns as obtained from the simulation presented in Gupta *et al.* [15]. Remarkably, the first sizable tension at the Al/Si interface occurs at 1.27 ns. The chopped stress pulse now acts as a new boundary condition for the Si surface.

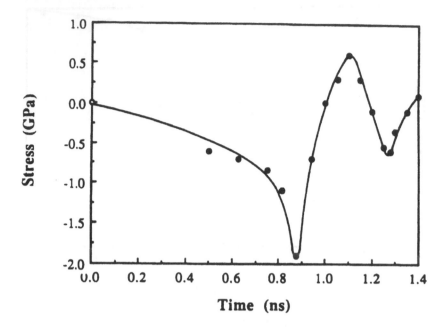

Figure 5. The Al/Si stress history at 1.4 ns as determined by the simulation given in Gupta *et al.* [15].

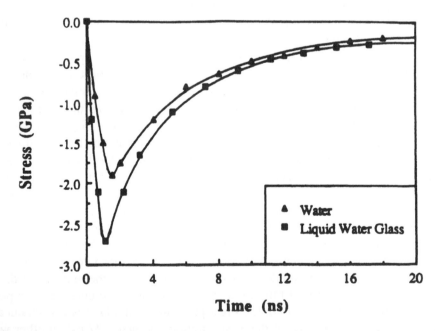

Figure 6. The stress pulse histories with different constraining media.

Figure 6 shows stress pulse profiles obtained with the same setup but with different constraining media, including a 25 mm-thick liquid column of water and liquid water glass. The details are given in Yuan *et al.* [18]. In all these cases, a typical asymptotic thermal tail at about 5% to 20% of the peak stress can be seen. No spallation of the

Al film was observed in any of these setups. Interestingly, the chopping of the stress pulse tail within the Si is analogous to the mechanism of stress pulse generation through an impact of a thin flyer plate in the plate impact experiment.

4. OPTICAL INTERFEROMETER FOR NON-SPECULAR SURFACES

To record velocity histories of rough surfaces, a new interferometer was developed [22]. The schematic of the interferometer is shown in Fig. 7. The beam from a frequency-stabilized He–Ne laser is brought to focus on the specimen surface by a convex lens. Light reflected from the diffuse surface is collected by the same lens and directed to a telescope, which condenses and collimates the beam. Next, a beam splitter directs the light to the two legs of the interferometer, and the returning beams are recombined at the Newport 877 photodiode with 200 ps rise time to produce the interference fringes. The fringes are recorded on a Tektronix SCD 1000 digitizer with a single-shot rise time of 5 ps. The key point of this optical setup is the second leg of the interferometer, which consists of two lenses (with focal length (f) equal to 50 cm each) and one mirror. It is possible to show that the system of two lenses, which are separated by the sum of their focal lengths, images the mirror M onto the apparent position M* (Fig. 7).The interferometer is so designed such that the distance of M* from the photodiode is exactly at the same distance as the other mirror M_2. But the actual difference in optical path between the two optical legs of the interferometer is equal to $8f/c$, which is about 13.3 ns in our case. The same apparent distance allows production of interference

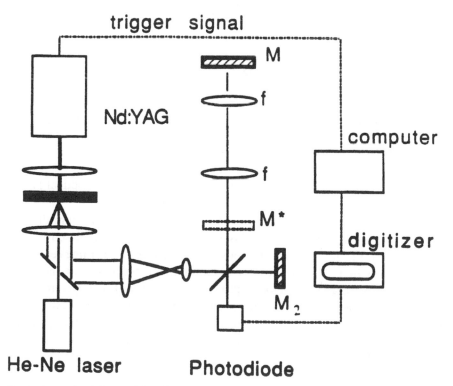

Figure 7. A schematic of the novel interferometer to record the displacement and velocity histories of rough surfaces.

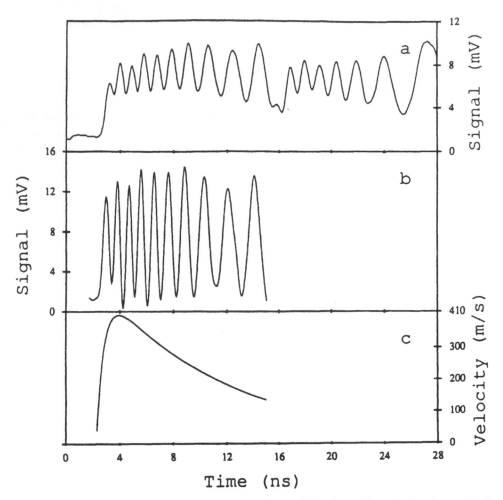

Figure 8. (a) The photodiode voltage output corresponding to the Si free surface obtained by the interferometer of Fig. 7, (b) the corrected fringe record without the effect of the photodiode rise time, (c) the stress pulse profile generated by the pulsed laser obtained from (b).

from spatially incoherent light, whereas, the difference in the actual optical paths allows the interference fringes to be related to the surface velocity.

To check the optical setup, we produced stress pulses in 25.5-mm-diameter and 1-mm-thick Si wafers with a Nd:YAG laser pulse in a manner discussed in Section 3. The (100) surface of the wafer provides a specular surface. The fringe record obtained directly is shown in Fig. 8(a) and that after removing the effect to the photodiode rise-time in Fig. 8(b), and the velocity profile associated with it is given in Fig. 8(c). Figure 8(a) shows two sets of the interference fringes. The time difference between these sets is equal to $8f/c$, which in our experiment is 13.3 ns. Here f is the focal length of the lenses, and c is the velocity of light. The signal of the ideal photodiode in the velocity interferometer is:

$$y(t) = A + B \cos \left\{ (u(t) - u(t - \tau)) \frac{4\pi}{\lambda} + \phi_0 \right\}, \qquad (4)$$

where $u(t)$ is the displacement of the moving surface, ϕ_0 is initial phase, λ is the laser wavelength, and A and B are constants. Before the arrival of the signal from the longer leg of the interferometer, $u(t - \tau)$ is constant, and the velocity interferometer within this period can be considered as a displacement interferometer. The time delay τ, which in our case is about 13.3 ns, is sufficient to obtain the information about the amplitude and even the shape of the stress wave pulse within the Si wafer. Clearly the fringes recorded after 13.3 ns in Fig. 8(a) provide the velocity information with the displacements averaged over 13.3 ns. The velocity profile of Fig. 8(c) was obtained from the displacement fringes obtained within the first 13.3 ns. The actual use of this interferometer is demonstrated on the Nb/alumina and diamond/alumina systems, where the Nb and diamond coatings assume the roughness of the substrate and makes the standard interferometer shown in Fig. 2, unusable.

5. APPLICATIONS OF THE LASER SPALLATION EXPERIMENT

Here, we summarize the interface strength measurements which we have made to date using both the old and the modified versions of the laser spallation experiment. The substrate discs were either 12.7 mm or 25.4 mm in diameter and 1 mm thick. Typical test coating thickness ranged from 1 μm to 3 μm. The laser beam was focused onto a 3 mm diameter spot. The incident laser fluence was controlled by changing the energy of the laser beam while keeping the illuminated area constant. All coatings could be successfully spalled except the diamond coatings from the diamond/alumina composites, where the failure was always within the substrate.

5.1. Interfaces between SiC coatings and Si and C substrates

The amorphous SiC coatings were deposited by the plasma-assisted chemical vapor deposition technique. The coatings had a thickness of about 1 to 3 μm. They were deposited in a nearly stress-free manner by maintaining the substrates at a certain temperature to prevent the entrapment of significant concentrations of hydrogen, that otherwise would have resulted in high residual compressive stresses and premature delamination in a manner described in detail by us in [2]. The substrates used in this study were Si single crystal wafers with (100) plane surfaces, Pyrolytic graphite (PG) platelets with the principal axis of layer normals lying in the plane of the platelet and Pitch-55 (P-55) type carbon ribbons of 600 μm width and 35 μm thickness and having a meso-phase morphology very similar to the Pitch-55 carbon fibers. Detailed description of the morphology of the Pitch-55 ribbon and PG can be found in [31].

The SiC/Si tensile strengths ranged from 3.7 to 10.5 GPa depending upon the structure of the coating, which, in turn, was controlled through control of deposition parameters. All spalled interfaces were flat with irregular coating edges, as shown by the micrograph in Fig. 9. Detailed Auger electron spectroscopy (AES) on the spalled spot indicated that the failure occurred at the interface. It was possible to delaminate successfully coatings of thickness ranging from 0.3 to 3 μm.

For the SiC/PG interface, SiC coatings of 2.1 μm thickness were successfully spalled at the interface. The substrate disc was 1.69 mm thick. Interestingly, the spalled patterns were elongated along the edges of the graphitic planes that terminate perpendicular to the surface. This is also the stiffest direction in that plane. For this case an average interface strength of 3.68 GPa was determined using the old version of the laser

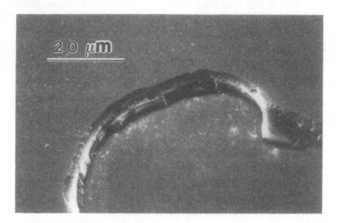

Figure 9. Spalled spot of SiC coating from a Si wafer.

spallation experiment. These calculations also include the anisotropic character of the PG substrate. Since the surface techniques available were not able to distinguish the carbon of the SiC from that of the PG substrate, the depth of the crater as determined by a mechanical profilometer (to an accuracy of 2.5 to 10 nm) compared remarkably well with the thickness of the deposited SiC coatings, thereby confirming the failure at the interface. Interface strength values ranging from 3.4 to 7.48 GPa were obtained for different SiC coatings.

Due to the presence of inhomogeneities and the weak (transverse) strength of the Pitch-55 ribbons, failure was predominantly observed within the ribbon in the SiC/P-55 interface experiments. However, in some cases, failure at the interface was also observed. An average interface strength value of 0.24 GPa was obtained via the old version of this technique. These calculations also include the anisotropic character of the ribbon material. A part of the same ribbon was tested without the coating in order to determine the transverse strength of the ribbon. A value of 0.26 GPa was obtained. As expected, this value is higher than the interface strength observed on the same coating/ribbon system. Interface strength values ranging from 0.22 to 0.24 GPa were obtained for the SiC/Pitch-55 ribbon system. The interface strength determined for this system was significantly lower than that obtained from the SiC/PG system. This is probably due to the inhomogeneities on the ribbon surface (see Gupta and Argon [31] for the structure of the ribbon), which act as sizable interface flaws and lead to lower strength values.

5.2. Interfaces between polymer and Cu coatings on passivated Si wafers for device applications

Strengths of interfaces between various polymer coatings and passivated surfaces of Si were successfully obtained. The average interface strength of 5.3-μm-thick polyimide coating on Si was determined to be 0.46 GPa, whereas the tensile strength of interfaces between another type of polyimide coating of 4.6-μm-thickness and oxidized and nitrided surfaces of Si were obtained to be 0.43 GPa and 0.55 GPa, respectively.

Similarly, the interface strength between 0.1-μm-thick Cu coatings on nitrided surfaces of Si was determined to be 41 MPa. Cu/Si samples were provided by Professor

C. V. Thompson's group at the Massachusetts Institute of Technology. Spallation of 0.1-μm-thick coatings shows the success of the short stress pulse generation technique, discussed in Section 3.

5.3. Metals/PG interfaces

All metal coatings were 3 μm thick, and deposited by a sputter system, which consisted of two 50.8 mm-diameter magnetron sputter guns, a 1000-W DC power supply, a 600-W RF power supply with a matching network, and a backsputter gun for cleaning the substrate surface prior to the deposition. The backfill argon pressure was 10 mTorr. Coatings of Sn, Sb, Cu, Nb, Al and Cr were chosen. The deposition rates were controlled to be 8 Å/second. All coatings were of polycrystalline microstructure. The modified version of the laser spallation experiment was used along with the strategy to produce short stress pulses. Since the PG substrates can be polished to a mirror surface, the displacement fringes were recorded on the free substrate surface by the standard Doppler displacement interferometer, shown in Fig. 2. The interface strengths for the various systems are given in Table 1.

Table 1 indicates that the Nb/PG interface has the highest strength (41.16 MPa), and the Sb/PG interface shows the lowest strength (5.99 MPa), while the rest of the interfaces have strengths close to 15 MPa. Detailed microscopy of the failed interfacial region revealed that all interface failures were brittle, with limited growth of voids without any coalescence. However, the deformation within the coating during the process of interfacial spallation was found to be quite different.

The micrographs of the spalled spots for some systems are shown in Fig. 10. The separation occurs at the interface in all systems. The most interesting feature is in the fracture mechanisms of the coatings. These fracture mechanisms can be divided into three categories: brittle fracture, quasi-brittle fracture, and ductile fracture. Sb, Cr and Nb coating display brittle fracture, here shown only for Sb in Fig. 10(a), in which the coating was spalled off directly from the substrate without any plastic deformation within the coating. Since the measured tensile strengths for the Cr and Nb interface of 15.47 MPa and 41.16 MPa are lower than their yield strengths of 362 MPa and 207 MPa, respectively, the coatings fracture in a brittle fashion. A similar conclusion follows for Sb, since the yield strength of 3.8 MPa as obtained from the conventional low strain tests is comparable to the measured strength of 5.99 MPa. Clearly at the typical strain rate of 10^6 s^{-1} in the spallation experiment, the yield strength of Sb must be much greater than 5.99 MPa, and should result in the observed brittle coating fracture. The

Table 1.

Interface strengths and intrinsic fracture energies for various metal/PG systems

Substrate	Coatings	Interface strength (MPa)	h (nm)	E_0 (GPa)	G_{co} (J/m^2)
Pyrolitic	Al	16.53	0.32	49	13×10^{-6}
graphite	Cu	15.46	0.30	79	7×10^{-6}
	Cr	15.47	0.26	153	3×10^{-6}
	Sb	5.99	0.33	41	2×10^{-6}
	Sn	18.53	0.41	39	27×10^{-6}
	Nb	41.16	0.28	65	54×10^{-6}

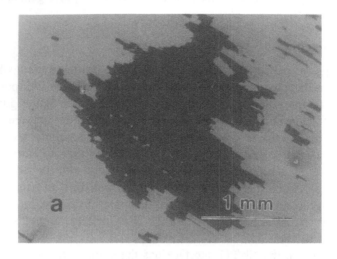

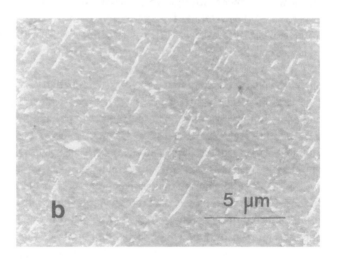

Figure 10(a–b). The micrographs of the spalled spots for metal/PG systems. (a) Sb, (b) a high magnification view of the Nb/PG interface towards the PG side.

failed coating patterns are aligned with the stiffest direction of PG, which is along the edges of the graphitic planes on the surface. The spalled interfaces were flat and clean. A high magnification view of the interface towards the PG side in Fig. 10(b) shows the marks left by the spalled coating. Since these striations are aligned with the edges of the graphitic planes, the coating adhered strongly along these sites. In conclusion, for the interfaces involving Sb, Cr, and NB coatings, both the interface and coating showed a brittle failure.

The ductile fracture was observed in the Al coating. In this fracture mode, shown in Fig. 10(c), the coating bulges out, indicating a complete interface separation, albeit with no complete spallation (i.e. the coating did not fracture from the edges) even at laser fluence levels greater than the threshold value. However, as shown in Fig. 10(c), some cracks were observed at the base of the bulge. The coating was carefully removed from

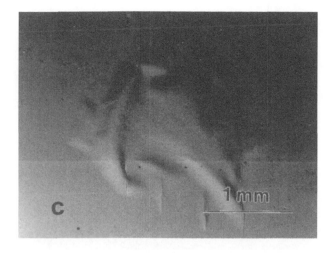

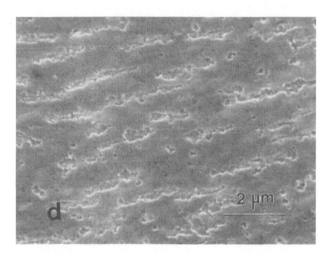

Figure 10(c–d). The micrographs of the spalled spots for metal/PG systems. (c) Al, (d) a high magnification view of the Al/PG interface towards the Al coating side.

the edges, and the surface toward the interface was examined under a scanning electron microscope. Interestingly, despite no plastic deformation within the coating some local plasticity in the form of hole growth was observed at the interface as depicted by the SEM micrograph of Fig. 10(d). Coalesced voids of about 1 μm diameter are arranged in slender pockets approximately 10 μm in length. These pockets are aligned along the direction of the graphitic plane edges on the PG surface. The PG surface showed identical features as shown in Fig. 10(b), indicating stronger interfacial bonds along the edges of the graphitic plane. Due to these stronger bonds, the local triaxial stress is elevated to several times the yield strength at the strain rate of the experiment of about 10^6 s^{-1} (the asymptotic analysis of Gupta [32] shows that values as high as 4 to 8 times the yield strength can be achieved). Due to the high triaxiality ratio, the voids within these slender pockets grow and coalesce as shown in Fig. 10(d). This

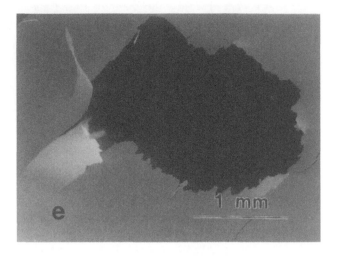

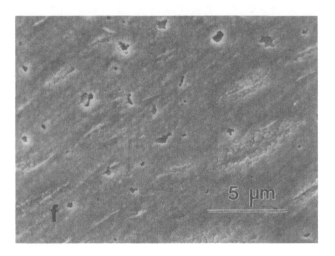

Figure 10(e−f). The micrographs of the spalled spots for metal/PG systems. (e) Cu, (f) a high magnification view of the Cu/PG interface towards the Cu coating side.

mechanism is feasible because the yield strength of Al is relatively insensitive to the strain rate; increasing from 15 MPa at 10^{-3} s^{-1} to only 55 MPa at 10^4 s^{-1}. Since the interface stress within the pockets is higher than the measured value of 16.53 MPa by 3 to 6 times, the voids can grow as the stress is sufficiently higher than the yield strength of Al, taken here as 55 MPa, albeit for a strain rate of 10^4 s^{-1}.

Finally, we discuss the quasi-brittle fracture mode observed in the failures of Cu and Sn coatings. Figure 10(e) shows a typical failure pattern involving Cu coatings. The coating is completely removed from the interface, except for some coating flaps which hang from the interface. The coating-flap was carefully removed and its surface toward the interface was observed under high magnification. Figure 10(f) depicts one such view and shows that the voids nucleated at the interface and grew only to 1 μm diameter before the cleavage strength of the interface was reached. PG surface showed markings similar to those observed for the Sb, Cr and Nb cases. Interestingly no void nucleation was seen on the PG surface. Figure 10(e) shows that the coating failure is brittle. The ductile tearing at the flap edge is not due to the stress pulse; instead, it is an after-effect caused by the coating inertia and occurs at very small strain rates. The same is true for the apparent ductile tearing of the Sn coating. A high magnification view of the coating interface showed a sparse distribution of partially grown voids of about 1 μm in diameter.

In conclusion, the Sb/PG and Nb/PG interfaces show the lowest and the highest interface strengths, respectively; however, they display the same fracture mechanism. Furthermore, Cu, Sn and Al have similar interface strengths, while they produce different fracture mechanisms within the coating, albeit with similar brittle interface failures. Therefore, it is concluded that the interface fracture mechanism is influenced by the mechanical properties of the coatings and is independent of the interface strength.

Next we discuss an application where due to the nature of the interface formation, the substrate was weaker than the interface.

5.4. Interfaces between diamond coatings and diamond/alumina composites

The samples were prepared by Professor Rishi Raj's group at Cornell University. Diamond/Al_2O_3 composites with 0, 5, 10 and 20 wt% diamond were prepared from the diamond powders with average particle sizes of 1, 2, 5 and 10 μm, and α-Al_2O_3 particles with 0.7 μm average size. The diamond powder was washed with 20% HF to remove any impurity. The two powders were stirred together in deionized water, sonicated, and finally blended in a shear blender. The slurry was dried on a hot plate while being stirred. Discs 12 mm in diameter and 1 mm thickness were hot pressed at 1500°C for 30 min in an argon atmosphere in a graphite die. The hot press pressure was 20 MPa. The disc was then polished by 320 and 600 grit SiC powder. After polishing, the diamond film was deposited in a microwave plasma system for 1 h. The deposition temperature was about 850°C and the CH_4 concentration in the CH_4–H_2 mixture was 5%. The SEM image and Raman spectra showed that the thickness of the diamond film was about 3 μm. A typical morphology of the diamond film is shown in Fig. 11.

The rough surface of the diamond coatings, as well as that of the alumina surface destroys the geometrical coherence of the He–Ne laser beam and limits the use of the regular interferometer. Here we used the novel interferometer discussed in Section 4 to record the coating's free-surface displacement velocities from the rough alumina surface. The success of the interferometer is shown in Fig. 12. Figure 12(a) shows a typical photodiode voltage output corresponding to the velocity of the Al_2O_3 free-surface. Figure 12(b) corresponds to the free-surface displacement obtained from the fringe record of Fig. 12(a) and, finally, the stress wave profile generated at the given laser fluence obtained from Fig. 12(b) is presented in Fig. 12(c). In addition, the maximum compressive stress was recorded and is plotted for the various laser fluences, shown in Fig. 13. Assuming a similar pulse profile for all the diamond composite substrates, the peak compressive stress at various fluence levels in these substrates was determined by multiplying the values given in Fig. 13 by the ratio $\rho_s c_s / \rho_a c_a$. Here, ρ and c represent the density and the longitudinal wave velocity respectively, and the subscripts 's' and 'a' denote, respectively, the composite and the alumina substrates. The stress wave

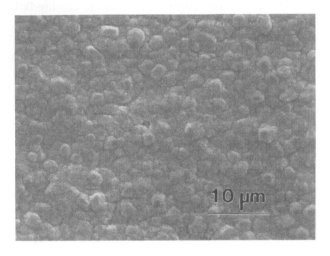

Figure 11. SEM micrograph showing the microstructure of the diamond coating.

profile of Fig. 12(c) was put into a computer program to calculate the maximum interface tensile stress. The ratios of the maximum interface tensile stress to the absolute value of the maximum compressive stress, defined as the transfer coefficient I, were determined to be 0.043, 0.042 and 0.038, respectively, for the 5, 10 and 20 wt% composites and a diamond coating thickness of 3 μm. In the simulation, the following parameters were assumed: density $\rho = 3720$ kg/m^3, Young's modulus $E = 303$ GPa, and Poisson's ratio $\nu = 0.23$ for the Al$_2$O$_3$, and $\rho = 3515$ kg/m^3, $E = 1050$ GPa, and $\nu = 0.104$ for the diamond film [33]. The composite modulus and density were determined by the rule of mixture, using the base density and Young's modulus values of Al$_2$O$_3$ to obtain the transfer coefficients.

In all of the samples the spallation was within the substrate, as shown in Fig. 14 for 5 wt% composite substrate. Since no interface failure was observed even for the highest threshold laser fluence, the adhesion strength should be larger than the interface tensile stress corresponding to the threshold laser fluence. The lower bound to the interface strength was obtained by multiplying the maximum compressive stress corresponding to the threshold fluence values, from Fig. 13 (changed appropriately for the composite substrates), and multiplying the result by the appropriate transfer coefficient. Interface tensile stresses of 0.126, 0.071 and 0.129 GPa, respectively, for the 5, 10 and 20 wt% composite with 1-μm diamond particle size were obtained. Similarly, the interface strengths should be greater than 0.116, 0.0979 and 0.0946 GPa for 2-μm particle size, and 0.1146, 0.0914 and 0.07486 MPa for the 5-μm particle size, all respectively, for 5, 10 and 20 wt% composites. The interface stress with pure alumina substrate yielded an interface stress value of 0.140 GPa.

The tensile strength of the substrate was also determined by calculating the stresses produced at the depth of the crater formed after spallation. For all composites, the crater depth as measured by a mechanical profilometer was found to be 46 μm. The tensile strength of various composites is given in Table 2. The missing values for 10-μm-size diamond particles indicate that the sample failed through a brittle exfoliation resulting in no spallation. Several other features in Table 2 are worth noting.

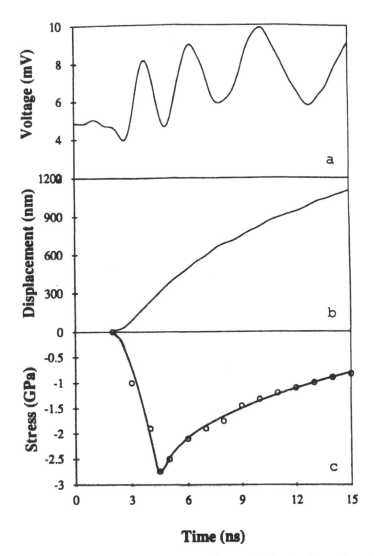

Figure 12. (a) The photodiode voltage output corresponding to the alumina free-surface obtained by the Doppler displacement interferometer, (b) the free-surface displacement obtained from the fringe record of (a), (c) the stress pulse profile generated by the pulsed laser obtained from (b).

First, the substrate without the diamond powder possesses the highest tensile strength of 1.76 GPa and, for a given wt% diamond, the strength decreases as the diamond particle size increases from 1 μm to 10 μm, except for an anomaly at 10 wt% and 1 μm particle size. For example, for 5 wt% composite the strength decreases from 1.61 GPa to 1.41 GPa with a corresponding increase of the particle size from 1 μm to 10 μm. This is because the weak diamond/alumina particle interfaces act as flaws and are fully separated to constitute the failure. Consequently, the tensile stress is lower for larger diamond particles. The failed surface is a locus of the diamond/alumina particle interfaces as shown in Fig. 14.

Similarly, the substrates with 5 wt% diamond showed higher tensile strength than those with 10 and 20 wt% diamond for all particle sizes. This is because the weak interparticle

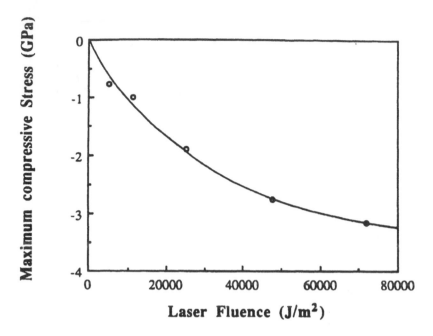

Figure 13. The peak compressive stress as a function of the laser fluence at the alumina free-surface.

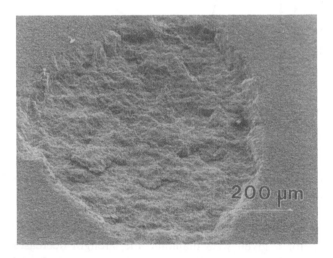

Figure 14. SEM micrograph showing the spalled diamond coating from the diamond/alumina composite surface.

interfaces become more closely spaced as the wt% increases, leading to lower strengths. For example, for the 5 μm particle size, the strength decreases from 1.46 GPa to 0.96 GPa with a corresponding wt% increase from 5 to 20.

In conclusion, although the addition of diamond particles increases the adhesion strength of the diamond film, it decreases the tensile strength of the composite substrate.

Table 2.
Tensile strength (in GPa) of various diamond/alumina composites as obtained by the laser spallation technique. The variation in the reported values is due to the difference in the size of the diamond particles

Diamond wt%	Diamond particle size (μm)				
	Without diamond	1	2	5	10
0	1.76	–	–	–	–
5	–	1.61	1.47	1.45	1.41
10	–	0.89	1.21	1.15	*
20	–	1.65	1.02	0.96	*

* Disintegration of the sample occurred.

5.5. Interfaces between Nb and SnO$_2$ coatings and alumina substrates

In this experiment, all the Nb coatings were 3 μm thick and were deposited using the sputter system discussed in Section 5.3. The modified version of the laser spallation experiment was used. In the as-deposited state, the interface strength was determined to be 0.28 GPa. Since in the laser experiment the stresses were generated only over a part of the interface, the coating in that part was completely removed at the interface, as shown in Fig. 15. The fracture occurs at the interface in a brittle mode similar to that found in the Nb/PG system. An average value of 320 MPa for the SnO$_2$ coating/alumina interface was measured [34].

To achieve a predetermined strength level at coating/substrate interface, and more specifically, at the fiber/coating interfaces in composites, it is not only necessary to measure the strength of the interface, but also to control the strength through control of interface structure and chemistry. For the Nb/alumina interface, the strength was controllably increased to 0.35 GPa by increasing the Cr interlayer thickness to 50 Å.

Figure 15. SEM micrograph showing the spalled Nb coating from the polycrystalline alumina surface.

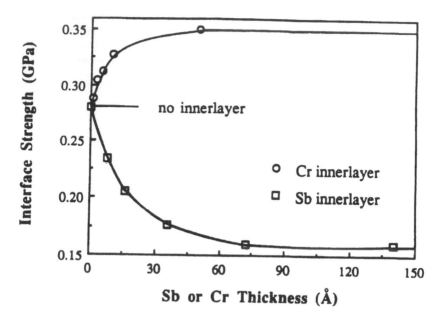

Figure 16. Control of Nb/alumina interface strength through control of Cr and Sb interlayer thickness.

The strength could be continuously decreased to 0.16 GPa by continuously increasing the Sb thickness to 70 Å [19]. This is shown in Fig. 16.

In all of the above applications, since the measured strengths were within the Hugoniot elastic limits for all materials, only an elastic wave mechanics simulation was used to relate the recorded free-surface velocities to the interface strength. As indicated above, the modified laser spallation experiment, along with the two developments of short stress pulses and a novel interferometer, has extended the potential of the technique. Next, we show how the interface strength can be related to the interface atomic structure and chemistry by taking an example of Nb/sapphire interface, with and without the inner layers of the Cr and Sb.

6. STRENGTH–STRUCTURE–CHEMISTRY RELATIONSHIP FOR Nb/SAPPHIRE INTERFACES

To control the interface strength through modification of interface structure and chemistry, we measured the strength of Nb/sapphire interfaces, with and without the inner layers (5–40 Å thick) of Cr and Sb. Spallation experiments were performed on Nb/sapphire, Nb/Cr/sapphire and Nb/Sb/sapphire interfaces in the as-deposited state, and also on samples with two different heat treatment cycles, viz. at 600°C for 24 h and at 1200°C for 10 min. The structure was obtained with HRTEM for all interfaces, here shown only for Nb/Cr/sapphire (Fig. 17). All interfaces were flat and free of macroscopic defects like cracks and voids except those which were annealed at 1200°C for 10 min. With such an annealing cycle, both the Nb/sapphire and Nb/Cr/sapphire interfaces showed interface voids, formed by the migration of vacancies from within Nb (arising from sputter deposition process) to the interface, (Fig. 18). In addition, the interfaces involving Cr showed the formation of the intermetallic Cr_2Nb (Fig. 19). Interestingly, the continuous layer of Cr is reduced to a uniform distribution of hemispherical-

Figure 17. High resolution transmission electron micrograph of the Nb/Cr/sapphire interface. The amorphous layer (A) formed due to the back-sputter Argon is about 25 Å thick.

shaped compound. All coatings and inner-layers were sputter-deposited using the RF-magnetron sputtering system. Interestingly, as shown in Fig. 17, back-sputter cleaning of the sapphire surface via flow of Argon ions, prior to coating deposition, forms a 10 to 25 Å thick disordered amorphous layer. This layer was present in all depositions.

This layer is re-crystallized at 1200 °C. The above structure and chemistry were related to the interface strength. Figure 20 shows various interfaces in ascending order of their strength, as measured by the modified laser spallation experiment. The highest strength was recorded for Nb/Cr/sapphire sample annealed at 1200 °C. The intermetallic compound Cr_2Nb acts as a strength enhancer. Finally, Fig. 21 shows how, by controlling the thickness of the Cr and Sb inner-layers, the interface strength can be remarkably tailored over a wide range, indicating the latitude in control of interface strength possible through control of interface chemistry. Sb can lower the strength by 90%, and Cr can elevate it by 50% over that for the clean interface.

Finally, the spalled samples were analyzed using XPS, Auger Spectroscopy and Atomic Force Microscopy to determine the locus of failure. All failures were at the interface adjacent to sapphire, except Nb/Cr/sapphire samples annealed at 1200 °C, where the failure was non-planar, following the Nb/sapphire interface, and within Nb at locations where the intermetallic compound Cr_2Nb was encountered. This was confirmed by an atomic force micrograph of the interface towards the sapphire which showed a uniform distribution of surface bumps (50–60 nm) corresponding to the anchored Cr_2Nb locations. Thus, it acts as a strength enhancer. All interface failures adjacent to the sapphire were in the amorphous layer, suggesting an interesting strength-degrading mechanism in metal–ceramic interfaces.

The above measured values when compared with those for the Nb/alumina system clearly indicate the influence of interface flaws on the Nb/alumina interface strength.

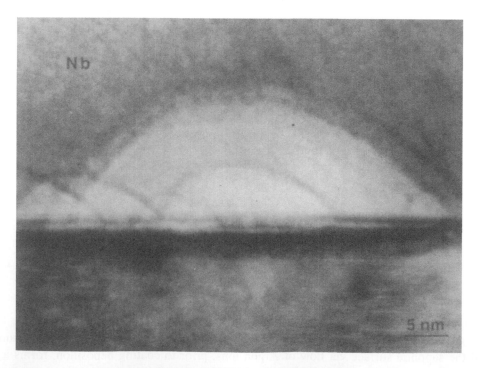

Figure 18. Interface voids formed at the Nb/sapphire interface after a 1200 °C annealing.

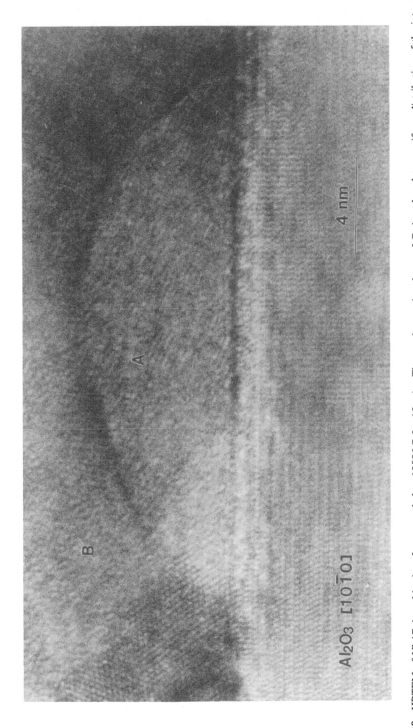

Figure 19. HRTEM of Nb/Cr/sapphire interface annealed at 1200°C for 10 min. The continuous interlayer of Cr is reduced to a uniform distribution of the intermetallic compound Cr_2Nb (A) at the interface between Nb (B) and sapphire.

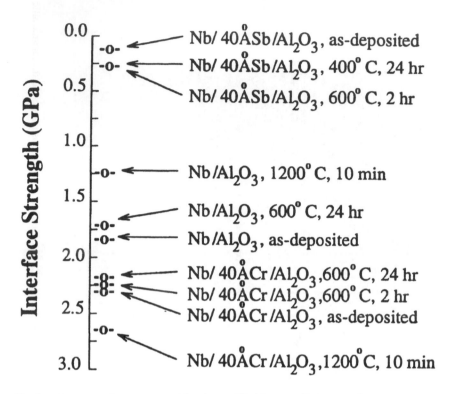

Figure 20. Interface strengths as measured by the modified laser spallation experiment.

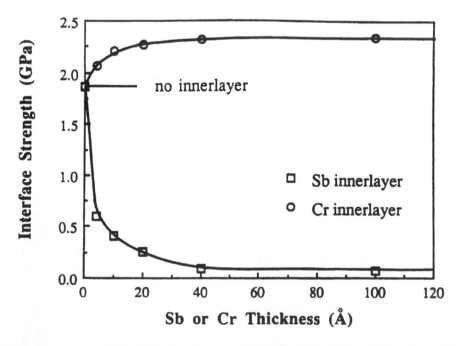

Figure 21. The control of Nb/sapphire interface strength through control of Cr and Sb interlayer thickness.

A more detailed discussion on the measured tensile strength values as provided by the laser spallation experiment is given next. Since in the laser spallation experiment the interface is loaded at a strain rate of almost 10^7 s^{-1}, all inelastic processes which usually accompany the interface decohesion process are largely avoided and, as shown in Gupta [35], the measured strengths can be related to the interface fracture energies, albeit for a given interface structure. This is discussed next.

7. INTERFACE STRENGTH–INTRINSIC TOUGHNESS RELATIONSHIP

7.1. Derivation of the strength–toughness relationship

It has been shown recently that the shape of the metallic binding energy–distance curves can be obtained from a simple two-parameter scaling of a universal function $E^*(a^*)$ [23],

$$E(a) = \Delta E\, E^*(a^*). \tag{5}$$

Here $E(a)$ is the total energy as a function of the interatomic separation distance a, ΔE is equilibrium binding energy, and $E^*(a^*)$ is a universal shape function which describes quantitatively all the first-principle calculations of bimetallic adhesion, molecular binding, cohesion, and chemisorption of gas molecules on metals. The coordinate a^* is a scaled length defined by

$$a^* = \frac{a - a_{\mathrm{m}}}{l}. \tag{6}$$

Here a_{m} is the equilibrium separation and l is a scaling length which is to be determined. Therefore, the total energy–distance curve for a given physical situation can be determined from the two scaling parameters ΔE and l once the general form of $E^*(a^*)$ is established. The basis for such a remarkable relationship stems from the fact that there is an accumulation of bond charge in the interstitial region (at the interface in bimetallic systems), and the forces involved are exponential in nature. Interestingly, such a relationship has also been shown to hold for metal/ceramic interfaces [36].

Using $E^*(a^*) = -(1 + a^*)e^{-a^*}$ as a fit to the universal energy function, the stress-separation curve or the σ vs. δ curve can be obtained by using $\sigma_{\mathrm{i}} = \partial E(a)/\partial a$ and equations (5) and (6):

$$\sigma_{\mathrm{i}}(\delta) = \frac{\Delta E}{l} a^* e^{-a^*}, \tag{7}$$

where e is a constant, $a^* = \delta/l$ and $\delta = (a - a_{\mathrm{m}})$ is defined as the excess separation over the unstressed separation distance $a_{\mathrm{m}} \equiv h$ at the interface. ΔE can the further expressed in terms of the initial modulus E_0 for one-dimensional tensile straining of the interface layer, using $[\partial^2 E(a)/\partial a^2]_{a=a_{\mathrm{m}}} = E_0/h$. With these simplifications equation (7) can be expressed as

$$\sigma_{\mathrm{i}}(\delta) = e\sigma_{\max} a^* e^{-a^*}, \tag{8}$$

where

$$\sigma_{\max} = \frac{E_0 l}{eh}. \tag{9}$$

The intrinsic toughness G_{co} or the ideal work of interfacial separation can be expressed in terms of the fundamental quantity by using equations (8) and (9) and the following expression [37]:

$$G_{co} = \int_0^\infty \sigma(\delta)\,d\delta,$$

to obtain

$$G_{co} \equiv \frac{E_0 l^2}{h}. \tag{10}$$

Using (9) and (10) we can relate the peak interface strength to the intrinsic toughness in either of the following forms:

$$\sigma_{max}^2 = \frac{G_{co} E_0}{e^2 h}, \tag{11}$$

or

$$\frac{G_{co}}{e\sigma_{max}} = l. \tag{12}$$

In deriving the above set of equations we have neglected the effect of shear parallel to the interface, which results from the dissimilarity in the elastic constants across the interface even when it is pulled apart by tensile forces acting perpendicular to the interface [38]. Equation (12) is the same equation as that derived by Rice and Wang [39].

Equation (11) is similar to that derived by Griffith [24] based on continuum arguments, albeit with constants that are different by a factor of 3.70. Equations (11) and (12) can be used to determine if the tensile strength values obtained from the spallation experiment are close to the expected flaw-free *intrinsic strength* of the interface. For this purpose, we use equation (11) to estimate a toughness value by using the measured tensile strengths, and compare them to the values of intrinsic toughness obtained from independent experiments. Since equation (11) should also hold for homogeneous materials, we also estimate the cleavage of Si through our measured strength values and compare it with that obtained by Gilman [40].

7.2. Evaluation of the measured interface strengths by using the strength–toughness relationship

In order to determine the intrinsic toughness of interfaces via equation (11) by using the experimentally measured values for the interface strength (Section 5), E_0 and h parameters are required. The equilibrium lattice spacing h at the interface can be estimated by using the transmission electron microscopy (TEM) at the interface. In cases involving amorphous coatings, as was the case for the SiC coatings, the inter-lattice spacing in the crystalline substrate within the interfacial region will define h. The other possibility arises when the interface is flat and both the materials are crystalline (e.g. see Nb/Al_2O_3 interface lattice images in [4]). In such cases, h can be taken as an average of the lattice spacings near the interface in the two materials.

The other parameter that needs to be estimated is E_0. Since the decohesion process relating to the intrinsic toughness should involve straining of material points that are located at most three to four lattice spacings on either side of the interface, E_0 can be taken as an average of the corresponding quantities in the two materials. Although E_0 for most substrates of interest is documented in Simmons and Wang [41], it is a largely unknown constant for thin films. This needs to be determined separately through other experiments. For the SiC coatings discussed here, E_0 was determined through the delamination experiments in a manner discussed in [2].

7.2.1. SiC/C interface: Consider an interface between a low modulus ($E = 116$ GPa [2]) SiC coating and a Pitch-55 carbon substrate. Taking the interface strength as determined from the laser experiment to be 6.91 GPa [15]; h equal to 0.9 nm (from TEM work of Li *et al.* [42]); and E_0 as an average for the two media $= 107$ GPa ($= [116$ GPa (E for SiC) + 98 GPa (effective E for anisotropic C fiber)]/2); we estimate a value of about 3 J/m^2 for the intrinsic toughness from equation (11). This value is close to the estimated interface toughness value of 3.28 J/m^2 [2]. The interface strength value used in the above calculations is the highest one recorded in the laser experiment so as to get the interface flaw-free value.

7.2.2. SiC/Si interface: For the SiC/Si interface, we take the interface strength to be 10 GPa as the highest recorded interface value for the low modulus SiC coating/Si interface; $h = 0.5$ nm and $E_0 = 111.5$ GPa ($= [116$ GPa (E for SiC) + 107 GPa (E for Si)]/2) and determine a value of 3.31 J/m^2 for the intrinsic toughness close to the estimated value of 2.7 J/m^2 [2]. Both of the above calculations confirm that the toughness values obtained from the spontaneous delamination experiment [2] ($= 5.5$ J/m^2 for the SiC/C case and 5.1 J/m^2 for the SiC/Si interface) still contain some dissipative component. However, it is clear that the measured interface values are closely related to the intrinsic toughness or the work of adhesion.

7.2.3. Silicon: The relationship of equation (11) should definitely hold for homogeneous solids. Here, we explore it for silicon. We measured the cleavage strength of Si from the laser spallation experiment by sending a stress pulse normal to the (111) planes to be 16.4 GPa. Taking $h = 0.32$ nm as the separation of (111) planes, and $E\langle 111\rangle = 185$ GPa [41] in equation (11) yields a value for the surface energy of 0.84 J/m^2. This value is close to 1.24 J/m^2 as measured by Gilman [40] in cleavage experiments. We attribute the difference to the fact that the silicon crystals used in our experiments contain dislocations and hence the measured cleavage strength is lower than the ideal strength.

For the above systems, the independently measured strength and toughness values satisfy the relationship in equation (11) satisfactorily. Next, we estimate the interface fracture energies using the measured interface strength values for the Nb/alumina and metals/PG interfaces.

7.2.4. Nb/Al$_2$O$_3$ interface: The ideal work of adhesion for this system can also be calculated from equation (11) using the measured value of the interface strength, provided that the non-instrinsic effects are negligible. Taking $\sigma_{max} = 0.28$ GPa, $h = 0.31$ nm and $E_0 = 203$ GPa ($= \{303$ GPa (E for Al$_2$O$_3$) + 103 GPa (E for Nb)}/2), a value of 0.0009 J/m^2 is obtained. Clearly this value is much smaller than the estimated value of 0.69 J/m^2 [19]. It cannot be the ideal work of adhesion but it is due to sizable interface flaws. Using 0.69 J/m^2 as G_{co} in equation (11) yields a flaw size of 0.68 μm when h is interpreted as the flaw size. Considering the rough surface of the polycrystalline Al$_2$O$_3$ in Fig. 15 such flaw sizes can be expected. Interestingly a much larger value of 80 ± 30 J/m^2 for the fracture energy was obtained by Evans *et al.* [43] using the notch-bend test geometry since in such setups the influence of roughness and plasticity becomes significant.

7.2.5. Metal/PG surfaces: The intrinsic toughness for various metal/PG interfaces calculated using equation (11), ranges between 2×10^{-6} to 54×10^{-6} J/m^2 (Table 1). Surprisingly, the chemically cleaned and polished PG surfaces yielded low intrinsic fracture energy values despite an additional back-sputtering surface treatment prior to the metal coating depositions. Although the influence of plasticity during interface separation is suppressed at high strain-rate loading in the spallation experiment, the extremely low surface energy values are clearly due to the interface flaws created during the deposition process.

In conclusion, although the laser spallation experiment measures the *actual* tensile strength of the interface, only for some interface systems the value is close to the expected flaw-free *intrinsic strength*.

8. DISCUSSION: MECHANICAL CHARACTERIZATION PARAMETER FOR THE BI-MATERIAL INTERFACES

While the interface tensile strength is an important quantity required to determine the critical initiation conditions in interface separation, in flawed interfaces or in cases where the propagation of existing cracks along the interface is the determining factor, the most relevant quantity required is the fracture toughness K_c or the critical energy release rate G_c for the propagation of the crack along the interface. If the separation is of cleavage-type, the plastic dissipative component of the total G_c scales directly with the intrinsic interface toughness, G_{co}. The plastic dissipative component depends on the specific ratio of K_I to K_{II} and is the primary source for the engineering $G_c - \Psi$ curves proposed recently in the literature (see for e.g. [43, 44]). For ceramics, it also includes other energy absorbing processes like frictional sliding on rough interfaces [43]. The parameter Ψ depends upon the remote loading and controls the local K_I to K_{II} ratio, and hence G_c. However, since G_{co} is the intrinsic interface property, it should be independent of Ψ. At the fundamental level, both the interface tensile strength and the intrinsic toughness G_{co} are related. Indeed, as shown in Section 7 the two are related, albeit only for flawless interfaces.

Recently compiled toughness values for various interfaces [43] as determined through different test configurations show that the plastic dissipative and the roughness-related shielding components form a significant part of these measurements. Since these components depend on the mixity of the local mode I to mode II stress field of the interface crack, the measured toughness values are not unique. However, the work ad adhesion should be independent of these effects and will be a unique property for a given bi-material interface. Since the strength determined from the laser spallation experiment relates directly to the work of adhesion (for flawless interfaces only), it appears that the bi-material interfaces can be uniquely characterized by their measured tensile strength values. Moreover, the work of adhesion (or equivalently the interface strength) directly controls the plastic dissipative [3, 6] and the roughness-related shielding components and should be the property of interest while studying the effect of interfacial segregants on the interface property. However, in order to determine the overall energy absorption related to any fracturing process of a composite member subjected to a given boundary load configuration, the $G_c - \Psi$ curves for the interfaces involved will be an immediate requirement. To characterize the interface fundamentally through the $G_c - \Psi$ curves [4, 44] will be very tedious and, at times, experimentally non-achievable. This is because most of the experiments suggested in the literature to determine the interface toughness utilize a non-homogeneous stress field in the vicinity of the interface and it is quite likely that the failure in the adjoining material may result even before the desired interface is stressed to failure. Therefore obtaining the complete $G_c - \Psi$ may not be possible for all bi-material systems. Nevertheless, information on both the $G_c - \Psi$ curves and the interface strength is collectively required in order to address the problems related to the mechanical properties of composites and the decohesion of films and coatings.

9. CONCLUSIONS

Recent developments in laser spallation to determine the tensile strengths of thin film interfaces have been summarized. A novel scheme is used to produce stress pulses with sub-nanosecond rise-times and amplitudes in excess of 3 GPa in the substrate. In contrast to pulse profiles assuming an asymptotic tail at about 5% to 10% of the peak stress, the pulses show, hitherto unreported, much sharper post-peak decays resulting in a zero stress at about 17 ns. Another important factor was the development of an optical interferometer which allowed us to record the displacement velocities of rough surfaces, which otherwise destroy the geometrical coherency of the reflected beam and make the regular interferometer unusable. This further expands the potential of the technique as coatings with rough surfaces (e.g. thermal spray coatings) can be tested.

The two ideas of short stress pulses and that of the novel interferometer allowed us to determine the tensile strengths of a variety of interfaces of interest to the microelectronic, composites and thermal spray industries. Noteworthy among them, with future potential, were the determination of the tensile strength of interfases between 0.1-μm-thick Cu films and nitrided surfaces of Si. An average value of 41 MPA was determined. Additionally, the strengths of interfaces between metallic coatings of Sn, Sb, Cu, Nb, Al, Cr, and substrates of pyrolytic graphite were measured in units of MPa to be 18.53, 5.99, 15.46, 41.66, 16.53 and 15.47, respectively. The associated failure mechanisms were also determined. Similarly an average value of 320 MPa for the SnO_2 coating/alumina interface was measured [34]. A value of 280 MPa was obtained for the Nb coating/alumina interface. In addition, the strength of the Nb/alumina interface was controllably increased to 0.35 GPa by increasing the Cr interlayer thickness to 50 Å. The strength could be continuously decreased to 0.16 GPa by continuously increasing the Sb thickness to 70 Å. The effect of interface flaws was evident when the above Nb/alumina values are compared with the Nb/sapphire interface strengths of 2.1 GPa.

With the interface characterization tool at hand, we are now determining the strength–structure–chemistry relationship for interfaces, an area of importance in interface science. This was exemplified here by taking an example of the Nb/sapphire interface.

Finally, it was shown how for some interface systems the measured strengths could be related to the intrinsic interface fracture energies using the concept of universal bonding correlation. This suggests that for interfaces with no flaws, the strength obtained by the laser spallation technique can be considered as a fundamental characterization parameter, and it can be related directly to the interface fracture energy.

Acknowledgements

The research leading to the above results has been supported initially by IST/SDIO through the ONR under Contract No. N00014-85-K-0645 and by ONR under Grant No. N00014-89-J-1609, and more recently through the United States Army Research Office under Contract No. DAAL03-91-G-0059. For support and continued keen interest in this research we are grateful to Dr S. Fishman of ONR, and Drs E. Chen and W. Simmons of ARO. The construction of the novel interferometer was accomplished as a part of the ONR program on carbon–carbon composites, for which we are grateful to Drs Y. D. S. Rajapakse and R. Barsoum of that agency. Finally, we also acknowledge the generosity of Tektronix Inc., who donated the SCD 1000 digitizer.

REFERENCES

1. A. S. Argon, V. Gupta, H. S. Landis and J. A. Cornie, *Mater. Sci. Engg.* **A107**, 41–47 (1989).
2. A. S. Argon, V. Gupta, H. S. Landis and J. A. Cornie, *J. Mater. Sci.* **24**, 1207–1218 (1989).
3. V. Gupta, A. S. Argon and J. A. Cornie, *J. Mater. Sci.* **24**, 2031–2040 (1989).
4. A. G. Evans, *Mater. Sci. Engg.* **A107**, 227–239 (1989).
5. A. Maheshwari, K. K. Chawla and T. A. Michalske, *Mater. Sci. Engg.* **A107**, 269–276 (1989).
6. A. G. Varias, N. P. O'Dowd, R. J. Asaro and C. F. Shih, *Mater. Sci. Engg.* **A126**, 65–93 (1990).
7. R. Jacobson, *Thin Solid Films* **34**, 191–199 (1976).
8. B. N. Chapman, *J. Vac. Sci. Technol.* **11**, 106–113 (1974).
9. S. S. Chiang, D. B. Marshall and A. G. Evans, in: *Surfaces and Interfaces in Ceramic and Ceramic–Metal Systems*, J. Pask and A. G. Evans (Eds), Vol. 14, pp. 603–617. Plenum Press, New York (1981).
10. A. Davutuglu and A. I. Aksay, in: *Surfaces and Interfaces in Ceramic and Ceramic–Metal Systems*, J. Pask and A. G. Evans (Eds), Vol. 14, pp. 641–649. Plenum Press, New York (1981).
11. T. S. Chow, C. A. Liu and R. C. Pennell, *J. Polym. Sci.* **14**, 1305–1310 (1976).
12. B. J. Dalgleish, K. P. Trumble and A. G. Evans, *Acta Metall.* **37**, 1923–1931 (1989).
13. V. Gupta, A. S. Argon, J. A. Cornie and D. M. Parks, *Mater. Sci. Engg.* **A126**, 105–117 (1990).
14. K. L. Mittal, *J. Adhesion Sci. Technol.* **1**, 247–259 (1987).
15. V. Gupta, A. S. Argon, J. A. Cornie and D. M. Parks, *J. Mech. Phys. Solids* **40** (1), 141–180 (1992).
16. J. Yuan and V. Gupta, *J. Appl. Phys.* **74** (4), 2388–2396 (1993).
17. V. Gupta and J. Yuan, *J. Appl. Phys.* **74** (4), 2397–2404 (1993).
18. J. Yuan, V. Gupta and A. Pronin, *J. Appl. Phys.* **74** (4), 2405–2410 (1993).
19. V. Gupta, J. Yuan and D. Martinez, *J. Amer. Ceram. Soc.* **76** (2), 305–315 (1993).
20. J. L. Vossen, in: *Adhesion Measurement of Thin Films, Thick Films, and Bulk Coatings*, K. L. Mittal (Ed.), STP-640, pp. 122–133. ASTM, Philadelphia (1978).
21. V. Gupta, J. Yuan and A. Pronin, *Rev. Sci. Instrum.* **64** (6), 1611–1613 (1993).
22. A. Pronin and V. Gupta, *Rev. Sci. Instrum.* **64** (8), 2233–2236 (1993).
23. J. H. Rose, J. R. Smith, F. Guinea and J. Ferrante, *Phys. Rev.* **B29**, 2963–2969 (1984).
24. A. A. Griffith, *Phil. Trans. Roy. Soc. (London)* **A221**, 163–198 (1920).
25. L. M. Barker, *Experimental Mechanics* **12**, 209–215 (1972).
26. L. M. Barker and R. F. Hollenbach, *J. Appl. Phys.* **41**, 4208–4226 (1970).
27. J.-P. Monchalin, *IEEE Trans. Ultrasonic, Ferroelectric, and Frequency Control* **5**, 485–499 (1986).
28. R. J. Von Gutfeld and R. L. Melcher, *Appl. Phys. Lett.* **30**, 257–259 (1977).
29. J. D. O'Keefe and C. H. Skeen, *Appl. Phys. Lett.* **21**, 464–466 (1972).
30. J. A. Fox, *Appl. Phys. Lett.* **24**, 340–343 (1974).
31. V. Gupta and A. S. Argon, *J. Mater. Sci.* **27**, 777–785 (1992).
32. V. Gupta, *J. Mech. Phys. Solids* **41** (6), 1035–1066 (1993).
33. *Metal Handbook*, 10th ed., Vol. 2, p. 1099, ASM International (1990).
34. A. Pronin, V. Gupta, J. Yuan, K. K. Chawla and R. U. Vaidya, *Scripta. Metall. Mater.* **28**, 1371–1376 (1993).
35. V. Gupta, *MRS Bulletin* **XVI** (4), 39–45 (1991).
36. T. J. Hong, J. R. Smith and D. J. Srolovitz, *Phys. Rev. Lett.* **70** (5), 615–619 (1993).
37. B. R. Lawn and T. R. Wilshaw, *Fracture of Brittle Solids*. Cambridge Univ. Press, Cambridge (1975).

38. J. R. Rice, *J. Appl. Mech.* **55**, 98–103 (1988).
39. J. R. Rice and J.-S. Wang, *Mater. Sci. Engg.* **A107**, 23–40 (1989).
40. J. J. Gilman, *J. Appl. Phys.* **31**, 2208–2218 (1960).
41. G. Simmons and H. Wang, *Single Crystal Elastic Constants and Calculated Aggregate Properties. A Handbook.* 2nd ed., MIT, Cambridge (1971).
42. Q. Li, J. Megusar, L. J. Masur and J. A. Cornie, *Mater. Sci. Engg.* **A117**, 199–205 (1989).
43. A. G. Evans, M. Rühle, B. Dalgleish and P. G. Charalambides, *Metall. Trans.* **A 21A**, 2419–2429 (1990).
44. J.-S. Wang and Z. Suo, *Acta Metall.* **38**, 1279–1290 (1990).

Adhesion Measurement of Films and Coatings, pp. 403–409
K. L. Mittal (Ed.)
© VSP 1995.

Assessment of adhesion of Ti(Y)N and Ti(La)N coatings by an *in situ* SEM constant-rate tensile test

ZHIMING YU,* CHANGQING LIU, LI YU and ZHUJING JIN

Corrosion Science Laboratory, Institute of Corrosion and Protection of Metals, Academia Sinica, Wencui Road 62, Shenyang 110015, China

Revised version received 3 January 1994

Abstract—A so-called constant-rate tensile test technique is proposed to assess the adhesion performance of ion-plated hard coatings to the substrate, in which the coated sample is elongated at a constant-rate in a scanning electron microscope, while the changes in the coating surface are observed and recorded. Critical elongation of the sample, which corresponds to the start of detachment of the coating from the substrate, has proved to be suitable for adhesion assessment. A comparative study shows that the proposed technique gives the same ranking for a number of coating/substrate systems as the classical scratch test does. It is shown that enhancement of the adhesion performance for the TiN coatings can be achieved by adding rare-earth elements such as Y and La and by introducing nitriding sublayers to the system. These beneficial effects are discussed in terms of the alteration of the microstructure of the transition zones in the systems.

Keywords: Adhesion assessment; coatings; constant-rate tensile test.

1. INTRODUCTION

Ion-plated hard coatings offer wide industrial applications due to their excellent characteristics. However, their service life is greatly influenced by the adhesion of the deposited coating to the substrate [1, 2]. It is well-known that, among other things, the adhesion of a coating to a substrate depends especially on the microstructure of the interface region. Thus, many efforts have been made to alter the interface structure, for example, by introducing rare-earth elements into the coating, or by the use of a nitriding layer as an intermediate layer.

However, there is no way to directly measure the adhesion strength of a coating to a substrate. For the assessment of adhesion performance, many techniques have been developed, such as the scratch test, the bending test, etc. [3, 4]. A disadvantage of these methods is the lack of a clear relation between the adhesion performance and the parameters obtained from the tests.

In this paper a new technique, the so-called constant-rate tensile test with simultaneous observation by scanning electron microscopy (SEM), is introduced, by which the adhesion performance of a coating to a substrate can be semi-quantitatively evaluated, and through SEM observation during the deformation process of the sample, a direct correlation between the parameters obtained from the test for the assessment of adhesion and the change in the coating morphology of the sample can be established.

---

*To whom correspondence should be addressed.

2. EXPERIMENTAL

2.1. Preparation of the coatings

A_3 steel sheet of 1.2 mm thickness was used as the substrate. The form and size of the sample tested are shown in Fig. 1. The substrates were ultrasonically cleaned for 10 min in acetone after mechanical polishing. Then Ti(Y)N and Ti(La)N coatings were deposited onto the substrate by the reactive ion-plating technique with DML-500A type ion-plating equipment. The deposition process consisted of two steps. A very thin Ti(Y) or Ti(La) layer (about 10 nm) was pre-deposited before a Ti(Y)N or Ti(La)N coating of about 5 μm thickness was deposited by introducing nitrogen gas into the vacuum chamber. The preparation of the N^++Ti(Y)N coating was performed firstly by the glow discharge nitriding process and then a Ti(Y)N coating of about 3 μm thickness was deposited by IPB30/30T type ion-plating equipment. Between the two processes, the sample surface was mechanically polished and ultrasonically cleaned in trichloroethylene and Freon-113 for 5 min each and then cleaned in Freon-113 vapor for 8 min in order to ensure that the sample surface remained highly active.

2.2. Assessment of adhesion

A scanning electron microscope with an attached miniature tensile tester was used for the investigation. The test sample was tensile-strained at a constant strain rate of 0.2 mm/min and during the deformation process, the change in the morphology of the coating surface was instantaneously observed and recorded. Critical elongation of the sample at which the coating under strain began to detach from the substrate was selected as the criterion for describing the adhesion performance of the coating to the substrate. The critical elongation is given by $\Delta L = L - L_0$, where L_0 is the original gauge length of the sample and L is the gauge length at the moment at which the coating begins to detach from the substrate. For comparison, the adhesion of the same coatings/substrate systems was also tested by a CSR-01 type scratch tester. The stylus of this tester is a diamond head of 0.2 mm radius and 120° in the spherical angle. The critical load was determined by combining the characteristic mechanics curve during the scratch with the observation by means of a microscope.

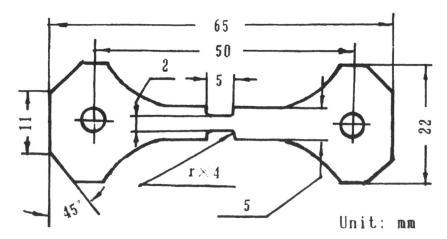

Figure 1.   The form and size of a constant-rate tensile specimen.

3. RESULTS AND DISCUSSION

3.1. Assessment of the adhesion of the coatings

Figure 2 shows the measured critical elongations of TiN, Ti(Y)N and Ti(La)N coatings. For TiN, $\Delta L = 0.8$ mm; for Ti(La)N, $\Delta L = 1.55$ mm; and for Ti(Y)N, $\Delta L = 1.75$ mm.

Figure 3 shows the SEM micrographs of the scratch on the coating surfaces. Some spallations at the edge of the scratch are visible for the TiN coating under a 150 g load; however, no damage to the Ti(Y)N coating can be observed even under a 300 g load. These results confirm that the adhesion of the TiN coating to the substrate is greatly improved by adding the rare-earth elements yttrium and lanthanum. Figure 4 shows that the rare-earth elements yttrium and lanthanum are concentrated in the transition zone between the TiN coating and the substrate as monitored by electron probe microanalysis (EPMA). We think that the concentration of rare-earth elements is induced by the prior evaporation of yttrium or lanthanum in the coating process.

Figure 5 shows schematically the structure of the cross-section of the Ti(Y)N/A$_3$ steel system, which was deduced from the observation of the cross-section by transmission electron microscopy (TEM). The transition zone consists of three sublayers: sublayer I is about 200 nm thick and is adjacent to the substrate. Y$_6$Fe$_{23}$ and α-Fe exist in this sublayer. Ti, Y, and FeTi phases exist in sublayer II, which is about 50 nm thick. Sublayer III is about 120 nm thick and contains YN, Ti$_x$N$_y$ and Ti$_2$N, TiN phases [5–7]. This shows that the adhesion of the TiN coating was enhanced by changing the structure of the transition zone by adding rare-earth elements.

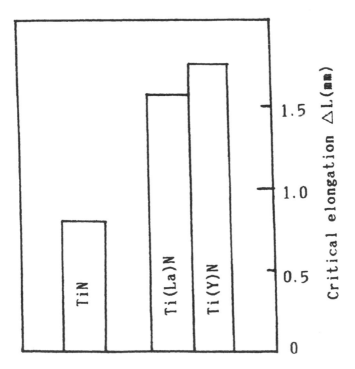

Figure 2. Comparison of the adhesion of TiN and Ti(Y)N, Ti(La)N coatings to A$_3$ steel.

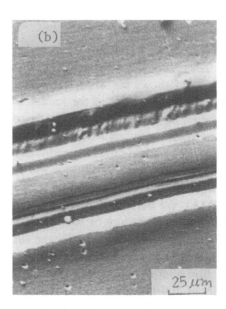

Figure 3. SEM micrographs of scratches on the coating surfaces: (a) TiN coating after the scratch test under a load of 150 g; (b) Ti(Y)N coating after the scratch test under a load of 300 g.

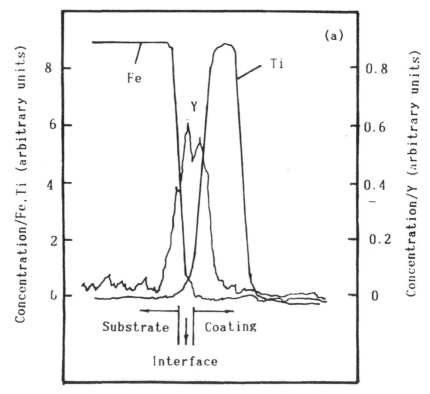

Figure 4a. The distribution of the rare-earth elements in the coating/substrate systems by EPMA. (a) Distribution of Y in the Ti(Y)N coating/A$_3$ steel system.

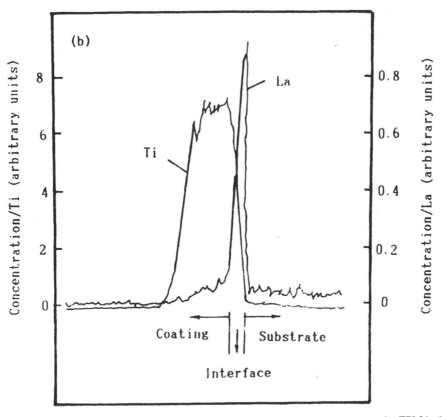

Figure 4b. The distribution of the rare-earth elements in the coating/substrate systems by EPMA. (b) Distribution of La in the Ti(La)N coating/A_3 steel system.

$TiN+Ti_2N$ Ti_xN_y+YN	Sublayer Ⅲ
$Ti+Y+FeTi$	Sublayer Ⅱ
$\alpha -Fe+Y_6Fe_{23}$	Sublayer I
A3 steel	Substrate

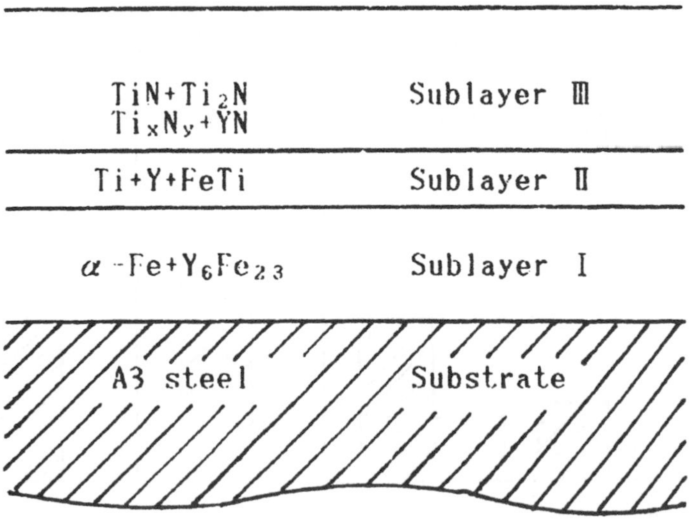

Figure 5. Schematic diagram of the structure of the transition zone in the Ti(Y)N coating/A_3 steel system.

Table 1.

Comparison of the adhesion of TiN, N$^+$+TiN, and N$^+$+Ti(Y)N coatings

Coating	TiN	TiN	Ti(Y)N
Substrate condition	Normal	Nitriding layer	Nitriding layer
Substrate hardness (Hv)	143.8	372.3	372.3
Critical load (g)	92.0	1640	2297

3.2. Assessment of the N$^+$+Ti(Y)N coating by the scratch test

The adhesion of the N$^+$+Ti(Y)N coating to the substrate was also assessed by the scratch test. The results are shown in Table 1. The adhesion of the N$^+$+TiN coating to the substrate was also significantly improved by adding the rare-earth element Y. We also observed that the TiN coating was partly detached from the substrate at some spots when the ε-phase exists in the transition zone of the sample. This is because the ε-phase is rather brittle and therefore may have some adverse effects on the adhesion of the coating to the nitriding layer. Therefore, the ε-phase was not present in the samples used in this study; this was achieved by changing the nitriding process.

3.3. Comparison of the two methods

For a comparative study, similar samples were divided into two groups, which were then assessed separately by the constant-rate tensile test and the scratch test. The test results are summarized in Fig. 6.

The critical elongations of the coatings were obtained by the constant-rate tensile test: for Ti(VP), $\Delta L = 0.14$ mm; for TiN–Ti(IP), $\Delta L = 0.30$ mm; for Ti(Y)N–Ti(Y)(IP), $\Delta L = 1.30$ mm; and for Ti(Y)N–Ti(Y)–Ti(IP), $\Delta L = 1.90$ mm.

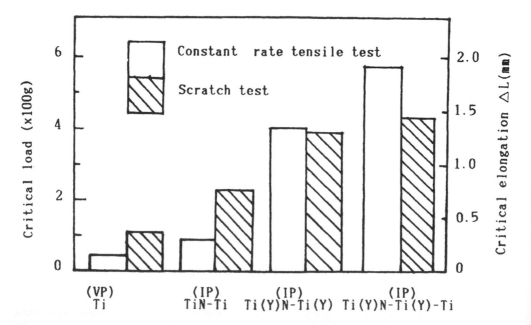

Figure 6.   Comparison of the results obtained by the scratch test and the constant-rate tensile test.

The critical loads (L_c) of the coatings were measured by the scratch test: for Ti(VP), $L_c = 110$ g; for TiN–Ti(IP), $L_c = 230$ g; for Ti(Y)N–Ti(Y)(IP), $L_c = 390$ g; and for Ti(Y)N–Ti(Y)–Ti(IP), $L_c = 430$ g. (VP indicates the vaporphase, i.e. evaporated; IP indicates ion-plated.)

From Fig. 6 it follows that both criteria, i.e. the critical elongation and the critical load, show the same tendency for the assessment of the adhesion performance for the same coating/substrate system.

From Fig. 6 it can also be seen that the critical load increases more slowly than the critical elongation does with the enhancement of adhesion of the coating/substrate systems. As a matter of fact, the critical load shows some dependence on the hardness of the coating [8, 9], whereas the increase of the critical elongation depends on the plasticity of the coating [10, 11].

4. CONCLUSIONS

(1) For the assessment of the adhesion of ion-plated coatings to a substrate, the constant-rate tensile test and the scratch test give the same ranking order. The critical elongation, at which the coating begins to detach from the substrate, can be used as a criterion for the adhesion assessment of coatings by the constant-rate tensile test.

(2) The adhesion of TiN coatings can be enhanced by adding rare-earth elements to the transition zone, or by introducing a nitriding layer between the coating and the substrate.

(3) It has also been shown that there is a clear relationship between the parameter used for the assessment of adhesion of the coating and the changes in the surface morphology of the deposited coatings.

REFERENCES

1. K. Inagawa, K. Watanabe, I. Tanaka and A. Itoh, *J. Nucl. Mater.* **128**, 925–928 (1984).
2. M. T. Laugier, *Thin Solid Films* **117**, 243 (1984).
3. K. L. Mittal, *J. Adhesion Sci. Technol.* **1**, 247 (1987).
4. K. L. Mittal (Ed.), *Adhesion Measurement of Thin Films, Thick Films and Bulk Coatings*. ASTM, Philadelphia (1978).
5. W. Wu, C. Liu, Z. Jin, Z. Yu and C. Shi, *Mater. Sci. Eng.* **A131**, 203–213 (1991).
6. Z. Jin, L. Yu, S. Dai and Z. Yu, *Acta Metall. Sinica (Engl. Ed.) Ser. B* **2** (4), 279–283 (1989).
7. C. Liu, W. Wu, Z. Yu and Z. Jin, *Thin Solid Films* **207**, 98–101 (1992).
8. S. J. Bull and D. S. Rickerby, *Surface Coatings Technol.* **42**, 149 (1990).
9. A. J. Perry, *Surface Eng.* **3**, 183 (1986).
10. C. Liu, Z. Jin, L. Yu, Z. Yu and W. Wu, *Acta Metall. Sinica (Chinese Ed.)* **28**, 173–178 (1992).
11. Z. Jin, Z. Yu, S. Dai and L. Yu, in: *Proc. Surface-Hardening Met. Mater.*, p. 59 (1987).

Adhesion Measurement of Films and Coatings, pp. 411–421
K. L. Mittal (Ed.)
© VSP 1995.

Adhesion studies of polyimide films using a surface acoustic wave sensor

D. W. GALIPEAU,[1,*] J. F. VETELINO[2] and C. FEGER[3]

[1] *Department of Electrical Engineering, South Dakota State University, Brookings, SD 57007, USA*
[2] *Laboratory for Surface Science and Technology and Department of Electrical Engineering, University of Maine, Orono, ME 04469, USA*
[3] *Room 39-114, IBM T. J. Watson Research Center, P.O. Box 218, Yorktown Heights, NY 10598, USA*

Revised version received 3 May 1993

Abstract—The feasibility of using a surface acoustic wave (SAW) sensor as a novel, nondestructive evaluation (NDE) technique for studying the relative adhesion of thin polyimide (PI) films on quartz (SiO_2) has been examined. PI films are of interest because of their widespread use in microelectronics, where there is a continuing need for improved film properties such as the dielectric constant and adhesion. A dual delay line SAW sensor was used to study the effect of humidity on the PI–quartz interface. The results show clear differences in the comparative SAW humidity response for films applied with and without an adhesion promoter and with and without a chromium intermediate layer. Temperature and humidity ageing was observed to have a greater effect on the SAW humidity response for films without adhesion-improving treatments. A theoretical analysis identified changes in the PI film properties as the physical mechanism responsible for the primary SAW sensor response. The properties of the PI film that change as a function of humidity are density, elastic constants, and stress. The more dominant of these factors appear to be density and elastic constant changes.

Keywords: Adhesion; thin films; polyimide; surface acoustic wave.

1. INTRODUCTION

Polyimide (PI) films are widely used in the microelectronics industry as intermetal dielectrics, thin film substrates, and protective layers. They possess several attractive features including high-temperature stability, a low dielectric constant, and excellent planarizing characteristics. Film adhesion and reliability, however, continue to be an important area of concern. The purpose of this work was to examine the feasibility of using a surface acoustic wave (SAW) sensor to study the adhesion of PI films nondestructively.

SAW sensors have been used extensively to study the vapor absorption and the thermomechanical properties of polymer films [1–9]. The sensing mechanism was attributed to small changes in the density, elasticity, and viscoelasticity of the films. The only work utilizing SAW sensors to study the relationship between the humidity response and the curing, surface properties, and adhesion of PI films has been done at the University of Maine [10, 11]. The purpose of the present work was to conduct SAW sensor studies on the effects of humidity on PI films applied on both delay lines of a dual delay line SAW sensor. This SAW sensor configuration allows for careful study of the response differences between two films prepared with different interface characteristics. Film density and elasticity are

---
*To whom correspondence should be addressed.

considered as sensing mechanisms since both are affected by humidity and temperature. In addition, this study also considers film stress as a sensing mechanism since it is well known that residual stress is present in cured PI films due to solvent loss and differences in the thermal expansion coefficients between the film and substrate [12, 13]. Humidity as well as temperature can affect the magnitude of these stresses [14], and since stress can affect the SAW response [15], it could therefore be a factor in adhesion studies.

Silane adhesion promotors such as Union Carbide A-1100 (γ-aminopropyltriethoxysilane) are commonly used to improve the adhesion of PI films [16, 17]. It is also known that a chromium intermediate layer can also improve film adhesion [18]. However, the overall effect of improved adhesion on the SAW sensor response is not clear. The relationship between improved adhesion and the sensor response to changes in film density, elasticity, and residual stress when an adhesion promoter or chromium is present is also not well understood. These topics will therefore be examined.

2. EXPERIMENTAL

The SAW sensor used in this study was the dual delay line type on a ST-quartz substrate as shown in Fig. 1. Quartz was used due to its similarity to SiO_2 in integrated circuits (ICs). The interdigital transducers (IDTs) consist of a patterned metal layer (copper or aluminium) approximately 1500 Å thick. In a typical configuration, one SAW delay path is covered with a PI film with adhesion promoter while the second path is covered with a PI film without adhesion promoter. The substrate was 2.52 cm by 3.15 cm and 0.9 mm thick. The operating frequency of the sensor was 80 MHz and the wavelength was 40 μm.

The PI and adhesion promoter used in this study were DuPont PMDA-ODA and Union Carbide A-1100, respectively. The adhesion promoter was prepared 2 h before application by mixing 0.1% (by weight) A-1100 with 99.9% distilled water. The solution was applied prior to the application of the PI by depositing a drop on the sensor path being treated. The drop was washed off after 30 s by

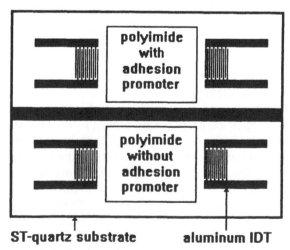

Figure 1. The dual delay line SAW adhesion sensor with polyimide films applied with and without an adhesion promoter.

applying distilled water and washing away from the untreated delay path. After several rinses, the substrate was dried at 80°C on a hot plate for 30 min. The thickness of the A-1100 layer was estimated to be 100 Å. Polyamic acid was then spin-coated onto the entire SAW substrate and dried at 80°C for 30 min. The PI was cured using standard procedures consisting of a four-step process of 30 min at 150°C, 30 min at 230°C, 30 min at 300°C, and 60 min at 400°C.

The final PI film pattern was obtained by covering the film to be left on the SAW delay path with a 0.1 μm copper mask. The copper mask was deposited using one of two methods. In the first method, the copper was evaporated through a Mylar contact mask. In the second method, a lift-off process was used in order to improve the pattern accuracy. In this method, the PI film was covered with photoresist and the area to be left was developed and removed with acetone. The PI was then sputter-cleaned with argon before copper was evaporated onto the entire sensor. The remaining copper and photoresist were then lifted off the sensor, leaving the copper mask only on the delay path. The unmasked PI was removed by oxygen reactive ion etching (RIE) for both methods. The copper mask was subsequently removed with standard photolithographic etching solution. For both methods, the final size of the PI film after etching was 6.0 mm square. Both processes provided the PI film pattern shown in Fig. 1. The cured PI had a final of thickness of approximately 1 μm.

The PI films were stored at room temperature and humidity (about 22°C and 30–50% relative humidity). Studies have shown that it is difficult to remove all moisture from PI films. Therefore, experiments intended to look at small differences in the humidity response were done using the dual delay line sensor. This ensured that both PI film samples had been exposed to the same environmental conditions.

The operation of the SAW sensor is as follows. SAWs are launched onto the propagation (delay) path of the sensor via the piezoelectric effect when an RF signal at the appropriate frequency is applied to the IDT on either end of the delay path. This creates a series of acoustic waves or SAWs which propagate along the delay path to the other IDT, where they are converted back to an electrical signal. A subtle change in a film's properties which is on the delay path can affect both the velocity and the attenuation of the SAW. These changes can then be monitored in one of two ways: first, by using the input signal as a reference to which the output signal of a single channel is compared (single delay line); and second, by using the output signal from one channel as a reference to which the output of a second channel is compared (dual delay line). Method 1 allows for small changes in film properties to be detected when it is perturbed. It was used to study the effect of humidity on a film. Method 2 allows for extremely small differences in the comparative response of two different films being studied. This second method was used when comparing the effect of humidity on films prepared with and without an adhesion promoter or a chromium layer.

A vector voltmeter was used to measure the phase difference change between the two SAWs as a function of temperature or humidity. This phase change is directly related to changes in the SAW velocity and can be measured with a precision of under 10 ppm. The electronic setup is shown in Fig. 2. The relative humidity (RH) was controlled by mixing dry compressed air and air which had been passed through a bubbler (approximately 100% RH) with rotameters to

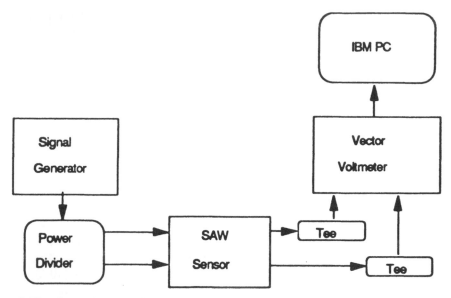

Figure 2. The electronic setup using a vector voltmeter.

provide the desired RH. The total air flow rate through the chamber was held constant at 500 cm^3/min (ccm), regardless of the RH. Although this humidity control system is not very precise, it was adequate since the measurements taken were examining either large responses of individual films or the comparative and simultaneous response of two films. The humidity in the test chamber was monitored by an Omega HX93 humidity sensor (3% accuracy, 90% response in 10 s). The environment control system is shown in Fig. 3.

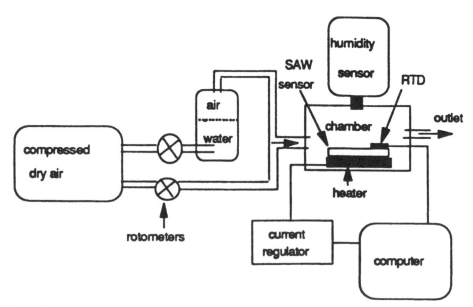

Figure 3. The environment control system.

3. RESULTS AND DISCUSSION

A SAW sensor with PI film with and without adhesion promoter (Fig. 1) was patterned using method 1 (unsputtered). Each PI film was first examined while being referenced to its input signal (single delay line). The RH was varied from 0% to 100% and back to 0%. RH steps of 25% were used in order to assess the linearity of the response. The results for both films were similar. The results for the film with promoter are shown in Fig. 4. The negative phase difference change corresponds to a decrease in the SAW velocity or 'loading' effect as the RH is increased. The output from the channel with a film and promoter was then referenced to the channel with a film and no promoter using the dual delay line sensor. This procedure allowed for the observation of only the differences in the response between the film with promoter and the film without, and therefore shows the effect of the adhesion promoter on the SAW response. The results are shown in Fig. 5. The positive change in phase difference indicates that the sample with promoter is exhibiting a smaller decrease in SAW velocity (negative phase difference change) than the sample without, or a smaller response to the increased RH.

A second dual delay line sensor with sputter-cleaned PI films, which were treated with KOH to eliminate the effect of the sputtering, was also studied. The response of a single film referenced to the input signal is shown in Fig. 6. This film had an overall response larger than that of the unsputtered film shown in Fig. 4 due to the effects of the KOH treatment. The output from the channel with the film and promoter was then referenced to the channel with the film and no promoter using the dual delay line procedure as done previously. The resulting phase shift difference was similar in shape and direction to the unsputtered films, although somewhat larger as shown in Fig. 7. The mechanism which causes the

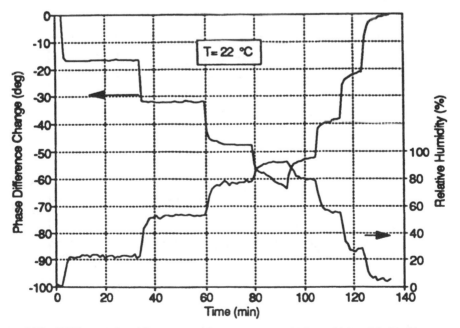

Figure 4. The SAW sensor humidity response for an unsputtered 1.2 μm thick polyimide film.

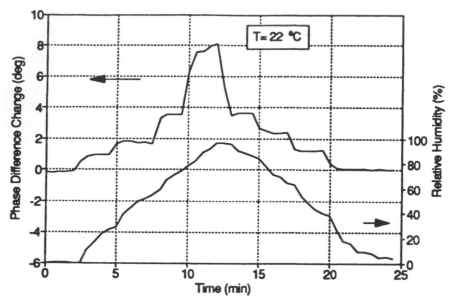

Figure 5. The dual delay line SAW sensor humidity difference response for an unsputtered polyimide film with promoter referenced to an unsputtered film without promoter.

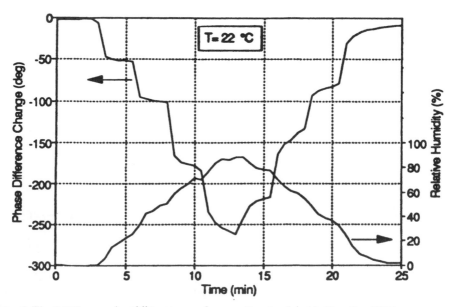

Figure 6. The SAW sensor humidity response for a sputtered polyimide film after KOH treatment.

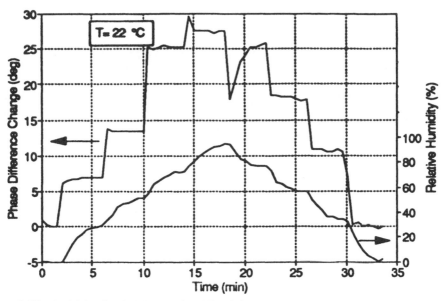

Figure 7. The dual delay line SAW sensor humidity difference response for a sputtered polyimide film with promoter referenced to a sputtered film without promoter.

different response between the two films could be the ability of water to penetrate a weaker interface layer. This would result in a larger humidity response for the film without promoter, which is what was observed. Also, if increased stress is present in the film with promoter, this stress could reduce the amount of water that can be absorbed and may also contribute to the different response. A relationship between film stress and film properties has been reported by others [19].

PI films with and without a chromium intermediate layer were also studied. Two different dual delay line sensor configurations were examined. Configuration 1 had a PI film with promoter on the upper channel while the lower channel had a PI film with a 300 Å chromium intermediate layer. Configuration 2 had a PI film without promoter on the upper channel while the lower channel had a PI film applied with promoter and chromium. The difference responses of the films with the chromium layer referenced to the films without the chromium were similar for both sensors. The response for configuration 2 is shown in Fig. 8. These results show that the chromium layer affects the SAW response in a fashion similar to the adhesion promoter in the prior experiments. Since the responses of both configurations were similar, this indicates that the effect of the chromium at the interface dominates any effect due to the adhesion promoter for these PI films.

Experiments were repeated on all dual delay line sensors after temperature and humidity ageing. Rothman [20] has shown that when PI films on SiO_2 are exposed to long periods of temperature and humidity, adhesion weakens for films without a promoter while PI films with a promoter or on metal show only small amounts of adhesion weakening. Figure 9 shows the humidity response difference for the same PI films used for Fig. 8 after they were exposed to 100% humidity at 22°C for 3 days. The results show over a 50% increase in the humidity response difference at the 50% and 75% RH steps. The other film samples showed a similar

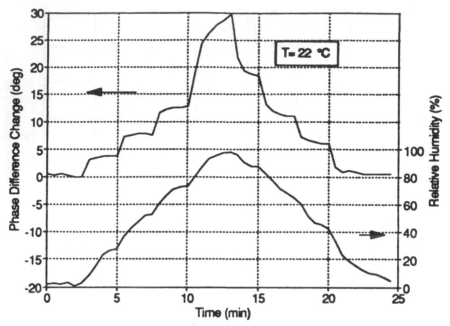

Figure 8. The dual delay line SAW sensor humidity difference response for a PI film on chromium referenced to a PI film on quartz before temperature and humidity ageing.

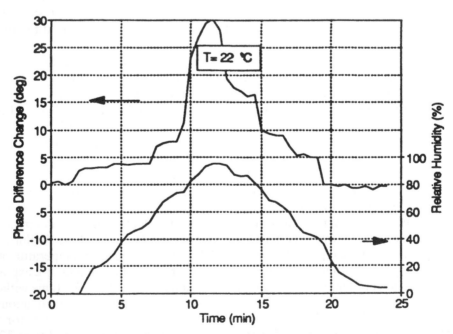

Figure 9. The dual delay line SAW sensor humidity difference response for a PI film with promoter and chromium referenced to a PI film without promoter and chromium after temperature and humidity ageing.

effect. These results suggest that the SAW sensor can detect changes in relative adhesion resulting from temperature and humidity.

Attempts to measure extremely weak adhesion and adhesion failure were not possible because the SAW became completely attenuated by either large amounts of water at the film–substrate interface, or partial and reversible localized detachment of the PI film for these limiting cases.

In order to better understand the mechanism behind the SAW-PI sensor humidity response and how adhesion affects the response, a theoretical analysis was done based on the Christoffel equations. Software developed at the University of Maine [21] can provide theoretical phase and velocity information for SAW propagation in a layered substrate. The program can predict the phase difference change for polymer layers as a function of layer parameters such as density, elastic constants, and film stress. Density and elastic constants are input parameters of the program. The stress effect is determined by calculating the stress dependence of the elastic constants, substrate density and path length. The stress results obtained were similar to those of Sinha and Locke [15]. Changes in the relative humidity can be correlated to changes in these input parameters by using other reported data [12, 14, 22]. The results shown in Table 1 compare the theoretical results with the phase difference change observed for the unsputtered PI film shown in Fig. 4. The combined theoretical value, which includes the approximate effect on phase caused by changes in the three film parameters, compares well with the experimental results. The dominant mechanisms for the SAW humidity response are density and elastic constant changes in the film caused by water uptake in the polyimide.

Differences in the SAW humidity response caused by film adhesion or the effect of temperature and humidity on adhesion were not included in the theoretical model. The most probable explanation for the PI-SAW sensor humidity response differences and the weakening of adhesion with humidity exposure is as follows: the amount of water that enters the interface region of the film is dependent on the strength of the interfacial bond. For films with good adhesion, only small amounts of water, if any, enter the interface area. For this case, the results for unsputtered films correlate well with theory. This explains the smaller phase change for films with an adhesion promoter or chromium. For films with poor adhesion, significant amounts of water can enter the interface area; this, in turn, can further weaken the adhesion over time as demonstrated by the temperature and humidity ageing experiments. This additional water causes a

Table 1.
The theoretical change in phase of the SAW sensor as a function of changes in density, elastic constants (ECs), and stress of the PI film with RH for a 1.2 μm thick PI film compared with experiment

% RH	Phase change due to changes in:			Total	Experimental
	Density	ECs	Stress		
0	0.0	0.0	0.0	0.0	0.0
25	− 10.0	− 6.0	− 0.8	− 16.8	− 16.0
50	− 19.0	− 12.0	− 1.6	− 32.6	− 32.0
75	− 28.0	− 18.0	− 2.4	− 48.4	− 47.0

phase change larger than that seen with films with good adhesion and also significant attentuation of the SAW due to the viscosity of the water at the interface. This model is in agreement with what is known about the chemistry of the PI–SiO$_2$ interface. That is, interfaces with an adhesion promoter form a chemical or acid–base type bond which would preclude water from entering the interface. Interfaces without a promoter have a much weaker hydrogen-type bond which would allow reversible substitution of water molecules at the PI–SiO$_2$ interface [16, 23].

4. CONCLUSIONS

These results indicate that the primary advantage of the SAW sensor in adhesion studies is that it provides a novel, nondestructive method to show differences in the adhesion strength of thin PI films. It appears that the SAW humidity response is related to the strength of adhesion at the interface and the corresponding water penetration of that interface. Additional PI samples with configurations similar to those already tested should be examined with the SAW sensor in order to expand the data base and ensure reproducibility of these results. Since variations in fabrication steps can significantly affect the experimental results, they need to be controlled closely. Applications of the PI-SAW sensor include adhesion difference studies between thin PI films and a standard with known adhesion properties. This will allow comparisons of either differences in adhesion due to modified interface properties or adhesion weakening due to film ageing and exposure to environmental conditions such as humidity or temperature. The sensor could also be used to compare the adhesion of test samples *in situ* for process control applications.

Acknowledgements

This work was supported, in part, by a grant from the IBM Corporation and a grant from the International Society for Hybrid Microelectronics (ISHM) Educational Foundation. Thanks are also due to J. Andle for the use of his 'Layers' software, and R. Falconer, W. Unertl, and R. Lec for many helpful discussions.

REFERENCES

1. See, for example, *IEEE Trans. Ultrason. Ferro-electrics Frequency Control* (*Special Issue Acoustic Sensors*) **34**, (2) (1987).
2. H. Wohltjen and R. Dessy, *Anal. Chem.* **51**, 1472–1475 (1979).
3. J. A. Groetsch III and R. E. Dessy, *J. Appl. Polym. Sci.* **28**, 161–178 (1983).
4. D. S. Ballantine, Jr. and H. Wohltjen, in: *Chemical Sensors and Microinstrumentation*, R. W. Murray (Ed.), pp. 222–236. American Chemical Society, Washington, DC (1989).
5. M. T. Fertsch, R. M. White and R. S. Muller, Paper presented at the Device Research Conf., Ithaca, NY (June 1980).
6. C. T. Chuang and R. M. White, *Proc. IEEE Ultrason. Symp.* Chicago, IL, pp. 159–162 (Oct. 1981).
7. S. G. Joshi and J. G. Brace, *Proc. IEEE Ultrason. Symp.* San Francisco, CA, pp. 600–604 (Oct. 1985).
8. S. J. Martin, G. C. Frye, A. J. Ricco and T. E. Zipperian, *Proc. IEEE Ultrason. Symp.* Denver, Colorado, pp. 563–567 (Oct. 1987).
9. J. G. Brace, T. S. Sanfelippo and S. G. Joshi, *Sensors Actuators* **14**, 47–68 (1988).

10. D. W. Galipeau, C. Feger, J. F. Vetelino and R. Lec, *Sensors Actuators B* **5**, 59–65 (1991).
11. D. W. Galipeau, C. Feger and J. F. Vetelino, *Int. J. Microcircuits Electron. Packaging* **15**, 53–59 (1992).
12. B. Han, C. Gryte, H. Tong and C. Feger, *Proc. Soc. Plastics Eng. ANTEC Conf.*, pp. 994–996 (1988).
13. S. C. Noe, J. Y. Pan and S. D. Senturia, *Proc. Soc. Plastics Eng. ANTEC Conf.*, Montreal, pp. 1598–1601 (May 1991).
14. J. Y. Pan and S. D. Senturia, *Proc. Soc. Plastics Eng. ANTEC Conf.*, Montreal, pp. 1618–1621 (May 1991).
15. B. K. Sinha and S. Locke, *IEEE Trans. Ultrason. Ferroelectrics Frequency Control* **36** (2), 231–241 (1989).
16. F. M. Fowkes, *J. Adhesion Sci. Technol.* **1**, 7–27 (1987).
17. D. Suryanarayana and K. L. Mittal, *J. Appl. Polym. Sci.* **29**, 2039–2043 (1984).
18. P. S. Ho, in: *Principles of Electronic Packaging*, D. P. Seraphim, R. C. Lasky and Che-Yu Li (Eds), pp. 809–839. McGraw-Hill, New York (1989).
19. C. L. Bauer and R. J. Farris, in: *Polyimides: Materials, Chemistry and Characterization*, C. Feger, M. M. Khojasteh and J. E. McGrath (Eds), pp. 549–562 Elsevier, Amsterdam (1989).
20. L. B. Rothman, *J. Electrochem. Soc.* **127**, 2216–2220 (1980).
21. J. Andle, Private communication (1991).
22. D. D. Denton, J. B. Camou and S. D. Senturia, in: *Moisture and Humidity*, Proc. 1985 Int. Symp. on Moisture and Humidity, pp. 505–513. Instrument Society of America, Research Triangle Park, North Carolina (1985).
23. L. P. Buchwalter, *J. Adhesion Sci. Technol.* **4**, 697–721 (1990).

Adhesion Measurement of Films and Coatings, pp. 423–434
K. L. Mittal (Ed.)
© VSP 1995.

Salt bath test for assessing the adhesion of silver to poly(ethylene terephthalate) web

J. M. GRACE,* V. BOTTICELLI, D. R. FREEMAN, W. KOSEL and R. G. SPAHN

Eastman Kodak Company, 1-81-RL-02022, Rochester, NY 14650-2022, USA

Revised version received 26 February 1993

Abstract—A time-resolved salt bath technique was employed for assessing the adhesion of silver coatings on poly(ethylene terephthalate) (PET) web. Furthermore, a basis for quantitative comparisons of adhesion was developed. The salt bath test was applied to samples of silver on PET web, made with a variety of web surface treatments. The results of the salt bath test demonstrate that the technique is capable of distinguishing levels of adhesion and, therefore, levels of treatment. It is thus shown that the salt bath test can be used to study the effects of treatment and coating process parameters on the adhesion of silver coatings to PET substrates. Although the test is specifically applied to the case of silver on PET, the concepts of a time-resolved soak test are generally applicable to systems that are known to exhibit delamination when exposed to certain chemical agents or harsh environments.

Keywords: Silver; poly(ethylene terephthalate); PET; adhesion; ageing; soak; humidity; salt.

1. INTRODUCTION

A problem endemic to coating technology is the poor adhesion of the coating to the substrate. Adhesion-related problems generally have severe ramifications for product longevity or even product yield. Solutions to such problems are often limited by the ability to measure adhesion in a practical manner — the measurement technique must be sufficiently quantitative to permit the study of the effects of process factors on adhesion for the coatings and substrates of interest (*in the products of interest*). It is not surprising, then, that the scientific literature contains a significant body of work in the area of assessing and improving the adhesion of coatings to substrates. For a comprehensive bibliography on adhesion measurement techniques for films and coatings, see Mittal [1].

For some coatings on polymer substrates, glow-discharge treatment of the substrate can improve adhesion [2]. Recent work has shown that glow-discharge treatment of ESTAR™ web (PET) prior to coating can markedly improve the adhesion of silver

---

*To whom correspondence should be addressed.

films to the web [2–4]. The surface modification of ESTAR™ (PET) by glow-discharge treatment has been studied by surface analysis techniques, and the results have been correlated with adhesion measurement by peel tests. (ESTAR™ is a PET web manufactured by Kodak; the term ESTAR™ is used here to refer to the web material, which is essentially PET.)

Some confirmation of the enhanced adhesion in silver/ESTAR™ with glow-discharge treatment has been made in a production-style system [5]. The optimization of the process, however, has been hindered by the difficulty in quantitatively assessing the silver/ESTAR™ adhesion over the range of process conditions of interest. This paper focuses on a technique recently devised to assess silver/ESTAR™ adhesion in a quantitative manner. The technique can discriminate sufficiently amongst various levels of treatment to enable process studies that will elucidate the effects of the various process factors on silver/ESTAR™ adhesion.

2. TIME-RESOLVED SALT BATH: MOTIVATION

As a first attempt to characterize the adhesion of silver to ESTAR™, one might attempt to measure the forces necessary to separate the silver coating from the web, either by peeling or pulling. In our case, this approach fails; only silver deposited on untreated web can be peeled from the web in a continuous strip. Any of the samples made with glow-discharge treatment of the web do not peel and, therefore, cannot be measured. Thus, over the range of process variables explored, only two values of adhesion are obtained: good ($= 1$) and bad ($= 0$).

It should be noted that standard peel tests (peel force measurements) have been used to demonstrate that adhesion of silver to polymeric substrates can be enhanced by glow-discharge treatment of the substrate prior to metallization [2, 3, 6]. In those cases, however, only samples with poor silver/polymer adhesion could be peeled without the use of epoxies; furthermore, the range of treatment conditions explored was significantly smaller than in our case. Because of concern about modifying the silver/ESTAR™ samples by use of epoxy, and because of concern about the potentially limited range of silver/ESTAR™ bonding over which a single epoxy might be useful, peel force measurements were ruled out for our purposes.

In the absence of any useful measurement scheme by mechanical methods, it would be practical to consider the environmental conditions that cause the silver/ESTAR™ interface to fail in the production application. (If we must modify the silver/ESTAR™ samples in order to assess adhesion, why not modify them in a way that is comparable to the way in which they are modified in their intended application?) Environmental conditions could be used to assess the influence of glow-discharge treatment on silver/ESTAR™ bond durability, thereby providing a practical measure of adhesion. Indeed, such practical measures of adhesion are often more relevant to product success than measurements of the work or forces required to separate coatings from substrates. For example, a coating might have good adhesion to its substrate when judged by peel force measurements, but the same coating/substrate system might fail when subjected to immersion in boiling water. If the coated substrate is likely to see such adverse conditions in its product application, an immersion test is far more useful than a peel force test.

This paper discusses an adhesion test based on the immersion of samples in a hot (80°C), saturated solution of sodium chloride in water. Suffice it to say that such a test is believed to be an accelerated environmental test, based on observed failures of Ag/ESTAR™ structures in more moderate conditions. Because the failure mode is essentially delamination of the silver from the ESTAR™, it can be described as adhesion failure. Assuming, then, that the durability in a hot salt bath is somehow a measure of silver/ESTAR™ adhesion, it remains to devise a quantitative adhesion test based on the exposure of the samples to the salt bath.

A good figure of merit for a salt bath test is the soak time required to cause complete failure of the silver/ESTAR™ bond (i.e. the time required for the silver to 'float off' the ESTAR™). The soak times before visible occurs may be quite long for silver on treated web. Furthermore, the logistics of monitoring a large number of samples for the 'float-off' phenomenon over long times become difficult at best. Nonetheless, the 'float-off' test can be implemented using a batch operation for soaking, with predetermined soak times.

A maximum soak time can be set and subdivided into smaller intervals (e.g. a maximum soak of 100 min in intervals of 10 min). All the samples can be immersed in the bath at once. At the end of each interval, a representative piece of each sample can be removed from the bath and analyzed for failure. For each sample, then, there must be at least as many pieces as there are soak intervals. In order to average over the variable web surface conditions and variable deposition conditions inherent in the course of some web-coating processes, several swatches of coated material can be taken from the length of processed web. Each of these pieces can then be used to provide samples for each soak interval. For a given sample of processed web, using a maximum soak of 100 min, one might take ten swatches, each cut into ten pieces. A piece from each swatch would be pulled out of the bath every 10 min, and the results after each interval would be averaged over the different swatches. A more convenient procedure for this time-resolved float-off test is presented in Section 3.

A simple extension of the time-resolved salt bath float-off test can be used to further differentiate between sampls that do not float off completely after the given soak interval. Recall that the initial motivation for the salt bath test was that a peel force measurement is impractical for many of the silver/ESTAR™ samples made with various glow-discharge treatments. An interesting question is whether or not the silver can be peeled from the ESTAR™ after the sample is soaked in the heated salt bath. Presumably, if the silver eventually floats off the ESTAR™, there must be some time interval of soak after which the silver can be peeled from the ESTAR™ with adhesive tape. Hence, the soaked samples can be tape-tested after soaking, in order to differentiate between those samples that minimally 'survive' the soak and those that 'survive' without fail.

By combining the float-off and tape tests described above, one should have the ability to distinguish between silver/ESTAR™ samples made with a variety of treatments. This ability to discern the effective degree of treatment permits the investigation of the treatment process itself, as well as interactions between treatment and coating processes. Thus, the salt bath tests can provide insight with which the treatment and coating processes may be examined.

3. SALT BATH TEST PROCEDURE

The most labor-intensive part of the salt bath test procedure is sample preparation. In our pilot coater, samples of silver/ESTAR™ are made by treating and coating 3–6 m long sections of 25.4 cm wide ESTAR™ base. In order to average out variability in the web surface and variability in the treatment and coating conditions over the length of a sample, ten pieces are taken from along the length of a given sample. From each of these pieces, ten 0.95 cm diameter disks are cut, using a punch. These disks are then glued to 3.8 × 6.4 cm tabs of 0.18 mm ESTAR™ web so that each tab has a disk from each of the ten pieces of the original sample. The disks are glued by applying RTV silicone adhesive to the ESTAR™ sides, leaving the silver surfaces facing outward from the tabs. For a given sample, then, there are ten tabs, each containing representative disks from each of ten pieces of the original sample. Each disk is identified by a sample number and a piece number.

After the samples have been mounted on the tabs, each tab is labeled to identify the sample and the soak interval. For example, sample #1 would have ten tabs labeled (#1, 10 min), (#1, 20 min), (#1, 30 min), ..., (#1, 100 min). The tabs are grouped by soak interval, and each group of tabs is suspended from a stiff copper wire, using binder clips. The samples are then ready to be tested in the salt bath.

The salt bath is prepared by heating water to roughly 80°C and adding salt (NaCl) until the solution is saturated. Actual bath temperatures used are in the range 80–84°C. A magnetic stirrer is used to keep the bath temperature and salt concentration uniform.

The samples are placed in the salt bath by laying the ten copper wires across the bath container, so that all the disks are submerged. Every 10 min, a wire is removed from the bath, and the samples suspended from it are rinsed in deionized water. The samples are patted dry with paper towels and laid out on a countertop to finish drying. After 100 min, all tabs have been removed from the bath, and the samples can be analyzed for float test results and subsequently tape-tested.

Analysis of the float test results is carried out by recording the percentage of failed silver/ESTAR™ disks for each sample at each soak interval. A disk is considered to fail if there is no silver left ('float-off') or if most of the surface is wrinkled, indicating separation of silver from the web. A disk is considered to pass if the silver surface is smooth and shows no visible signs of separation from the web. Anything between fail and pass is considered to be half way. (For one of ten disks, such a disk would contribute 5% fail to the score.)

After the float test results are recorded, a tape test is performed. Strips of Scotch Magic™ tape are placed over the silvered disks and pressed down firmly. The tape strips are then peeled, and the peeled tape, with any silver stuck to it, is mounted on an appropriately labeled index card. For each tab, the percentage of area of the ten disks still covered by silver is recorded as 'percent remaining' (= 100−% fail).

The 'percent remaining' figures for each soak interval, combined with the percent fail from the float test results, are used to calculate an adhesion score, as described below.

4. METHOD FOR SCORING THE SALT BATH TEST

The scheme for assigning numbers to the salt bath test results is based on the premise that the soak time required for float-off or tape test failure is a practical measure of the adhesion of silver to ESTAR™. For the purposes of determining a suitable assignment scheme, it is useful to consider the dependences of float-off and tape test failure on the soak time. First, we examine the float-off test.

In considering the float-off test, it is important to realize that the bond between silver and ESTAR™ is attacked by the salt bath. In principle, the silver coating, if thick enough, remains intact, but eventually its bond with the ESTAR™ web is weakened to the point that the silver floats away from the ESTAR™. In practice, the silver coating may float off as described, but it may also remain on the ESTAR™ in a severely wrinkled form. If all silver/ESTAR™ disks from a given batch were completely identical, the soak time required for float-off or severe wrinkling would be the same for the entire batch. Hence, one might expect a sharp rise in the failure rate at a critical soak time, as depicted in Fig. 1.

In practice, there is a distribution of critical soak times as a result of variability in the web surface conditions and process parameters for the treatment and coating of the web. Measured curves of the failure rate vs. the soak time are shown in Fig. 2. For untreated samples, the critical soak time is so small that the failure rate rises very quickly, almost in an exponential manner. For moderately treated samples, there is clearly a range of soak times for which the failure rate is low, but eventually the silver debonds from the ESTAR™. At higher treatment levels, the failure rate remains low for the duration of the salt bath test.

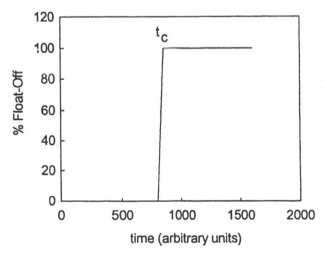

Figure 1. Float-off failure as a function of the soak time for an ideal sample. After a time t_c, the silver/ESTAR™ bond fails to the point that the silver coating floats away from the ESTAR™ web.

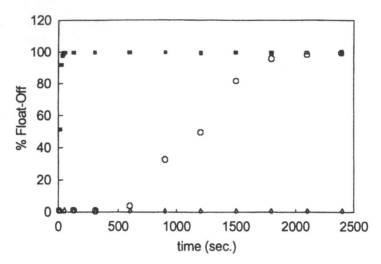

Figure 2. Float-off failure as a function of the soak time for actual samples. The data are for three samples: silver on untreated (■), moderately treated (○), and well-treated (△) ESTAR™.

In contrast to the float-off test, the tape test essentially measures the amount of silver still appreciably bonded to the ESTAR™ as a function of the soak time. Prior to the soak time for which a silver/ESTAR™ disk is likely to exhibit float-off, portions of the disk still have appreciable silver/ESTAR™ bonding. When the tape is applied to and then peeled from such disks, measurable amounts of the silver coating are left behind. As the silver/ESTAR™ bond is destroyed by chemical activity in the salt bath, one might expect the failure rate to increase to 100% with some time constant:

$$\% \, F(t) = 100(1 - e^{-kt}). \tag{1}$$

In equation (1), $\% \, F(t)$ is the percentage of disks exhibiting failure as a function of time t, and k is the decay rate. Ideally, one would like to measure k, as it should relate inversely to the adhesion of the silver to the ESTAR™. The larger the decay rate, the worse the adhesion and, consequently, the shorter the time required to have all the disks exhibit failure.

As shown in Fig. 3, some of the samples tested exhibit the behavior described in equation (1). When the quantity $\ln(100 - \% \, F)$ is plotted vs. the soak time, a line is obtained. In principle, one can plot the failure rates for each sample as a function of the soak time and fit the curves to equation (1). A somewhat simpler assessment of k can be made by weighting the failure rates inversely with their respective soak times and averaging the weighted failure rates as in equation (2) below:

$$k_{\text{eff}} = \frac{1}{10} \sum_i \left[\% \, F_i / 10i \right]. \tag{2}$$

Here, k_{eff} is an effective value of the decay rate; $\% \, F_i$ is the percentage of failed disks after the ith soak interval; and $10i$ is the duration (in min) of the ith soak interval. In the limit $t \ll 1/k$, k_{eff} approximates an average value of the decay rate k, as can

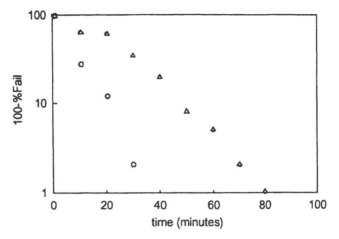

Figure 3. Tape test failure as a function of the soak time. Results are shown for two samples: silver on moderately treated (O) and well-treated (△) ESTAR™. Silver on untreated ESTAR™ failed 100% after the first soak interval and therefore would give a line downward along the vertical axis in the plot.

be inferred from equation (1). For time scales comparable to or longer than $1/k$, k_{eff} underestimates k. Nonetheless, k_{eff} provides a simple way to quantify salt bath tape test results.

The concept of k_{eff} may also be applied to the float test. Although the float test results do not obey equation (1), the fact that a range of critical soak times is observed (see Figs 1 and 2) makes the approach of equation (2) useful for analyzing the float test results as well. Samples for which failure occurs at shorter soak times will still have higher values for k_{eff} than those that fail at longer times. Hence the use of k_{eff} to analyze float test results provides a workable scheme for quantifying them.

Finally, we must convert k_{eff} into an adhesion score. As is evident from above, two failure rates are monitored as a function of the soak time. First, the float-off failure rate (% FO) is measured. Second, the tape test is performed, and the 'peel-off' failure rate [% PO = (100 − % Remaining)] is measured. (Note that 'peel-off' failure is used here to describe failure in a *tape* test performed on samples *after* exposure to the salt bath.) From the discussion above, we see that both of these tests can produce a k_{eff}. The k_{eff} for each test may be used to rate the level of adhesion:

$$\text{FLOAT} = \frac{\sum_i (\% \, FO_i)/(10i)}{2.928968}; \quad \text{PEEL} = \frac{\sum_i (\% \, PO_i)/(10i)}{2.928968}. \quad (3)$$

In equation (3), FLOAT and PEEL are adhesion scores, proportional to their respective values of k_{eff}; the number 2.928968 is a normalization factor. If all disks failed at all soak intervals, the sum in the numerator would be 29.28968. Thus, the normalization factor in the denominator makes the FLOAT and PEEL scores fall between 0 and 10, lower scores signifying better adhesion. According to this scheme, a sample for which all disks fail at all soak times gets a score of 10, whereas a sample for which all disks pass at all soak intervals gets a score of 0. A total score can be made by adding the FLOAT and PEEL scores, giving a range of values between 0 and 20.

The adhesion scores obtained from equation (3) are convenient for quantifying adhesion; because they are based on k_{eff}, however, they are not necessarily equivalent to physical quantities. As mentioned earlier, k_{eff} approximates an actual failure rate constant, provided that the failure rate follows an exponential rise [as in equation (1)] *and* the soak times are significantly shorter than $1/k$. For situations where the soak times are significantly greater than $1/k$, any rise in the failure rate with time is not discernible, as all disks will probably fail, and extraction of a true failure rate constant from the data is impractical. In the rare event that the maximum soak time is of the order $1/k$ (the maximum failure rate is appreciably less than 100%), a true failure rate cosntant can be calculated by using equation (1) for each soak time. Thus, it is evident that even if the failure rate follows equation (1), it is difficult to extract a true rate constant except in special circumstances. Furthermore, for salt bath test data that do not obey equation (1) (FLOAT test data, in particular), a true rate constant cannot be calculated. Nonetheless, by the use of k_{eff} as a loose quantifier for salt bath failure rates, equation (3) provides a simple means of quantifying adhesion.

5. TEST OF THE TECHNIQUE (RESULTS)

Initial tests of the salt bath technique were carried out by comparing the soak times required for float-off and peel-off failure of some untreated silver/ESTAR™ and some archived samples made with glow-discharge-treated web. The untreated product exhibited float-off failure within minutes of immersion in the 80°C salt bath, whereas the aged treated samples withstood an hour in the bath with no float-off failure and only moderate amounts of peel-off failure. Some other archived samples with various treatments showed behaviors between that of the untreated and original archived samples tested.

With some indication that the salt bath test could discriminate between levels of treatment, a full salt bath test was carried out on 34 fresh samples. (These samples were part of an experiment designed to study the effects of treatment process parameters on adhesion; we focus here on the salt bath test technique only.) For all 34 samples, the same ESTAR™ thickness was used, the silver deposition technique was the same, and the silver thickness was roughly constant. Among the 34 samples were four replicate runs with all the process conditions at mid-range and two pairs of replicates made without treatment of the ESTAR™ web. In addition, six of the 34 samples were analyzed twice in the salt bath. The range of values obtained for FLOAT and PEEL, as well as the results for the replicate groups are shown in Figs 4 and 5.

The results for the FLOAT and PEEL adhesion scores, as compared with the peel force test, are shown in Fig. 4. Plotted on the vertical axis is the sum of the FLOAT and PEEL scores, while on the horizontal axis are the sample numbers, ordered to have descending values of PEEL + FLOAT from left to right. For comparison, the results of a peel force test are indicated by the open circles. As the peel force test gave only two results (see Section 2), we may map the 'good' and 'bad' peel force results to adhesion values of 0 and 20, respectively, on the FLOAT + PEEL scale. It

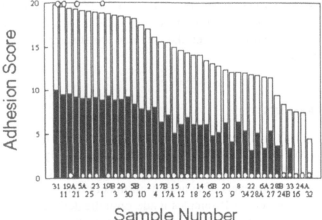

Figure 4. Ordered list of adhesion scores. The contributions of FLOAT (PEEL) to the adhesion scores are indicated by dark (light) bars. Adhesion values are arranged left-to-right from worst (highest) to best (lowest). The sample numbers denote silver depositions on webs with a variety of treatments, including deposition on untreated web. For comparison, results of a peel force test are shown as open circles. Samples that could be peeled without soaking were all made with untreated ESTAR™ and are given a score of 20; those that could not be peeled are given a score of 0.

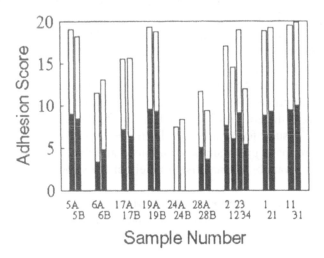

Figure 5. Adhesion scores for groups of replicates. As in Fig. 4, contributions of FLOAT (PEEL) are indicated by dark (light) bars. Sample numbers with letter suffixes indicate replicate measurements on samples from the same batch. Different sample numbers grouped together indicate batches made under replicate process conditions.

is interesting to note that several samples that did not peel before soaking performed quite poorly in the salt bath test.

In Fig. 5, the adhesion values for the groups of replicates are shown. As can be seen, there is some variability in the adhesion values obtained from repeated salt bath tests of the same sample material; there is also variability in the adhesion values obtained from salt bath tests on different samples made under nominally the same processing conditions. Possible sources of variability are discussed in the next section. For now it suffices to note that the variability in the salt bath test results is not large enough to prevent the discrimination between the best and worst samples, as well as those of intermediate quality.

6. ANALYSIS OF RESULTS

The results of the salt bath test on the 34 samples (see Figs 4 and 5) show that the test can discriminate between different levels of adhesion. By comparison, the peel force test (without the use of epoxy) can only be carried out on untreated samples. The ability to distinguish levels of adhesion ultimately determines the ability to distinguish levels of treatment. All samples made on untreated web exhibit comparably low peel forces; the rest cannot be peeled and would be considered to have good adhesion. Essentially, a peel force measurement (without the use of epoxy) can only distinguish one process condition (silver on untreated web) from the rest. In contrast, there is a range of salt bath adhesion scores obtained for samples made with a range of web treatments. Furthermore, the differences measured by the salt bath test are quite relevant to the product application, whereas the peel force test does not probe the environmental stability of the product. Based on the peel force test, one might conclude that 30 of the 34 samples have adequate adhesion of silver to ESTAR^{TM}. Examination of the salt bath test results, however, might result in a significantly different conclusion, depending on what degree of environmental degradation is considered acceptable for the product.

Having established that the salt bath technique can measure adhesion for practical purposes, it is appropriate to discuss the degree of variability in the measurements and the possible sources thereof. As illustrated in Fig. 5, the salt bath test gives similar, but not identical, results for samples made in the same batch. Furthermore, when several batches are made under similar conditions and then tested, there is variability in the results. Such variability can arise from either the sample preparation or the test procedure.

There are two obvious sources of variability with regard to sample preparation. First, there is the issue of sample uniformity. Some factors affecting sample uniformity are

• uniformity of the web surface conditions (particulates, chemical residue, local stress, etc.);

• uniformity of the process parameters over the course of making a batch (are treatment and coating parameters held constant over the length of coated web?); and

• uniformity of the treatment and coating conditions over the width of the web.

A second source of sample variability arises from the difficulty in reproducing a given set of process conditions for separate bathces. Such difficulty may arise from problems in maintaining uniform process conditions for a single batch, depending on the nature of the jitter and drift in the process settings. In addition, the materials used from batch to batch may differ in quality. The web surface conditions, in particular, might differ from roll to roll, thereby causing variability of the results.

The salt bath test may also introduce variability in the results. Both the salt concentration and the bath temperature strongly affect the float-off and peel-off phenomena. If these test conditions are not properly controlled, measurement reproducibility will be degraded. Although variability is evident in comparison of the results for similar samples in different baths, the magnitude of this variability is comparable to the variability seen among sets of disks from the same sample in the same bath. We may therefore conclude that the reproducibility in test results is not limited by the ability to control bath conditions. (Typically, the bath temperature is kept in the range 80–84 °C, and the salt content is kept high enough to saturate the solution.)

From available information on the variability of treatment and coating process conditions, it appears that most of the coating process conditions are controllable and repeatable. The glow-discharge power, however, is particularly difficult to stabilize and is difficult to set repeatably. (It is likely that the power instability is related to pressure fluctuations caused by variation in the gap between the web and glow box. Such variation can be reduced by mounting the glow box near a web transport roller and by changing the shape of the glow box appropriately.) The effective variability in the glow-discharge power likely degrades both sample uniformity and batch-to-batch reproducibility. An additional factor that is likely to cause such variability is the nature of the web surface, as described above. With no further diagnostics performed, we can only speculate that the key contributors to the variability in the salt bath test results are the lack of adequate control of a key treatment parameter and the variability of the web surface condition. Both of these sources may cause variations within a given roll or from roll to roll of product.

An interesting extension of the concept of measurement variability is that of measurement tunability. As mentioned above, the bath conditions strongly affect the test results, but the conditions are adequately controlled. Given the conditions used in this study, a range of sample behaviors can be observed, as is evident from Fig. 4. Suppose, however, that all the samples studied were of extremely good or extremely bad quality and that there was a reason to believe that the samples were significantly different fron one another. For example, if all samples had adhesion scores close to 0, how could they be compared with one another? If all samples had scores close to 10, how could they be compared? In the former case, the bath temperature or the soak interval could be increased to see if any of the samples would eventually fail. In the latter case, the soak interval could be reduced, the bath temperature could be reduced, or the salt concentration could be substantially reduced; these changes would be made to see if any samples could survive. In either case, a set of bath conditions could be found to give adequate time resolution of float-off and peel-off failure. The new conditions could then be used to see if any of these samples behaved differently from one another.

The salt bath test results establish that different degrees of web treatment produce different degrees of adhesion. Thus, a capability to distinguish levels of adhesion is established. Furthermore, this capability should be tunable to different ranges of levels of adhesion.

Of more general interest, however, is the result that time-resolved soak tests can provide quantitative assessment of practical adhesion. The coating and substrate in question need not be silver and ESTAR™. There may be many situations in which a coating delaminates from a substrate under exposure to specific environments. Using the ability to quantify adhesion failure in such environments (or extreme cases thereof), processes can be found that enhance the robustness of the coated structures in adverse conditions. Although the salt bath may not be a relevant environment for the adhesion testing of other systems, the basic methodology of the time-resolved soak test, as well as the analysis of failure rates, should be readily applicable to tackling problems of coating delamination in harsh environments.

7. CONCLUSION

The time-resolved salt bath technique provides a way to characterize the adhesion of silver coatings to ESTAR™ web. This technique may be tuned for analysis of samples having extreme adhesion values. Tunability of the salt bath test can be achieved by manipulating the salt bath conditions and the soak interval. The limiting factors in test reproducibility appear to be related to sample uniformity and the control of some process parameters rather than the bath conditions. Because this technique can discriminate between samples made under a variety of treatment and coating conditions, it can be used to optimize the treatment and coating processes.

As previous attempts to characterize adhesion could give only two results (good or bad), we now have the ability to see differences that previously could not be seen. We can use this newfound vision to assess the importance of the various deposition/glow-treatment parameters on the adhesion of silver to ESTAR™ and other PET webs. In addition, the concepts of the time-resolved soak test and analysis based on the failure rate are more widely applicable to problems of coating delamination in harsh environments.

REFERENCES

1. K. L. Mittal, *J. Adhesion Sci. Technol.* **1**, 247 (1987).
2. R. W. Burger and L. J. Gerenser, in: *Metallized Plastics 3: Fundamental and Applied Aspects*, K. L. Mittal (Ed.), pp. 179–193. Plenum Press, New York (1992).
3. R. W. Burger and L. J. Gerenser, *Soc. Vacuum Coaters 34th Annu. Tech. Conf. Proc.*, p. 162 (1991).
4. L. J. Gerenser, *J. Vac. Sci. Technol.* **A8**, 3682 (1990).
5. J. Roth and R. G. Spahn, Eastman Kodak Company, private communication (1992).
6. L. J. Gerenser and K. E. Goppert-Berarducci in ref. 2, pp. 163–178.

Adhesion Measurement of Films and Coatings, pp. 435–456
K. L. Mittal (Ed.)
© VSP 1995.

Testing the adhesion of paint films to metals by swelling in *N*-methyl pyrrolidone

W. J. van OOIJ,* R. A. EDWARDS, A. SABATA and J. ZAPPIA

Armco Research and Technology, Middletown, OH 45044-3999, USA

Revised version received 15 March 1993

Abstract—A new test for estimating the adhesion of paints to metal substrates is presented. Small painted disks are immersed in *N*-methyl pyrrolidone (NMP) at 60°C and the time for the paint film to delaminate completely and intact from the substrate is recorded. This time, termed the NMP paint retention time or NMPRT, is shown to be sensitive to changes in the metal pretreatment, the type of paint, the cure conditions, and the presence of water at the interface. As the strong swelling of the paint in NMP induces interfacial shear stresses, the NMPRT value is believed to be a measure of the number or strength of the interfacial bonds. Two examples are discussed in some detail: (i) delamination of automotive epoxy-based electrocoat systems from phosphated cold-rolled steel (CRS) and electrogalvanized steel (EGS), and (ii) modification of the interface between stainless steels (types 301 and 409) and epoxy or polyester powder paint systems by means of organofunctional silanes. In both examples it is shown how the NMP method can be used to optimize paint performance on metals.

Keywords: Paints; adhesion; metals; silanes; testing; interfaces; swelling; NMP.

1. INTRODUCTION

The most widely used adhesion tests for paints, especially in the automotive industry, are crosshatch tests [1]. For instance, in the General Motors test, a pattern of 100 small squares 1.5×1.5 mm^2, is cut through the paint into the base metal using a specially designed tool. The hatched area is then tested by adhesive tape and the number of paint squares that have come off is recorded. Other tests that exist are the nickel scratch test, impact tests, bend tests, pull-off tests, and the like. It is not the purpose of this paper to review any of these tests in detail. However, the general comment that can be made is that in all existing tests for paints, the mechanical properties of the paint play an important and sometimes dominant role. For instance, in impact testing of painted automotive steels, failure is often observed to occur at the interface between the phosphate crystals and the metal, and not between the paint and the phosphate crystals. The mechanical properties of the paint are important here as

---

*To whom correspondence should be addressed.

the paint absorbs the impact energy and transmits part of it to the phosphate–metal interface.

Other paint adhesion tests that have been described are the blister test [2] and tests in which acoustic emission from blistering or mud-cracking paint coatings are recorded [3]. The latter are complicated and there is still controversy as to the origin of the acoustic signals that are sometimes recorded when paint coatings are subjected to cathodic delamination conditions. Further, no signals are observed when the system is at rest, so this technique cannot be regarded as a general adhesion test. The blister test, on the other hand, is an excellent test for measuring the actual fracture energy, Γ, of a polymer–metal system. Recently, for instance, the effect of substrate surface roughness and the roughness profile for polyurethane coatings on stainless steels was investigated [2]. The fracture energy or peeling energy was found to decrease as the roughness increased.

For paint systems on structural steels, adhesion is often estimated by tensile tests. Studs of a certain standard diameter, e.g. 1 cm^2, are glued to the paint surface and then removed in a 90° pull-off arrangement with or without the application of a toroidal force. The work required to remove the paint is recorded [4]. Here mechanical properties such as the cohesive strength of the adhesive or the paint determine the maximum adhesion level that can be measured. Since paints do not have high cohesive strengths, only relatively low levels of adhesion can be measured. In the event of covalent interfacial bonding, failure in the tensile test will frequently occur cohesively in the paint.

Minor differences in interfacial chemistry cannot be detected by the current adhesion test methods. An example is the effect of a final chromate rinse of phosphate conversion coatings. While it has been firmly established that such rinses improve the corrosion performance of the painted system, the common tests do not detect any effect of the chromate rinse on the adhesion of the paint system. Zinc phosphate systems on CRS or EGS are known to be partly dehydrated from the tetrahydrate to the dihydrate hopeite structure [5]. During exposure of the painted system in the field or in accelerated corrosion tests, rehydration to the tetrahydrate occurs. Since it has also been postulated that bonding between phosphate crystals and epoxy paints occurs by hydrogen bonding (or acid–base interactions) involving the water of hydration and the oxygen functionalities in the paint vehicle [6], this rehydration should be expected to have a strengthening effect on the interface. So the dry adhesion will be improved because the number of hydrogen bonds has increased. No adhesion tests have been developed that are capable of demonstrating this effect.

As a result, it is difficult to establish correlations between the adhesion of a particular system and its corrosion performance in the field or in accelerated tests. Knowledge of such correlations is important for the further improvement of metal pretreatments or paint systems.

Another important effect that cannot be established easily by the current tests is the effect of water on the interfacial adhesion and chemistry. Wet adhesion is expected to be improved if covalent bonds can be established across the interface. Such bonds will improve the durability of the system. A test method that would provide more detailed information on the paint–metal adhesion in a wet environment would be extremely useful in, for instance, the development of metal pretreatments by organofunctional silanes. Such silanes are reported to improve the hydrolytic stability of polymer–metal interfaces [7]. However, their performance as adhesion promoters when directly applied to metals is strongly dependent on the cleanliness of the metal, the silane application conditions, and other factors. Hence an *in situ* adhesion test that would provide information on the strength of the interface in the dry and wet states would be very useful for the development and optimization of metal pretreatments by functional silanes.

Clearly there is a need for an additional adhesion test that provides more information on the chemistry at the interface and the type of interfacial bonding, e.g. by being able to measure the effects of interfacial water or of interfacial changes occurring during exposure to a corrosive environment. Ideally, in such a test, contributions to the measured peel or fracture energy due to the paint mechanical properties, stresses, and presence of stress raisers in the paint coating should be eleminated. What remains is the contribution of the interfacial bonds only.

In our ongoing studies of corrosion mechanisms in painted metals, we have frequently used a method to remove paint systems from metal substrates, without damaging the paint, interfacial corrosion products, or phosphate crystals. This method consists of immersing a small piece of the painted metal N-methyl pyrrolidone (NMP). This is a colorless liquid (b.p. 203 °C) with the composition C_5H_9NO and structure:

$$
\begin{array}{c}
CH_3 \\
|\\
N \\
\langle\;_H\;\rangle = O
\end{array}
$$

This solvent is a swelling solvent, i.e. as a result of its high polarity, it tends to swell many polymers without actually dissolving them. The swelling increases the dimensions of the paint film markedly and it delaminates rapidly from the metal. The delaminated paint film can be dried, which takes about 24 h. In this process, it shrinks back to its original dimensions.

After complete drying, both the metal surface and the backside of the delaminated paint can be analyzed by techniques such as SEM, EDX, XPS, and even TOFSIMS [8]. This procedure has proven to be extremely useful for analytical studies of paint–metal interfaces. No traces of NMP are detected at the interface, not even by sensitive techniques such as TOFSIMS [8]. When this technique is used to remove paint films from corrosion products, e.g. in painted automotive steels following exposure in a corrosion test, corrosion products are frequently observed on both sides of the interface with the original unattacked phosphate crystals firmly adhering to the paint film. In regions where no corrosion has taken place, delamination is always exactly at the interface between the paint and the phosphate crystals, as could be established by interfacial analysis using techniques such as SEM, EDX, XPS, and TOFSIMS.

In using this paint removal method, it was observed that the paint delamination time was reproducible and, further, certain trends were observed. For instance, it took more time to remove an overcured paint than a paint which was not cured properly. Paints were difficult to remove from metal substrates which had been pretreated with certain silanes, and a consistent and reproducible difference was noticed between systems with and without chromate rinse of the phosphate conversion coating. Paints could not be removed at all from certain regions of the corrosion products propagating under the paint from a defect (scribe), indicating that the paint in such regions was no longer swellable. Some paints that were found to be removable intact in a reproducible way using NMP were epoxies and epoxy-urethanes (cataphoretic electrocoat systems and powder paints), acrylic and melamine-based automotive topcoats, thick PVC coatings, polyester powder paints, unsaturated polyesters and modified polyester solvent-based appliance paints, and various blends and hybrids. Paints that cannot be removed by this solvent are generally those that are highly crosslinked and hence not swellable. Further, the method does not work with paints that are highly filled with metal powders or flakes. Here, too, the paint resists swelling, probably because of the bond that exists between the metal particles and the paint resin. The paint removal times observed for different paint systems can be quite different, even if these systems have approximately identical thicknesses. It is, therefore, obvious that a quantitative method based on the paint removal time cannot be used to compare different systems, but only for detecting changes in a system in which the resin, filler, filler content, etc. are kept constant.

The objective of the study described in this paper was to establish, on a more quantitative basis, the relationship between the paint removal time and certain parameters of the paint and substrate, such as curing conditions, type of substrate metal, pretreatment of the metal, and ageing of the painted metal sample. In other words, it was investigated whether the test could be quantified so that it could be proposed as an adhesion test for the determination of subtle changes at the paint–metal interface. The examples that are presented and discussed are:

(1) catophoretic epoxy-based automotive paints on phosphated CRS and EGS. Among the variables here are the type of substrate, metal pretreatment, paint cure, and exposure in a standard corrosion test. Information on how such parameters affect adhesion is considered necessary for the further improvement of the corrosion of painted automotive steels. The exact mechanism of cosmetic corrosion of painted automotive steels and the role of the paint adhesion is, despite many studies, still not well known [9];

(2) powder paint systems (both epoxies and polyesters) on stainless steels whose surfaces were modified by organofunctional silanes. Painting of stainless steels is known to be problematic because of the presence of Cr_2O_3 in the oxide film. Although stainless steels have an inherently good corrosion resistance, there is interest in improving the paintability of such steels, primarily for decorative purposes. An improved pretreatment would enable various stainless steel grades to be used more extensively in the appliance industries [10].

2. EXPERIMENTAL

2.1. Materials and painting procedures

The substrates in this study were cold-rolled steel (CRS) sheets of standard automotive grade and thickness 0.7 mm, and electrogalvanized steel (EGS) sheets of 70 g/m^2 coating weight, both supplied by Armco Steel Company in Middletown, Ohio. Also used were sheets of grade 409 stainless steel (12% Cr, denoted by SS4) of thickness 0.7 mm, and 301 stainless steel (17% Cr, denoted by SS3) of thickness 1.5 mm. Both stainless steel substrates were of 2D surface finish and were supplied by Armco Advanced Materials Corporation in Butler, Pennsylvania.

N-Methyl pyrrolidone (NMP) was reagent grade (>99%) obtained from J. T. Baker.

The silanes used in the surface preparation of the stainless steels were all obtained from Dow Corning, namely glycidoxypropyltrimethoxy silane (GPS), styrylaminoethylaminopropyltrimethoxy silane (SAAPS), and 1,2-bis(trimethoxysilyl)-ethane (TMSE) crosslinker. GPS and the crosslinker were obtained as pure liquids (>98% purity); SAAPS was received as a 40% solution in methanol.

Both the CRS and the EGS substrates were zinc phosphated using the Bonderite 958 immersion process from Parker-Amchem (Madison Heights, Michigan). This process produces fine-grained Mn- and Ni-containing zinc phosphate crystals (hopeite) on EGS and predominantly zinc iron phosphate crystals (phosphophyllite) on CRS. The process involves an alkaline precleaning step and a final rinse in Parcolene 60 from Parker-Amchem which contains hexavalent chromium. On some panels, the final Cr(VI) rinse was omitted. In the second experiment with phosphated CRS and EGS, standard phosphated CRS and EGS (60 g/m^2) panels from ACT in Hillsdale, Michigan were used (Parker panels). These panels were zinc phosphated using the Bonderite 952 process, which is very similar to the Bonderite 958 process but instead it is applied by spraying rather than dipping. Here, too, panels were prepared with and without the standard final hexavalent Cr rinse. The phosphate coating weight on all panels was approximately 2–3 g/m^2.

All CRS and EGS panels were first cathodically electrocoated for 2–3 min at 175 V using the epoxy-based ED-11 primer from PPG. Most E-coated panels were phosphated as described above, but in some cases non-phosphated CRS and EGS panels were E-coated as well. The E-coats were cured by curing in air at 175°C for 30 min, which is the standard cure. Some panels were undercured at 150°C and others were overcured at 185°C. The thickness of the cured E-coat film was 25–30 μm. The cured primer is essentially an epoxy-urethane (epoxy crosslinked by a blocked diisocyanate system). Following priming, the panels were painted by spraying using a standard automotive topcoat system consisting of a basecoat (UBC-8554 White) and a clearcoat (UR-1000), both from PPG. Both of these paints are of an acrylic-melamine type. The total paint thickness after baking for 30 min at 121°C was 80–85 μm.

The powder paint used for painting the stainless steel substrates was either an epoxy paint (HK 002U from Interpon, Houston, Texas) or a polyester, 9W125 from Glidden Company in Cleveland, Ohio. They were electrostatically sprayed using a hand-held spray gun and then baked in a hot air oven for 20 min (for the polyester) or 10 min (for the epoxy) at 193°C. The cured film thickness was approximately 40 μm.

Some stainless steel panels were pretreated with Firstcoat MP (from Martin International in Columbus, Ohio), a chromate-containing epoxy-based primer and adhesion enhancer for paints on steel. It was applied by roll coating.

2.2. Metal pretreatments by etching and organofunctional silanes

The powder paints were applied to the stainless steel substrates with or without etching. First, all 409 and 301 samples were alkaline-cleaned in an industrial cleaner (Chemkleen from Chemfil, Troy, Michigan). Some of the 409 samples were then etched for 5–10 s in Vilella's reagent (5 ml of concentrated HCl, 1 g of picric acid, 100 ml of methanol). Some of the 301 stainless samples were first etched in a 2 : 1 mixture of nitric and acetic acids for 10 s and then for 5 s in a 10% solution of oxalic acid.

The organofunctional silanes were, prior to application, hydrolyzed in water which was acidified to pH 4 with acetic acid. Their concentration during hydrolysis was around 300 vol%. Hydrolysis of the silanes was considered complete when the solutions had become clear. The solutions were then diluted with deionized water to 1 vol% concentration. The metal samples were dipped in these freshly hydrolyzed solutions for 5 s and then blown dry. When the silane crosslinker TMSE was used, the silane and crosslinker were hydrolyzed separately and then mixed just prior to application to the metals. In all cases, the mixtures were 0.8 vol% in functional silane and 0.2 vol% in crosslinker.

Prior to dipping in the silane solutions, some 409 and 301 samples were rinsed for 30 s in a silicate solution which was 0.005 M in waterglass (sodium silicate) and 0.005 M in barium nitrate. The pH of this solution was adjusted to 12 and the temperature of the application was 55 °C. Following the rinse, the samples were dried in air for not more than a few minutes and then dipped in the silane solutions as described above. In the Results and Discussion sections of this paper, the silane treatment will be referred to as SL rinse. The combined silicate and silane rinse treatment is termed SSL rinse, and the combined silicate + silane + crosslinker treatment is termed SSLX. Applications of these treatments on cold-rolled steels to improve their adhesion and corrosion performance have been described [11].

2.3. Corrosion and paint adhesion testing

The water soak test was carried out by keeping the painted panels with sealed edges immersed in water at 70 °C for a period ranging from 3 days (CRS and EGS panels) to 3 weeks (most painted SS panels).

The standard GM 9071P crosshatch test was performed by cutting patterns of 100 squares of 1.5×1.5 mm^2 in the paint film and down into the base metal using a special tool. When this test was applied after the water soak test, the panels were dried only superficially using blotting paper. Immediately following cross-hatching, the area was tested several times with adhesive tape (type 898 from 3M) until no more paint could be removed. The test results as presented in the tables are expressed as the number of squares in the crosshatched area from which the paint had been removed in the tape test.

In the NMP paint delamination test, disks of 16 mm diameter (2 cm^2 area) were punched out of painted panels and then immersed in NMP held at 60°C. The paint always delaminated from the edges inwards. The time when the paint had completely come off was recorded. This experiment was done three or five times per panel and the average was calculated as the NMPRT for a particular paint–metal combination. Each punched disk was always treated in fresh solvent.

The GM scab corrosion test applied to the painted CRS and EGS substrates was the 9540P test [12]. This is an accelerated indoor corrosion test in which 24 h cycles of temperature, salt exposure, and humidity are applied for 5 days, followed by a different cycle during the weekend. The percentage of the time that the panels are wet in this test is about 95%. The test usually runs for one or two period of 4 weeks. Normally, in corrosion tests painted panels are shot-blasted and scribed down to the base metal in order to create artificial defects where underfilm corrosion will initiate. The panels used in this study were not scribed, however, since it was our objective to determine interfacial changes without interference from corrosion products.

2.4. Experimental design of NMP test of painted CRS and EGS substrates

A total of 72 10 × 15 cm^2 panels were prepared using various levels of seven different variables. Three variables were given three levels and four variables were given two levels. In Table 1 the variables and their levels are summarized.

In this scenario, a total of 432 ($3^3 \times 2^4$) treatment combinations are possible. In order to be able to estimate and test the effect of each of these variables on paint adhesion (NMPRT value), including interactions between them, it was necessary to use only a fraction (one-sixth) of the 432 possible combinations, i.e. $3^2 \times 2^3$ (= 72). Modular arithmetic was used to choose the appropriate fractional design. Nine of the 27 three-level treatments were chosen by representing the level of each variable as a 0, 1, or 2, summing across the three variables, and selecting only those sums which when divided by 3 had a remainder of 2 (i.e. sum = 2 or 5). Eight of the 16 two-level treatments were chosen by representing the level of each variable as a 0 or 1, summing across the four variables, and selecting only those sums which when divided by 2 had a remainder of 1 (i.e. sum = 1, 3, or 5).

Panels were prepared according to this design and the NMPRT was measured for five punched disks per panel. These samples were taken from areas close to the corners and one from the center of each panel. The NMPRT data reported are averaged for these five samples. The data were then evaluating using an ANOVA (analysis of variance) and Duncan's multiple range test.

3. RESULTS

3.1. Effect of organofunctional silanes on the NMPRT of E-coat primers on EGS

Some preliminary results with the NMP test have been published previously [13]. As a typical example of the effects that can be observed when the metal pretreatment is varied, the NMPRT values for a set of EGS samples are shown in Table 2.

Table 1.

Variables in the NMP adhesion test of painted CRS and EGS

1.	Substrate	
	CRS	Cold-rolled steel
	EGS	Electrogalvanized steel (70 g/m^2 Zn)
2.	Degree of wetness	
	Dry	Tested as such
	Wet	Water soak 24 h, air-dry 4 h, then test
	NaCl	Dip in 5% NaCl, air-dry 4 h, then test
3.	Phosphate	
	No	No phosphate application beneath paint
	Yes	Standard B952 phosphate process
	Prebaked	Heating phosphated panel at 200°C for 30 min
4.	Paint cure	
	165°C	Paint cured for 30 min at 165°C (undercure)
	175°C	Paint cured for 30 min at 175°C (standard cure)
	185°C	Paint cured for 30 min at 185°C (overcure)
5.	Post rinse	
	No	Cr(IV) final rinse omitted
	Yes	Standard Cr(IV) final rinse applied
6.	Topcoat	
	No	Panels E-coated only
	yes	Basecoat/clearcoat system applied + cured
7.	GM test	
	No	No GM test applied to painted panel
	Yes	Panel 4 weeks in GM test (unscribed)

Table 2.

NMPRT values for phosphated and E-coated EGS. Comparison of standard Cr(VI) post rinse with silicate–silane[a] rinses

Post rinse	NMPRT (min)[b]
None	4
Cr(VI)	25
Silicate + γ-APS	5
Silicate + MPS	8
Silicate + GPS	60
Silicate + SAAPS	> 180

[a] See Section 2 for desription of silanes and silicate–silane process.

[b] Average of three punchings per panel.

In this experiment, the final Cr(VI) rinse of the zinc phosphating process had been replaced with the combined silicate–silane rinse, as has been described in the Experimental section for the SS materials [14]. Two additional silanes were used here, namely γ-aminopropyltriethoxy silane (APS) and γ-methacryloxypropyltrimethoxy silane (MPS).

Remarkable differences can be noted between the paint removal times, depending on the type of silane used. Also, a difference is observed between the systems with and without the standard Cr(VI) rinse. Both the crosshatch and the nickel scratch test were applied to the same panels, but they failed to reveal any difference between the systems.

The differences shown in the table were reproducible for a given set of panels. However, there was some variation between the NMPRT values obtained in different experiments using the same substrates. This variability may be related to differences in paint cure conditions, changes in paint and phosphate compositions, or to differences in silane performance. It has recently been shown that the performance of silanes on metal substrates is critically dependent on the metal pretreatment and also on the application conditions of the silane. This phenomenon is related to the orientation of the silane molecules and the thickness of the silane film formed on the metal surface [15]. These are important parameters, but they are difficult to control on substrates of industrial origin.

The cause of the wide variation in NMPRT with type of silane has not been investigated in any detail. The observed effect can, at least in part, be attributed to different reactivities of the functional groups in the silanes with respect to the functional groups in the paint polymer, which are primarily hydroxyl and isocyanate groups.

Recent studies, which are currently being published elsewhere, have demonstrated that the ranking of the silanes in Table 2, in terms of increasing NMPRT, parallels the performance of the same systems in corrosion tests [14]. Both accelerated laboratory tests, such as the GM scab test, and long-term outdoor exposure show the same trends: omission of the Cr(VI) rinse deteriorates the corrosion performance and several silanes outperform Cr(VI) rinse. The best silane for use on EGS was SAAPS in all tests.

3.2. Designed experiment with automotive paint systems on CRS and EGS substrates

The results for the materials and variables as described in the Experimental section are presented in Figs 1 and 2. Figure 1 shows the average NMPRT values and their error bars (2σ limits) for two three-level variables and four two-level variables. Variable No. 4 of Table 1 is not plotted because its effect on the NMPRT was negligible. It thus seems that in the narrow temperature range used here, the paint cure does not affect the rate of paint removal. No significant variations or trends were found between the NMPRT data obtained with the five different sample locations.

Figure 2 shows the three interactions that were found, namely the variable 1∗3 (substrate∗phosphate), 1∗5 (substrate∗post rinse), and 3∗5 (phosphate∗post rinse).

The figures indicate the following trends. The variables are discussed here in decreasing order of importance:

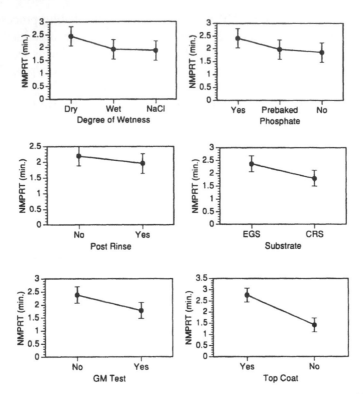

Figure 1. Effect charts with 2σ limits of six variables in the designed experiment with NMP paint delamination from CRS and EGS substrates.

- the presence of a topcoat is the most important factor; it increases the NMPRT by 1.34 min;
- after the GM test, the NMPRT has reduced by 0.60 min;
- the retention time is higher for EGS than for CRS by 0.56 min;
- phosphated substrates have a better paint retention than non-phosphated panels, and prebaking has a negative effect on the retention time;
- the paint retention is deteriorated after immersion in water, regardless of whether pure water or a salt solution is used; when interpreting these results, it should be noted that NMP is readily miscible in all proportions with water. Therefore, NMP can be expected to diffuse easily through water-saturated paints;
- the Cr(VI) post rinse has a slight negative effect on the NMPRT.

As for the interaction between variables (Fig. 2), the following conclusions can be drawn:

- the phosphate∗substrate interaction is evident in that EGS behaves better than CRS with treatment (i.e. phosphate, with or without the prebake) but it is worse than CRS without any pretreatment;
- the Cr(VI) post rinse makes CRS worse by 0.70 min, but has little effect on the EGS material (post rinse∗substrate interaction);

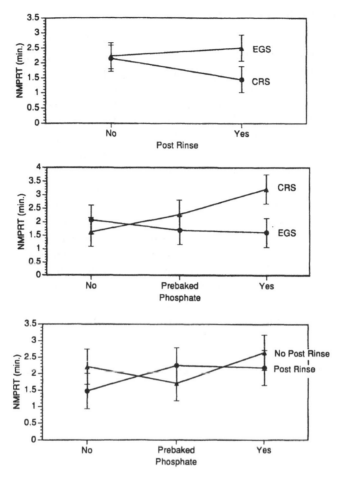

Figure 2. Interactions chart with 2σ limits in the designed experiment with NMP paint delamination from CRS and EGS substrates.

- the post rinse∗phosphate interaction is demonstrated by the observation that the post rinse improved the NMPRT by 0.54 min for prebaked panels but worsened the phosphated non-prebaked materials by 0.46 min.

The significance of these observations and how they may be related to the interfacial chemistry of the systems will be discussed later. It should be commented here, however, that most of the above conclusions from the experimental data, e.g. those on the effects of phosphating and of water immersion, are in accordance with what one would intuitively expect. Wet adhesion and adhesion after corrosion testing in the GM scab test should deteriorate and phosphate conversion coatings should improve adhesion to paints. The commonly used adhesion tests in the automotive industry (cross-hatch, nickel, tape, etc.) will not show differences until after gross corrosion has occurred. Hence the NMP method seems to be capable of detecting changes at the interface at a much earlier stage.

3.3. Additional NMP testing of automotive materials

Another experiment conducted with similar automotive CRS and EGS panel but supplied by a different source. A difference with respect to the experiment of the previous section is that a more rigorous wet adhesion test was performed here. The E-coated panels were continuously immersed for 65 h in water heated at 70°C. This water soak is known to cause severe blistering of paints on steel if the metal pretreatment is inadequate, or if the paint adhesion is poor for other reasons. The NMPRT was determined here on each panel immediately following the water soak, i.e. without a drying period as in the previous experiment. Another variable in this experiment was again the Cr(VI) post rinse of the phosphate coating. The spray phosphating process was standard and somewhat similar to the process used in the previous experiment. The cure temperature of the paint was also varied in this experiment, i.e. the standard cure temperature was compared with samples that were severely undercured.

This was a simple experiment with four two-level variables (i.e. substrate, Cr(VI) rinse, wet/dry, undercure/normal cure). The NMPRT values were determined in triplicate using separate 10×15 cm^2 panels. The results shown in Table 3 are the average values. The NMPRT adhesion data were complemented with those obtained from the standard crosshatch test of identical panels. This test was performed on wet panels and on the same panels after drying. In all the systems shown in Table 3 the paint retention in this crosshatch test was 100%, i.e. no paint was removed, thus detecting no differences between systems.

Very interesting trends in this experiment are observed. These can be summarized as follows:

- all EGS substrates have a higher NMPRT than the CRS systems. This effect was also observed in the designed experiment, but the effects are more pronounced here;
- there is now a considerably greater difference between the paints cured at different temperatures, which can undoubtedly be attributed to the greater deviation from the standard cure temperature here as compared with the designed experiment. The same effect is observed for all panels and the magnitude of the effect is approximately the same for EGS and CRS;
- for all systems, the Cr(VI) rinse improves the NMPRT markedly in this experiment. This is different from the results seen in the designed experiment. Apparently, the

Table 3.

NMPRT values (in min) for E-coated automotive CRS and EGS[a]

	Dry		Wet[b]	
Substrate	149°C cure	176°C cure	149°C cure	176°C cure
CRS–no post rinse	0.21	0.67	0.30	0.78
CRS–post rinse	0.42	0.95	0.58	1.00
EGS–no post rinse	0.28	1.35	0.73	2.43
EGS–post rinse	3.08	4.48	15.78	19.47

[a] Averaged for three panels.
[b] 65 h soaked in water of 70°C.

two phosphate systems behave differently in this respect. However, the effect is considerably greater for all EGS systems than for the CRS systems and this observation is in agreement with that of the designed experiment;

- the wet adhesion is in this experiment in all cases higher than the dry adhesion, which is opposite to the effect observed in the previous experiment. In the Discussion section a possible explanation for this effect will be given;

- these results again are not in disagreement with intuition, especially the lower interfacial strength of severely undercured paints, which is what should be expected.

3.4. Powder paint adhesion to silane-modified stainless steel substrates

It is well known in the industry that stainless steel (SS) exhibit notoriously poor paint adhesion. This is usually attributed to the presence of a passive Cr_2O_3-containing oxide film. This passivity is also the underlying cause for the high corrosion resistance of such steels. Therefore, most of the treatments that have been proposed in the open and patent literature to improve the paintability of SS involve an etching step. In such an etch, the Cr_2O_3 film is removed by a strong acid. However, an acid treatment is not attractive in large-scale painting operations of SS sheets. In our laboratory we have therefore developed a pretreatment for painting SS which does not require an acid etching step [10]. In the optimization of this treatment, the NMP adhesion test was extremely useful, because it differentiated much better and faster between substrates and treatments than any other test method.

Table 4 shows the NMPRT data for an epoxy powder paint applied to SS4 samples following a wide range of treatments employing two types of organofunctional silane. Details on the silanes and all application procedures (i.e. SL, SSL, and SSLX) have been given in the Experimental section. For comparison, data for CRS are also included. CRS was compared with and without phosphate coating in this experiment, because phosphating of CRS is done primarily to improve paint adhesion. A similar adhesion promotion process for SS does not exist, because under normal conditions, stainless steels cannot be phosphated. Therefore, the objective of this experiment was to obtain similar adhesion levels (i.e. NMPRT values) to those obtained for phosphated CRS.

Since certain powder paints did not always delaminate as an intact film, but rather as a powdery material, a new aspect of the NMP test is introduced here. The % area listed in Table 4 is the percentage of the substrate surface area from which, upon visual inspection, the paint had become delaminated after the NMPRT. This concept also enabled us to abort the experiment when the NMPRT became excessively long.

It is beyond the scope of this paper to discuss the chemistry of silanes at polymer–metal interfaces in any detail. Such discussions will be published elsewhere, and the behavior of silanes at surfaces and interfaces has also been covered extensively in a recent symposium proceedings [16]. In terms of the treatments of the NMPRT values, we can draw the following conclusions:

- the powder paint delaminates quickly from the control SS4 and etched SS4 substrates. The etching does not improve the paint adherence to SS at all. The

Table 4.

NMPRT values (in min) for epoxy powder coatings on silane-modified 409 stainless steel (SS4) (averaged for three punching; see Section 2 for abbreviations)

Treatment	Dry		Wet[a]	
	NMPRT	% area[b]	NMPRT	% area[b]
Control[c]	2.7	100	1.2	100
Firstcoat MP	2	100	0	100
SL SAAPS	10	50	10	90
SL GPS	6	70	1.5	100
SSL SAAPS	15	20	5	90
SSL GPS	8	80	3.1	100
SSLX SAAPS	25	10	30	10
SSLX GPS	15	60	2.1	100
Etched control	3	100	1.5	100
E[d] + SL SAAPS	30	50	30	50
E + SL GPS	35	50	35	50
E + SSL SAAPS	52	0	50	0
E + SSL GPS	50	0	30	0
E + SSLX SAAPS	70	0	50	0
E + SSLX GPS	65	0	40	0
CRS[c]	3	100	—	—
CRS + phosphate	50	0	—	—

[a]21 days soaked in water of 70°C.
[b]Degree of paint delamination in percent of the area tested.
[c]Solvent-cleaned only.
[d]E = Etched.

commercial primer Firstcoat MP does not offer any improvement for dry adhesion. In the water soak test, spontaneous delamination had occurred. The paint adhesion of the bare CRS surface is also very poor. The phosphate treatment of CRS has a tremendous effect on the NMPRT. In fact, this sample never delaminated;

- both silanes improve the NMPRT value markedly. This value goes up further for the SSL treatments and are even higher for the SSLX treatments, i.e. when a silane crosslinker is used;

- there is a rather dramatic effect of the two silanes (in all SL, SSL, and SSLX treatments) on the NMPRT of the etched samples. It can be speculated that now there is a better interaction with the SS oxide which contains an appreciable amount of iron oxide after etching. The NMPRT values for these (non-phosphated) substrates are now equivalent to that of phosphated CRS;

- a pronounced deteriorating effect on the NMPRT is observed when the samples are soaked in hot water. This effect is largely absent, however, in the samples that had been etched prior to silane treatment. The silane GPS is much worse after water soak than SAAPS for those samples that are not etched. After etching, the silanes SAAPS and GPS are equivalent;

- the best overall treatment, for practical purposes, is SSLX SAAPS. The NMPRT value of this system does not deteriorate in the water soak (as opposed to SL SAAPS and SSL SAAPS or SSLX GPS), and this treatment does not require the etching step. Its performance is close to that of the phosphated CRS material and may be optimized further, for instance, by varying the amount of crosslinker in the silane solution;

- although not tested extensively, the NMPRT data in the table parallel those observed in corrosion testing of the same or similar systems. Here, too, the SSLX materials performed best. The common corrosion test for painted SS is continuous salt spray exposure.

The 300 series of SS are even more difficult to paint than the 400 series because of the higher Cr content. Therefore, an experiment similar to that of Table 4 was carried out with SS3. Only those systems that showed good performance with SS4 were prepared. This experiment was carried out with a different paint, namely a polyester powder paint. For comparison, the SSL and SSLX treatments were also extended to non-phosphated CRS substrates. Another comparison was with an SS4 substrate which had been treated in the same phosphate process normally used for CRS. The NMP results for CRS in this experiment were compared with crosshatch data. The results are summarized in Table 5, from which we can draw the following conclusions:

- the NMPRT values for the controls, unetched and etched SS3, are considerably higher than in the previous experiment. Since the value for non-phosphated CRS is

Table 5.
NMPRT values (in min) for polyester powder coatings on silane-modified 304 stainless steel (SS3) and CRS (averaged for three punchings; see Section 2 for abbreviations)

Treatment	Dry		Wet[a]		
	NMPRT	% area[b]	NMPRT	% area[b]	% crosshatch
Control[c]	16	95			
SSLX SAAPS	60	0			
Etched control	33	75			
E[d] + SSL SAAPS	60	0			
E + SSLX SAAPS	60	0			
CRS[c]	30	90	1.33	100	100
CRS + phosphate	60	0	10	10	0
CRS + SSL SAAPS	60	0	2.5	80	100
CRS + SSLX SAAPS	60	0	10	80	0
SS4 + phosphate	34	85			

[a] 21 days soaked in water of 70°C.
[b] Degree of paint delamination in percent of the area tested.
[c] Solvent-cleaned only.
[d] E = Etched.

now also considerably higher, these differences must be related to the paint system and not to the different substrates used in these two experiments;

- the preferred treatment for SS4, namely SSLX SAAPS, works very well here on both etched and non-etched SS3 substrates. Even without the crosslinker (SSL SAAPS treatment) the paint adherence to etched SS3 is very high;

- the phosphated SS4 material has a considerably lower NMP performance than the phosphated CRS. This is, of course, to be expected, because SS4 does not form phosphate crystals properly;

- the SSL SAAPS treatment works well on bare CRS. For dry systems the behavior is identical to that of phosphated CRS. After water soak, however, all systems have deteriorated in their NMP adhesion behavior. The effects seen here are in accordance with expectation, i.e. non-phosphated CRS is the worst and phosphated CRS is the best. The SSL SAAPS treatment gives some improvement and the CRS with the SSLX SAAPS treatment begins to approach the phosphated CRS. It is to be noted here that the trend after the water soak treatment of both substrates, CRS and SS, is that the NMPRT values decrease. In the experiments with the E-coated automotive materials (Section 3.3, Table 3), the trend was that the NMPRT value went up after water soak. The exact mechanism for this discrepancy is not known at this point, but there are two differences between the experiments of Tables 3 and 5, namely (i) the length of the exposure to the hot water (3 days vs. 21 days) and (ii) the paint systems, namely a cathodic electrocoat primer (an epoxy-urethane) vs. a polyester powder paint. It is thus entirely reasonable that different interfacial reactions occurred in the two experiments;

- the crosshatch data for the wet CRS materials show much less discrimination than the NMPRT results, but a correlation can be noted, i.e. both tests identify the phosphated and the SSLX SAAPS materials as better than the other two systems.

4. DISCUSSION

4.1. *General aspects of the new adhesion test*

The results described above for widely differing systems indicate that the NMPRT is a reproducible number, specific for a certain combination of metal, surface pretreatment, and paint system. It has been demonstrated that certain changes in the system, such as metal pretreatment, paint cure, or the presence of water at the interface, result in statistically significant variations in the NMPRT value.

The behavior of the paint in the solvent demonstrates that the mechanism of delamination must be by shear stresses which develop at the interface. The paint becomes very soft in the test. It seems to be completely plasticized by the solvent. It is thus reasonable to assume that the effects of mechanical aspects such as stresses and strains, stress raisers, cracks, voids, and other defects are largely, if not completely, eliminated. Viscoelastic energy dissipation during fracture is also minimized or eliminated. This aspects distinguishes the NMP test sharply from existing tests in which paints are removed mechanically.

The interfacial stresses result in a destruction of the interfacial polymer–metal bonds, and the delamination rate is therefore a function of the number of polymer–metal bonds and also of the strength of such bonds. It is clear from the foregoing that if a secondary type of interaction prevails, e.g. hydrogen bonding or acid–base interactions, the paint will delaminate. The only type of bond that resists delamination is covalent bonding, such as that obtained by functional silanes, if applied properly. Evidence for this mechanism is the following series of observations:

The delaminated interface is clean, i.e. analysis of the metal and paint sides after delamination in NMP has shown that the separation is exactly at the interface, even on a surface analysis scale [8]. There are no traces of metal in the paint and virtually no paint residues remaining on the metal side, except for certain low-molecular-weight components or additives (crosslinkers, plasticizers, antioxidants, etc.) that are strongly adsorbed on the metal surfaces. These are usually very polar materials, such as quaternary amines in the case of cathodic E-coats. In the case of phosphated metals, the separation is sharply at the paint–phosphate interface. There are no phosphate crystals that remain on the paint and the impressions of the phosphate crystals can, after drying, be observed by SEM on the backside of the paint film. An example is given in Fig. 3.

The NMPRT value is not strongly dependent on the thickness of the paint film. Experiments with powder coatings such as those described in Section 3.4 have shown that the delamination rate for a two-coat system consisting of 50 μm epoxy paint + 50 μm polyester paint on hot-dip galvanized steel is almost identical to the NMPRT for a single coating of 50 μm epoxy on the same substrate [17], although it is sometimes observed that the delamination occurs at the epoxy–polyester interface first. Whether or not this occurs was found to depend on the cure conditions of the epoxy primer layer. The conditions and geometry of the test and the test sample are such that diffusion of NMP throughout the entire paint is extremely fact, i.e. faster than the rate of delamination. When a delaminated paint film of 50–100 μm thickness is, after complete drying, immersed into NMP again, it will swell to about 1.5–2 times its diameter in less than 1 min. These observations thus demonstrate that the diffusion rate of NMP in the paint polymer itself is not rate-determining for the delamination process. This diffusion rate is generally musch higher than the rate of delamination.

Paint delamination from a microrough surface, such as a phosphated metal or a shot-blasted or etched substrate, is not necessarily slower than from a polished substrate. It all depends on how the metal is pretreated. Polished metals can resist delamination completely, for instance if they are treated properly with functional silanes. This again demonstrates that the NMP diffusion rate through the paint or along the interface is not the rate-determining step, but this step is the rate at which existing hydrogen bonds are replaced by NMP.

The rate of delamination from a given substrate is in this mechanism, of cource, also strongly dependent on the type of paint. Important factors here are the pigment concentration, the interactions between the pigment particles and the resin, the polarity of the paint polymer, and the type of functional groups in the polymer. All of these parameters will determine the degree of swelling of the paint in NMP, and hence the

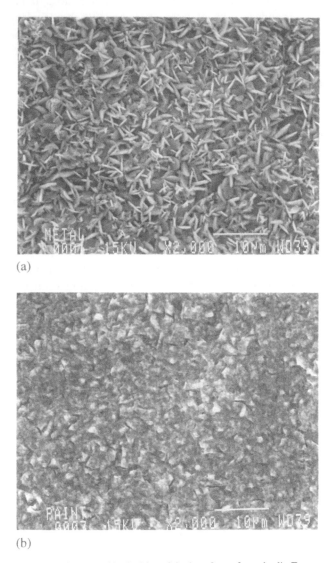

Figure 3. Secondary electron images of both sides of the interface of a cathodic E-coat on zinc phosphated CRS substrate following paint removal by NMP. (a) CRS side; (b) paint side after drying.

rate of paint delamination. It should therefore be concluded that it will be impossible to convert the NMPRT values to actual adhesion data, and further, different paint systems cannot be compared strictly on the basis of their NMPRT results. Where the method seems to be most applicable and useful is for the comparison of different metal pretreatments and their effects of the interfacial strength. Also, interfacial changes as a result of corrosion processes can be investigated.

An important aspect of the proposed test is the capability of detecting changes in the system upon immersion in water. The examples cited in the foregoing indicated in some cases an increased NMPRT in the wet state (e.g. some phosphated automotive systems); in other systems (all powder paints), a drop of the NMPRT value is observed. This variability suggests that the results are system-dependent and that the effects are not caused by a reduced swelling capability of NMP in the presence of water. It is known that paints on stainless steels lose their adhesion quickly when immersed in water. This effect is caused by the absence of strong paint–metal bonds and because cathodic sites are set up which destroy these rather weak bonds.

The NMP molecule itself can be expected to be a strong hydrogen bond former. This is probably the underlying reason why it is such a strong swelling agent. It will thus swell and delaminate paints that are weaker hydrogen bonders (or weaker bases) than NMP. The paint delamination process can thus also be viewed as a displacemnt from the interface of hydrogen bonding paint molecules by molecules of the stronger hydrogen bonder NMP.

NMP is not the only swelling agent that can be used to study interfacial phenomena. For paints that cannot be displaced by NMP because they do not swell appreciably, it should, of course, be possible to find other swelling solvents or to use mixtures of solvents. Also, the swelling capability of NMP is dependent on the temperature. Paints come off faster at higher temperatures. The temperature of $60\,^{\circ}$C was chosen rather arbitrarily. The delamination times are conveniently short. The swelling capability of NMP can also be increased by dissolving some LiF in it. This increases the rate of paint delamination, but it makes the method less useful for interfacial analysis by surface techniques, because all samples will show high concentrations of Li and F.

4.2. Automotive paint systems

The effect of the automotive topcoat on the NMPRT of CRS and EGS (Section 3.2) seems to contradict the above conclusions about the paint thickness. However, it should be pointed out that the E-coat is very thin (20–30 μm) and that, following the application of the clearcoat and basecoat layers, the entire system is baked again. These additional baking steps cause an excess curing of the E-coat film, as compared with systems which have been E-coated only [6]. Also, diffusion of topcoat constituents, such as curing agents, into the E-coat layer may occur. It is thus plausible that the E-coat film properties may be modified by the much thicker topcoat system.

The increased NMPRT value for painted automotive steels and the greater effect for EGS, as compared with CRS, can be explained by an acid–base mechanism [18]. It has been argued recently that the mechanism of adhesion between epoxy-urethane E-coats and phosphate crystals is by hydrogen bonding, involving the oxygen functionalities in the paint and the water molecules in the phosphate crystals [6]. Baking of the electrodeposited primer and the application of a second and a third coating (basecoat and clearcoat) followed by curing generate large stresses in the paint film [19]. This is caused, in part, by the fact that the phosphate crystals are dehydrated from a tetrahydrate to a dihydrate during the paint cure [5]. During exposure of the painted system to moisture, rehydration of the phosphate crystals takes place from $Zn_3(PO_4)_2 \cdot 2 H_2O$ to $Zn_3(PO_4)_2 \cdot 4 H_2O$ (hopeite) [19, 20]. This phenomenon causes expansion of the phosphate crystals and hence an increase of the shear stresses between the phosphate layer and the paint film, except in systems where the phosphate coating contains a large amount of Ni^{2+} ions which retard this rehydration reaction. Upon immersion of the painted phosphate crystals in water, a hydrolysis reaction of $Zn_3(PO_4)_2$ occurs. The surface of the crystals is converted to $Zn(OH)_2$ and $Zn(H_2PO_4)_2$. The $Zn(OH)_2$ then dehydrates easily to form ZnO, which is believed to weaken the phosphate–paint bond [19]. Again for zinc phosphate systems containing Ni^{2+} ions, this effect is suppressed, i.e. $Zn(OH)_2$ is stabilized. This mechanism is believed to be the underlying cause for the improved wet adhesion of cataphoretic paints to modern zinc phosphate systems containing Ni, which has been reported [19].

The increased NMPRT value for EGS upon rinsing with the (acidic) Cr(VI) solution and upon subsequent exposure of the painted system to moisture is in agreement with the hydrogen bonding (acid–base) concept. The Cr(VI) rinse will make the zinc phosphate crystals more acidic and the exposure to water will hydrolyze the zinc phosphate surface to $Zn(H_2PO_4)_2$, which is also more acidic. The PO_4-rich surface of hopeite after water exposure has actually been detected by TOFSIMS [8]. The increased hydrogen bonding of the hydrolyzed phosphate surface manifests itself as a higher NMPRT value. The observation that the paint still comes off can be interpreted as an indication that the mechanism in all cases is by secondary bonding. Primary (covalent) bonds, as may be formed by silanes, or in some cases by saponified paint residues [6], are not displaced by NMP.

The striking difference between EGS and CRS (Table 3) can be easily explained by the same model. The phosphate formed on CRS is predominantly phosphophyllite, $ZnFe_2(PO_4)_2 \cdot 4 H_2O$. This phosphate also dehydrates to a dihydrate during baking, but it also becomes amorphous and does not rehydrate or hydrolyze during subsequent exposure to water [21]. Hence the increase in the number of hydrogen bonds, such as occurs in the rehydration process of $Zn_3(PO_4)_2 \cdot 2 H_2O$ on EGS, does not occur here. As a result, the increase of NMPRT upon rehydration is for CRS much less then for EGS.

4.3. Silanes on stainless steels

The results presented for the stainless steel modification with organofunctional silanes illustrate that the NMP technique can detect interfacial changes at paint–metal interfaces with much greater sensitivity than the existing techniques based on mechanical paint removal. At the same time, the results seem to confirm that paints which are covalently bonded to the model (oxide) surface can no longer be removed by NMP. The silane, together with a suitable crosslinker, can be expected to form a three-dimensional interpenetrating interfacial network with the paint [7, 22]. Such a network is assumed to be hydrolytically stable. Hence the NMPRT after water soak is still excellent for the optimized systems.

The results also tend to show that the structure of hydrolytic stability of the interfacial network may be different for the silanes SAAPS and GPS. The cause of this difference is not known and a discussion of it is beyond the scope of this paper. However, it can be speculated that the interaction between the functional groups in GPS (an oxirane group) or in SAAPS (a vinyl group) and those of the paint (hydroxyl and urethane groups) may be quite different, resulting in different hydrolytic stabilities of the interphase.

The remarkable effect on the interphase hydrolytic stability of etching of the SS4 substrate in the case of GPS, whereas SAAPS performs well on etched or non-etched substrates, suggests that GPS is adsorbed in a different way on Cr_2O_3 than on Fe_2O_3. This is plausible because of the high polarity of the oxirane group which may interact differently with the different oxides. For SAAPS such an effect should be absent because the functional group is non-polar. It should adsorb in the same way on different metals, as indeed has been demonstrated for SAAPS (and also γ-APS) adsorbed on EGS and CRS [15]. However, more work is needed to confirm this also for GPS and SAAPS on SS materials.

5. CONCLUSIONS

- A new adhesion test for paints on metals has been presented which consists of immersing small punched disks in N-methyl pyrrolidone of 60°C. The paint comes off intact and the time required for complete delamination is termed the NMPRT. This number is reproducible for a given system.

- This approach is useful for studying interfacial changes, for instance, those that occur when the paint cure is varied, the metal pretreatment is changed, or when the painted metal is exposed to a wet environment. The changes observed are generally in agreement with expectation, i.e. a higher cure temperature or better surface preparation results in higher NMPRT values; phosphating improves the interfacial strength; water may increase or decrase the NMPRT, depending on the system; chromate rinsing of phosphates usually retards the paint delamination.

- The mechanism of paint delamination in NMP appears to be by inducing interfacial stresses which build up because the paint swells strongly in NMP. At the interface, the polymer molecules adsorbed on the metal are replaced by NMP, a strong hydrogen bond former.

- The method to work well in cases where the interfacial bonding is of secondary nature, such as hydrogen bonds or, more generally, acid–base interactions. Paints that form covalent bonds or paints with silane-modified interfaces cannot be removed by NMP. The method can thus be used to optimize metal pretreatments, e.g. by silanes.

- The NMPRT value agrees in many cases with the corrosion performance of the same system.

- Silane/crosslinker mixtures can be used to improve paint adhesion to stainless steel types 409 and 301.

REFERENCES

1. General Motors Engineering Standards, GM9071P; General Motors, Warren, MI (1989).
2. B. J. Briscoe and S. S. Panesar, *J. Phys. D: Appl. Phys.* **25**, A20–A27 (1992).
3. L. M. Callow and J. D. Scantlebury, in: *Polymeric Materials for Corrosion Control*, R. A. Dickie and F. L. Floyd (Eds), ACS Symposium Series No. 322, pp. 115–122. American Chemical Society, Washington, DC (1986).
4. A. F. Skenazi, D. Davin, V. Leroy and D. Coutsouradis, *Metall* **38**, 1187–1192.
5. Y. Miyoshi, J. Oka and S. Maeda, *Trans. ISIJ* **23**, 974 (1983).
6. W. J. van Ooij, A. Sabata and A. D. Appelhans, *Surface Interface Anal.* **17**, 403–420 (1991).
7. E. P. Plueddemann, *Silane Coupling Agents*, 2nd edn. Plenum Press, New York (1991).
8. W. J. van Ooij and A. Sabata, *Surface Interface Anal.* **19**, 101–113 (1992).
9. M. Ström, G. Ström, W. J. Van Ooij, A. Sabata, B. A. Knueppell and A. C. Ramamurthy, *Proc. GALVATECH '92*. Amsterdam, 8–10 September, pp. 521–527 (1992).
10. A. Sabata and W. J. van Ooij, to Armco, Inc., US Patent Appl. (2 Dec. 1992).
11. W. J. van Ooij and A. Sabata, to Armco Steel Company, US Patent 5,108,793 (28 April 1992).
12. J. W. Davis, *SAE Special Publications SP-612*, pp. 53–59. Society of Automotive Engineers, Warrendale, PA (1985).
13. W. J. van Ooij and A. Sabata, *Corrosion/91 (NACE)*, Cincinnati, OH, 1–5 March, Paper No. 517, pp. 1–17. NACE, Houston, TX (1991).
14. A. Sabata, W. J. van Ooij and R. J. Koch, *J. Adhesion Sci. Technol.* (in press).
15. W. J. van Ooij and A. Sabata, *J. Adhesion Sci. Technol.* **5**, 843–863 (1991).
16. K. L. Mittal (Ed.), *Silanes and Other Coupling Agents*. VSP, Zeist (1992).
17. W. J. van Ooij and A. Sabata, unpublished work.
18. K. L. Mittal and H. R. Anderson, Jr (Eds), *Acid–Base Interactions: Relevance to Adhesion Science and Technology*. VSP, Zeist (1991).
19. S. Maeda, T. Asai, M. Yamamoto, H. Asano and H. Okada, in: *Organic Coatings Science and Technology*, Vol. 7, G. D. Parfitt and A. V. Patsis (Eds), p. 223. Marcel Dekker, New York (1984).
20. W. J. van Ooij and A. Sabata, *J. Coatings Technol.* **61**, 51–65 (1989).
21. J. P. Servais, B. Schmitz and V. Leroy, *Corrosion/88 (NACE)*, St. Louis, MO, 21–25 March, Paper No. 41. NACE, Houston, TX (1988).
22. J. P. Bell, R. G. Schmidt, A. Malofsky and D. Mancini, *J. Adhesion Sci. Technol.* **5**, 927–944 (1991).